# Topics in Applied Physics
## Volume 90

Available  at
## Springer Link.com

Topics in Applied Physics is part of [SpringerLink] service. For all customers with standing orders for Topics in Applied Physics we offer the full text in electronic form via [SpringerLink] free of charge. Please contact your librarian who can receive a password for free access to the full articles by registration at:

springerlink.com → Orders

If you do not have a standing order you can nevertheless browse through the table of contents of the volumes and the abstracts of each article at:

springerlink.com → Browse Publications

There you will also find more information about the series.

## Springer
*Berlin*
*Heidelberg*
*New York*
*Hong Kong*
*London*
*Milan*
*Paris*
*Tokyo*

**Physics and Astronomy**

ONLINE LIBRARY

springeronline.com

# Topics in Applied Physics

Topics in Applied Physics is a well-established series of review books, each of which presents a comprehensive survey of a selected topic within the broad area of applied physics. Edited and written by leading research scientists in the field concerned, each volume contains review contributions covering the various aspects of the topic. Together these provide an overview of the state of the art in the respective field, extending from an introduction to the subject right up to the frontiers of contemporary research.

Topics in Applied Physics is addressed to all scientists at universities and in industry who wish to obtain an overview and to keep abreast of advances in applied physics. The series also provides easy but comprehensive access to the fields for newcomers starting research.

Contributions are specially commissioned. The Managing Editors are open to any suggestions for topics coming from the community of applied physicists no matter what the field and encourage prospective editors to approach them with ideas.

See also: springeronline.com

## Managing Editors

### Dr. Claus E. Ascheron

Springer-Verlag Heidelberg
Topics in Applied Physics
Tiergartenstr. 17
69121 Heidelberg
Germany
Email: ascheron@springer.de

### Dr. Hans J. Koelsch

Springer-Verlag New York, Inc.
Topics in Applied Physics
175 Fifth Avenue
New York, NY 10010-7858
USA
Email: hkoelsch@springer-ny.com

## Assistant Editor

### Dr. Werner Skolaut

Springer-Verlag Heidelberg
Topics in Applied Physics
Tiergartenstr. 17
69121 Heidelberg
Germany
Email: skolaut@springer.de

Peter Michler (Ed.)

# Single Quantum Dots

Fundamentals, Applications,
and New Concepts

With 181 Figures

 Springer

Prof. Dr. Peter Michler

Universität Stuttgart
5. Physikalisches Institut
Pfaffenwaldring 57
70550 Stuttgart
Germany
p.michler@physik.uni-stuttgart.de

Library of Congress Cataloging in Publication Data
Single quantum dots: fundamentals, applications and new concepts/ Peter Michler (ed.). p. cm. – (Topics in applied physics, ISSN 0303-4216; v. 90) Includes bibliographical references and index. ISBN 3-540-14022-0 (alk. paper)
1. Quantum dots. I. Michler, Peter, 1963– II. Series.   TK7874.88.S56 2003   621.3815'2–dc22   2003066675

Physics and Astronomy Classification Scheme (PACS):
78.67.Hc, 71.35.Cc, 42.50.Dv, 42.50.Pq

ISSN print edition: 0303-4216
ISSN electronic edition: 1437-0859
ISBN 3-540-14022-0 Springer-Verlag Berlin Heidelberg New York

Springer-Verlag is a part of Springer Science+Business Media

springeronline.com

© Springer-Verlag Berlin Heidelberg 2003
Printed in Germany

Typesetting: DA-TEX · Gerd Blumenstein · www.da-tex.de
Cover design: *design & production* GmbH, Heidelberg

Printed on acid-free paper      56/3141/mf      5 4 3 2 1 0

SD
1/28/04
TR.

# Preface

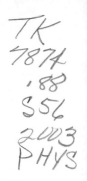
Semiconductor quantum dots (QDs) are nanometer-sized three-dimensional structures which confine electrons and holes in all three directions giving rise to a discrete energy spectrum. Many types of QDs have already been developed such as electrostatically and lithographically defined QDs, chemically synthesized QDs, naturally formed QDs by interface fluctuations, and epitaxially grown self-assembled QDs. The latter possess the advantage that their size, shape, composition and location can be tailored to a large extent by modern growth techniques. In addition they can be naturally embedded during growth into more complex structures making them attractive for different kind of electronic and photonic devices.

The growth of self-assembled QDs has been presently performed on a wide variety of semiconductor alloys such as Ge/Si, InAs/GaAs, (In,Ga)As/GaAs, (Ga,In)P/InP, CdSe/Zn(S,Se), and (In,Ga)N/GaN. The spectral region reaches from the near infrared ($\sim 1.5\,\mu m$) to the blue-green region ($\sim 430$–$500\,nm$). Self-assembled dot growth leads to a distribution of dot sizes, and hence a distribution of energy levels. On one hand, the inhomogeneous broadening reduces the advantage expected from the zero-dimensional systems, on the other hand the broad distribution of energy levels can be used for different types of applications such as wavelength selective storage devices. Nowadays, different companies are moving towards commercialization of devices based on quantum dot ensembles. For example current QD diode lasers can outperform quantum well lasers with respect to threshold density, temperature insensitivity, and gain bandwidth.

Within the last few years single quantum dot spectroscopy moved more and more into the focus of many research groups to study the exciting physics of *individual* and *coupled pairs* of QDs and to exploit their tremendous potential for device applications. In fact, the full advantage of their superior properties can be utilized if *single* quantum dots are used, e. g., as active elements for single-electron and single-photon devices. Recently, a single-photon source based on a single QD embedded in a photonic nanostructure has been demonstrated. The interest for such nonclassical light sources rapidly increases since the generation of photons on demand is an important prerequisite towards many novel technological applications in the field of quantum information science. More recently, a coherently driven single-QD photodiode has been fabricated which works as an optically triggered single-electron source which offers single electrons (or holes) on demand. Such a device combines *coherent* optical excitations and electric currents and could be one of the building blocks of a new generation of optoelectronic devices. Furthermore, a cavity-quantum electrodynamics effect, i.e. the Purcell effect, has been demonstrated in the coupled quantum-dot cavity system and reaching the strong coupling regime seems to be within reach during the next few years. These exciting developments encouraged the editor to ask leading scientists for articles for this book on self-assembled quantum dots with special emphasis on *single* quantum dot properties.

The book is organized as follows: *Pierre Petroff* gives an introduction into the crystal growth processes of self-assembled quantum dots and presents some newer methods for controlling the ordering and positioning of the quantum dots. Novel quantum dot based devices like a quantum dot memory device and a spin polarized light emitting diode are also discussed in his contribution.

Readers will find a fundamental introduction into the theory of the electronic and optical properties of self-assembled quantum dots in the contribution by *Pawel Hawrylak and Marek Korkusiński*. Especially the spectra of a single exciton, the exciton in the presence of many electrons, and many-exciton complexes are discussed in detail.

*Manfred Bayer* presents extensive spectroscopic studies of single In(Ga)As/GaAs self-assembled quantum dots. The dependence of photoluminescence (PL) and PL excitation spectra on the quantum dot symmetry and in magnetic fields is discussed. Special emphasis is put on fine structure effects and on the exciton-phonon interaction in his contribution. The electronic structure and coherent coupling of the electronic states of quantum dot molecules are also discussed in his contribution.

Quantum dot eigenstates can be manipulated in a well-controlled way by applying either external electric or magnetic fields. In addition, II–VI semiconductor quantum dots offer the possibility to incorporate magnetic ions isoelectronically into the crystal matrix giving access to magnetic semiconductors. *Gerd Bacher* describes in detail giant magneto-optical effects like huge effective g-factors and the formation of quasi-zero-dimensional magnetic polarons. In addition, the Stark effect for lateral and vertical electrical fields is discussed.

As the quantum dot laser is one of the most important quantum dot based devices so far, the contribution by *Heinz Schweizer, Michael Jetter and Ferdinand Scholz* has been solely and comprehensively devoted to the fabrication of quantum dot lasers and their static and dynamic properties. Strong emphasis is put on carrier dynamics, i.e., transport, recombination and relaxation processes which determine the modulation dynamics and the spectral properties of the device.

*Paola Borri and Wolfgang Langbein* review the dynamics of the optically-induced interband polarization in (In,Ga)As/(Ga,Al)As quantum dots. Radiative recombination, exciton–phonon interaction and Coulomb interaction of excitons with other carriers are discussed as mechanisms responsible for the excitonic dephasing. Optical Rabi oscillations of the excitonic population in quantum dots are also reviewed.

Cavity-quantum electrodynamics effects such as the enhancement or inhibition of the spontaneous emission rate, the funneling of photons into a single mode and the control of spontaneous emission processes on the single photon level can be investigated by use of self-assembled quantum dots embedded in high-quality solid-state microcavities. These effects are particularly attractive to improve optoelectronic devices with respect to low threshold or high modulation response or develop novel ones, e.g. a single quantum dot laser. The progress on these topics is reviewed by *Jean-Michel Gérard*.

The last chapter of this book is written by *myself* where I review the recent progress in the new fascinating field of nonclassical light genereration using semiconductor quantum dots. The generation of triggered single photon emission, correlated photon pair emission, and the prospects for the generation of entangled photons are discussed. Such light sources are required for the rapidly evolving field of quantum information technology.

Finally, I would like to thank all my colleagues for writing the various chapters and for spending much of their free time for our common book. Very special thanks to my wife Silke for her unconditional support and the appreciation for my absence during numerous weekends.

Stuttgart, September 2003                                    *Peter Michler*

# Contents

**Epitaxial Growth and Electronic Structure of Self-Assembled Quantum Dots**
Pierre M. Petroff ....................................................... 1

1. Introduction ...................................................... 1
2. Formation of Quantum Dots, Quantum Rings and Strain Effects
   on Island Nucleation .............................................. 2
3. Positioning and Ordering Quantum Dots ........................... 4
   3.1. Island Nucleation and Quantum Dot Lattices ................. 6
4. Strain Effects and Growth Characteristics of Quantum Dots ........ 8
5. Fundamental Electronic Properties of Self-Assembled
   Quantum Dots ..................................................... 10
   5.1. The "Atom-Like" Shell Model for the Quantum Dots ......... 10
   5.2. Charged and Neutral Quantum Dots .......................... 11
6. Quantum Dot Based Devices ....................................... 16
   6.1. Information Storage using Quantum Dots .................... 16
   6.2. A Spin Polarized Quantum Dot Light Emitting Diode ........ 18
7. Conclusions ...................................................... 20
References ........................................................... 21

**Electronic Properties of Self-Assembled Quantum Dots**
Pawel Hawrylak and Marek Korkusiński ............................... 25

1. Introduction ...................................................... 25
2. Single-Particle States ............................................ 26
   2.1. Strain Effects .............................................. 27
   2.2. Effective Mass Adiabatic Models ............................ 32
   2.3. $k \cdot p$ Models ......................................... 38
   2.4. Tight-binding Models ....................................... 42
   2.5. First Principle Methods .................................... 43
   2.6. Coupled Dots ............................................... 45
3. Excitons in Quantum Dots ........................................ 49
   3.1. Excitons in Jacobi Coordinates ............................. 49
   3.2. Magneto-Excitons ........................................... 53
   3.3. Exciton in a Double Quantum Dot Molecule .................. 54
   3.4. Exciton, Biexciton, and Charged Exciton .................... 57

    3.5.   Exciton–Phonon Interaction and Raman Scattering . . . . . . . . . . 59

    3.6.   Exciton and Biexciton in an Optical Cavity . . . . . . . . . . . . . . . . . 61

4.   Multi-Exciton Complexes . . . . . . . . . . . . . . . . . . . . . . . . . . . . . . . . . . . . 62

    4.1.   Exact Diagonalization Technique . . . . . . . . . . . . . . . . . . . . . . . . . . . 64

    4.2.   Hidden Symmetries . . . . . . . . . . . . . . . . . . . . . . . . . . . . . . . . . . . . . . 66

    4.3.   Excited States and Emission Spectra . . . . . . . . . . . . . . . . . . . . . . 71

5.   Charged Quantum Dots . . . . . . . . . . . . . . . . . . . . . . . . . . . . . . . . . . . . . 73

    5.1.   Ground State of Charged Dots . . . . . . . . . . . . . . . . . . . . . . . . . . . . 74

    5.2.   Emission . . . . . . . . . . . . . . . . . . . . . . . . . . . . . . . . . . . . . . . . . . . . . . 77

    5.3.   Absorption . . . . . . . . . . . . . . . . . . . . . . . . . . . . . . . . . . . . . . . . . . . . 81

    5.4.   Far-Infrared Spectroscopy . . . . . . . . . . . . . . . . . . . . . . . . . . . . . . . 85

    5.5.   Photocurrent Spectroscopy . . . . . . . . . . . . . . . . . . . . . . . . . . . . . . 85

References . . . . . . . . . . . . . . . . . . . . . . . . . . . . . . . . . . . . . . . . . . . . . . . . . . . 86

## Exciton Complexes in Self-Assembled In(Ga)As/GaAs Quantum Dots

Manfred Bayer . . . . . . . . . . . . . . . . . . . . . . . . . . . . . . . . . . . . . . . . . . . . . . . 93

1.   Introduction . . . . . . . . . . . . . . . . . . . . . . . . . . . . . . . . . . . . . . . . . . . . . . . 93

2.   Experimental Technique . . . . . . . . . . . . . . . . . . . . . . . . . . . . . . . . . . . . 99

3.   Excitonic Absorption in Quantum Dots . . . . . . . . . . . . . . . . . . . . . . 102

4.   Dephasing of the Ground State Exciton . . . . . . . . . . . . . . . . . . . . . . 112

5.   Fine Structure of Excitons and Excitonic Complexes . . . . . . . . . . . . 114

    5.1.   Faraday Configuration . . . . . . . . . . . . . . . . . . . . . . . . . . . . . . . . . . 117

    5.2.   Voigt Configuration . . . . . . . . . . . . . . . . . . . . . . . . . . . . . . . . . . . . 122

    5.3.   Fine Structure of the Biexciton . . . . . . . . . . . . . . . . . . . . . . . . . . 125

    5.4.   Charged Excitons . . . . . . . . . . . . . . . . . . . . . . . . . . . . . . . . . . . . . . 128

6.   Quantum Dot Molecules . . . . . . . . . . . . . . . . . . . . . . . . . . . . . . . . . . . . 135

7.   Conclusions . . . . . . . . . . . . . . . . . . . . . . . . . . . . . . . . . . . . . . . . . . . . . . . 143

References . . . . . . . . . . . . . . . . . . . . . . . . . . . . . . . . . . . . . . . . . . . . . . . . . . . 145

## Optical Spectroscopy on Epitaxially Grown II–VI Single Quantum Dots

Gerd Bacher . . . . . . . . . . . . . . . . . . . . . . . . . . . . . . . . . . . . . . . . . . . . . . . . 147

1.   Introduction . . . . . . . . . . . . . . . . . . . . . . . . . . . . . . . . . . . . . . . . . . . . . . 147

2.   II–VI Single Quantum Dots – Intrinsic Properties . . . . . . . . . . . . . . 149

    2.1.   Single Excitons in a Quantum Dot . . . . . . . . . . . . . . . . . . . . . . . . 149

    2.2.   Biexcitons and Charged Excitons . . . . . . . . . . . . . . . . . . . . . . . . . 157

3.   Eigenstate Manipulation by External Fields . . . . . . . . . . . . . . . . . . . 164

    3.1.   Applying Electrical Fields – The Stark Effect . . . . . . . . . . . . . . . 164

    3.2.   Applying Magnetic Fields – The Zeeman Effect . . . . . . . . . . . . . 167

4.   Single Magnetic Semiconductor Quantum Dots . . . . . . . . . . . . . . . . . 171

    4.1.   Giant Magneto-Optical Effects in a Zero-Dimensional System 171

    4.2.   Statistical Fluctuations of the Nano-Scale Magnetization . . . . . 174

5.   Summary and Outlook . . . . . . . . . . . . . . . . . . . . . . . . . . . . . . . . . . . . . . 178

References . . . . . . . . . . . . . . . . . . . . . . . . . . . . . . . . . . . . . . . . . . . . . . . . . . . 179

**Quantum-Dot Lasers**
Heinz Schweizer, Michael Jetter and Ferdinand Scholz ................. 185

1.  Introduction ..................................................... 185
    1.1.  History of Dot Lasers ....................................... 185
    1.2.  Motivation/Applications/Material Systems .................. 187
    1.3.  Artificial versus Self Assembly ............................ 188
2.  Epitaxially Fabricated Quantum Dot Lasers ....................... 189
    2.1.  Growth Considerations for Self-assembled Quantum Dot
          Laser Structures .......................................... 191
    2.2.  Specific Problems of Self-assembled Quantum Dot
          Laser Structures .......................................... 191
    2.3.  InAs-GaAs Quantum Dot Lasers ............................. 193
    2.4.  Red Light Emitting Dot Lasers ............................. 200
    2.5.  Nitride Quantum Dots for Laser Action ..................... 204
    2.6.  ZnSe Based Quantum Dot Lasers ............................ 205
3.  Lithography Based Quantum Dot Laser Fabrication .............. 206
    3.1.  Electron Beam Lithography ................................. 206
    3.2.  Structurization Technology ................................ 207
4.  Basic Considerations ............................................ 208
    4.1.  Electronic States and Statistics ........................... 208
    4.2.  Optical Transitions ....................................... 210
    4.3.  Dot Laser Threshold and Laser Dynamics ................... 211
5.  Devices and Applications ........................................ 219
    5.1.  Dot- and Wire-DFB-Lasers ................................. 219
    5.2.  Modulation Dynamics ...................................... 221
    5.3.  Chirp ..................................................... 222
    5.4.  High Power Lasers ........................................ 223
6.  Conclusions ..................................................... 225
References ........................................................... 226

**Dephasing Processes and Carrier Dynamics in (In,Ga)As
Quantum Dots**
Paola Borri and Wolfgang Langbein ................................. 237

1.  Introduction ..................................................... 237
2.  Coherent Spectroscopy of Semiconductor Quantum Dots .......... 238
    2.1.  Theoretical Background .................................... 238
    2.2.  Optical Experiments ....................................... 246
3.  Dephasing of Excitons in (In,Ga)As Quantum Dots .............. 248
    3.1.  Radiative Processes ....................................... 248
    3.2.  Exciton–Phonon Interaction ............................... 250
    3.3.  Coulomb Interactions ..................................... 255
4.  Optical Rabi Oscillations in (In,Ga)As Quantum Dots ........... 258
5.  Summary and Outlook ............................................ 264
References ........................................................... 265

## Solid-State Cavity-Quantum Electrodynamics with Self-Assembled Quantum Dots

Jean-Michel Gérard .................................................. 269

1. Introduction .................................................. 269
2. Some Assets of Self-Assembled QDs for Quantum Optics .......... 270
3. Weak and Strong Coupling Regimes: Some Basics ................. 272
   3.1. The Strong Coupling Regime ............................... 273
   3.2. Weak Coupling Regime: The Purcell Effect .................. 275
4. Presentation of Available 0D Semiconductor Microcavities ........ 278
   4.1. Pillar Microcavities ...................................... 279
   4.2. Microdisks ................................................ 283
   4.3. Photonic Crystal Microcavities ........................... 284
   4.4. Conclusion ................................................ 285
5. Single QDs and the Strong Coupling Regime ...................... 285
6. CQED Effects in the Weak Coupling Regime ....................... 288
   6.1. InAs QDs in Planar Cavities ............................... 288
   6.2. Purcell Effect for QD Arrays in 0D Microcavities .......... 289
   6.3. Purcell Effect for Single QDs in 0D Microcavities ......... 294
   6.4. "Nearly" Single-Mode Spontaneous Emission ................ 297
7. Application Prospects .......................................... 299
   7.1. QD-Microcavity Lasers: Toward Single QD-Lasers ........... 299
   7.2. The Single-Mode Single Photon Source
        and the Photon Collection Issue .......................... 303
   7.3. Generation of Indistinguishable Single Photons
        and Entangled Photon Pairs ............................... 308
8. Conclusion .................................................... 309
References ........................................................ 309

## Nonclassical Light from Single Semiconductor Quantum Dots

Peter Michler ..................................................... 315

1. Introduction .................................................. 315
2. Theoretical Background ........................................ 316
   2.1. Photon Statistics ........................................ 316
   2.2. Photon Cascade from a Quantum Dot ........................ 320
3. Experimental: Photon Statistic Measurements ................... 322
4. Photon Antibunching in Single Quantum Dot Photoluminescence .. 323
5. A Quantum Dot Single-Photon Source ............................ 326
   5.1. Triggered Single Photons from a Single Quantum Dot ....... 327
   5.2. Single Photons from Quantum Dots
        Coupled to a Microcavity ................................. 329
   5.3. Electrically Driven Single-Photon Source ................. 336
6. Correlated Photon Pairs from a Single Quantum Dot ............. 338
   6.1. Cross-Correlation Measurements ........................... 339
7. Summary and Outlook ........................................... 344
References ........................................................ 346

Index ............................................................. 349

# Epitaxial Growth and Electronic Structure of Self-Assembled Quantum Dots

Pierre M. Petroff

Materials Department, University of California,
Santa Barbara, CA 93103, USA
petroff@engineering.ucsb.edu

**Abstract.** Semiconductor self assembled quantum dots have emerged as one of the simplest means of exploring and exploiting the physics and device applications of carriers and excitons in the three dimensional confinement regime. This chapter covers the epitaxial growth processes involved in the formation of self-assembled quantum dots. A number of approaches for self-ordering and positioning quantum dots using direct epitaxial growth are also developed. The electronic structures of quantum dots using capacitance and luminescence spectroscopy techniques are presented to demonstrate the analogy between the quantum dots and "artificial atoms". To emphasize the difference with the properties of the isolated artificial atom, the role of the quantum dot coupling to the outside environment and the importance of many-body effects on the electronic properties of the quantum dot are also discussed. Finally we give examples of several quantum dot devices that exploit some of the quantum dot electronic and optical properties. Some of these quantum dot devices could lead to the implementation of quantum computing or quantum cryptography applications.

## 1 Introduction

The use of strain induced islands as a process for forming self assembled quantum dots (QDs) [1] in III–V semiconductors using epitaxy has been part of a large effort to exploit the novel quantum properties which arise from the three dimensional (3D) quantum confinement of carriers and excitons. The other methods for producing self assembled QDs are based on: a) the monolayer thickness fluctuations of a quantum well [2]; b) the electrostatically defined quantum dots in a two dimensional electron gas [3]; or c) the self-assembly of colloidal clusters using solution chemistry [4]. Each method has its strength and weaknesses and it is safe to say that presently, the epitaxial self-assembly method provides the best means of incorporating quantum dots into a wide variety of devices and exploit some of their unique properties [5,6,7,8]. It should also be mentioned that the carrier confinement provided by the self-assembled quantum dots is much larger than that provided by the electrostatically defined quantum dots or by the quantum well-quantum dots [2]. These self-assembled QDs have also provided a convenient "laboratory bench" for studies of carrier confinement and many-body effects in semiconductors. Because of the possibility of band gap engineering

P. Michler (Ed.): Single Quantum Dots, Topics Appl. Phys. **90**, 1–24 (2003)
© Springer-Verlag Berlin Heidelberg 2003

the self assembled QD structures, it has been possible to make a wide variety of solid state devices such as electrically pumped QD lasers [6,9] or infrared detectors [10] some of which are now near the industrial production stage.

The role of self assembled islands as an efficient strain relieving process was first identified in the investigation of SiGe/Si strained layers [11] and the growth of self assembled islands as a useful method to fabricate QDs [1] was also recognized later. The direct crystal growth of self-assembled QDs has been widely applied to a variety of strained layer semiconductor systems: several III–V compounds systems as well as II–VI and group IV heterostructures systems [5,6,7] and wide band gap III–V compounds [12] allow a range of wavelengths between 0.3 µm and 1.5 µm to be spanned.

The formation of QDs using epitaxy is based on the deposition of an epitaxial film on a lattice mismatched substrate. The elastic strain energy during the film deposition builds up as the square of the lattice strain. The total energy in the film, including strain energy, interfacial energy and surface energy, will be minimized during growth through the formation of coherently strained islands at the surface. These islands are transformed into QDs by embedding them into a larger band gap material.

This chapter covers the crystal growth processes involved in the QD formation using epitaxy as well as some of the newer methods for producing self-ordered QD lattices. We review some of the QD electronic properties relevant to the shell atom model which applies to the self-assembled QD electronic structure. We then use examples of QD devices which may be involved in the implementation of quantum computing and quantum cryptography. Throughout this contribution, we will take the InGaAs/GaAs QD system as a model system since it has been the most extensively studied.

## 2   Formation of Quantum Dots, Quantum Rings and Strain Effects on Island Nucleation

Thermodynamics and kinetics are both involved in the formation of self-assembled quantum dots. To discuss the nucleation and growth of the islands, we will take the InAs/GaAs system since this is the one which has been the most studied. For this system, the InAs is in compression to accommodate the strain associated with the smaller lattice constant of the GaAs. The initial stage in the QD formation is the formation of islands during deposition of an epitaxial layer with a small lattice mismatch (a few percent) on a crystalline substrate. As illustrated in Fig. 1 for the case of III–V compounds semiconductors deposited by molecular beam epitaxy, In and As atoms falling on a clean GaAs substrate held at high temperature will self assemble into smooth defect free epitaxial atomic layers if the lattice mismatch between the material deposited and the substrate is not too large. The diffusion length of some of the group III elements deposited by molecular beam epitaxy (MBE) are sufficiently large to insure the layer by layer growth until a build up in

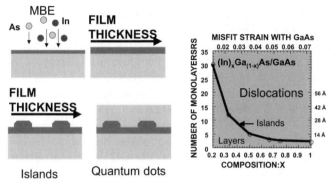

**Fig. 1.** Schematic of the island growth process illustrated for the deposition of InAs on GaAs. Capping of the islands will transform the islands into quantum dots. An increase in the film thickness beyond the island formation stage will produce island coarsening and the introduction of misfit dislocations. The *graph* shows the critical thickness for island formation measured experimentally as a function of the In content $x$ in the $In_x Ga_x As$ layer. The three fields in this plot distinguish the three growth regimes: layer-by-layer growth, island growth and layer growth with dislocations. The lower field region characterizes the thickness of the wetting layer connecting the quantum dots

the strain and surface energy of the epitaxial film switches the growth to the island growth mode. This change in the surface morphology is induced by the minimization of the total film energy. This interplay between the strain and surface energy of the film can be used, as we will see later, to control the island nucleation and promote self-ordering of islands.

Figure 1 also shows that increasing the film thickness, for a given composition of the $In_x Ga_{1-x} As$ alloy, beyond the island formation stage will introduce misfit dislocations. The formation of a thin wetting layer is inherent to the island formation process. However as indicated in Fig. 1, the wetting layer can become very thick (up to 30 monolayers) for low In content in the film. In this case, the wetting layer thickness approaches that of the QDs and will play an important role in the carrier confinement and relaxation processes and device characteristics since the active layer of the device becomes a corrugated quantum well or quantum well coupled to a QD layer.

The growth of self-assembled quantum dots is achieved by covering the smaller band gap material of the islands with a wider band gap epitaxial film. The island covering step will drastically change the island composition, shape and dimensions.

A listing of some of the general structural characteristics for this type of QD is informative as a background for designing QD devices and understanding some of their physical properties:

a) The islands and QDs are all deposited "in situ" in a reactor and are coherently strained and their growth remains essentially defect and impurity free. This view is supported by the very high internal quantum efficiency for optically pumped QDs.

b) The epitaxial QDs can easily be incorporated within an epitaxial structure thus making it possible to use band gap engineering techniques for electrical or optical injection and developing novel devices.

c) The QD size distribution is poorly controlled. Their size dispersion is in general $\approx 10\%$ and is a function of the substrate deposition temperature and surface growth kinetics. Although the initial nucleation of islands takes place preferentially at surface step edges, random nucleation events on terraces will eventually dominate as the QD density is increasing. Random island nucleation and coarsening kinetics during growth are controlling the size distribution.

d) The exact shape, dimensions and composition of QDs are not known accurately. Although the island shape and dimensions are well characterized using atomic force microscopy (AFM) [13,14], they will dramatically change during the capping by the wider band gap film. Surface exchange reactions and diffusion processes are playing an important role on the final shape, dimensions and composition of the QDs. Detailed studies of the QD composition in the InAs/GaAs system using X-ray diffuse scattering and AFM have shown the presence of an In concentration gradient within the QDs [15]. The existence of this gradient and its magnitude should strongly depend on the deposition conditions. In many instances the shape of the island is not a hemispherical cap and is more often a truncated pyramid. The elongated base of this pyramid reflects the crystallographic orientation dependence of the diffusion coefficient of the species during growth.

Several of these characteristics can affect the properties of quantum dot devices and a great deal of effort has been directed to address some of these problems.

## 3   Positioning and Ordering Quantum Dots

The absorption coefficient of QDs is small and for some device applications, a very high local QD density would be desirable. This would be the case for QD laser or far infrared (FIR) photodetector devices [9,10]. The optimization of coupled QD-microcavity devices will also rely heavily on the controlled positioning and ordering of QDs. In fact, controlling the island nucleation sites and density are difficult problems which involve both the thermodynamics and kinetics of growth. We discuss in this section two techniques which are yielding promising results.

The first method uses the directed migration of adatoms to a localized surface site that will promote preferential nucleation. This can be achieved by engineering the surface chemical potential [16,17,18] using a pre-patterned

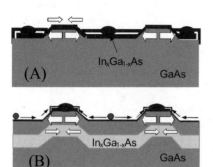

**Fig. 2.** Schematics of two possible configurations for controlled nucleation on a patterned substrate. (**a**) The strain distribution in the structure is illustrated by arrows and the InGaAs layer growth will preferentially occur in the valleys between the mesas. In (**b**) a coherently strained InGaAs stressor layer is used to enhance the strain on the mesa edge tops. The growth of islands will preferentially occur on the mesa tops where the strain build up is the largest

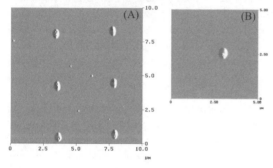

**Fig. 3.** AFM pictures of the GaAs {100} surface with surface depressions after a GaAs buffer layer regrowth and an InAs thin layer deposition. In (**a**) the In flux was adjusted to grow between on average 2 InAs islands per depression. The features observed between the depressions have been identified as surface defects and not QDs. In (**b**), the statistical distribution of In allows the formation of a single InAs island in some of the surface depressions [19]

substrate surface. A lowering of the surface chemical potential due to capillarity effects induces a driving force for adatom diffusion towards a concave surface site and causes island nucleation before the critical thickness for islanding is reached in other parts of the surface. Figure 2a illustrates a patterned surface and the preferential growth of the $In_x Ga_{1-x}As$ QDs taking place preferentially in the concave surface sites. This preferential accumulation of $In_x Ga_{1-x}As$ requires a fast In diffusion on the surface and is energetically favored [19]. The local stress build up induced by a thicker $In_x Ga_{1-x}As$ layer will trigger the InAs island nucleation first in these areas. An illustration of the technique is shown in the atomic force microscopy (AFM) pictures (Fig. 3). In this example, the periodicity between the surface depressions formed by e-beam lithography prior to the InAs deposition is 4 µm.

The InAs islands are found to form only in the surface depressions. The measured In surface diffusion using this technique is $\approx 15$ µm.

One of the drawbacks of this approach is associated with the formation of facets in the surface depression which takes place during the regrowth of a thin GaAs buffer layer following the pre-patterning step. The formation of an In rich wire like region at the junction of facets can take place prior to the formation of the QDs and the resulting coupling between the wire like region and the QDs may be problematic for some devices.

## 3.1  Island Nucleation and Quantum Dot Lattices

By controlling the island nucleation process to well defined areas, it is possible to minimize random nucleation and improve the island size uniformity. The interaction of strain fields between nucleating islands can promote nearly identical growth rates for all islands. This method was initially introduced to order islands on a mesa ridge where the diffusion kinetics favored a build up of strain on one edge of the mesa [16]. Using this method, strings of InAs islands were produced in a patterned GaAs surface and Monte Carlo simulations [17] support the view that enhanced island interactions will induce a significant narrowing of the island size distribution.

The thermodynamics and diffusion kinetics of the In atoms are changed during growth by introducing a local sub-surface strain field. As shown in Fig. 2b this can be achieved by growing a coherently strained film of InGaAs below the surface. In this case, the InAs film will grow more rapidly on the mesa tops and induce a preferential growth on InAs islands on top of the subsurface stressors [20].

In this method, a substrate is first patterned using optical holography and chemical etching on a GaAs surface to form a lattice of mesas. After desorption of the native oxide from the GaAs patterned surface, a thin GaAs buffer layer and an InGaAs stressor layer are deposited by molecular beam epitaxy (MBE) (see Fig. 2b). The InGaAs islands are then deposited after a thin GaAs layer is deposited on top of the InGaAs stressor layer. As shown in the AFM pictures (Fig. 4), the lattice orientation can be adjusted by the orientation of the mesa lattice on the pre-patterned substrate. The number of islands in the lattice basis is found to depend on the mesa shape and width and on the In flux.

A finite element calculation of the strain distribution in these structures indicates that the InAs islands nucleate at the surface sites where the stress is highest, i.e. the mesa edges and end points of the mesa ridges [20]. This effect is supported by experimental results shown in Fig. 5a. InAs islands lattice with a basis of one, two or three islands is deposited using a reduced In flux compared to that shown in Fig. 4. The strain distribution on the mesa surface controls the island nucleation sites. The number of islands per mesa will greatly depend on the shape of the mesas and on the In flux. In Fig. 5b, a double row of islands is observed on each mesa since the mesa top is larger than $\approx 60\,\text{nm}$.

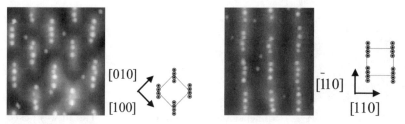

**Fig. 4.** Examples of ordered InAs island lattices [7]. The InAs islands are deposited on a GaAs patterned substrate and the schematic of the unit cell of these two lattices corresponds to each of the atomic force micrographs. The mesa lattice parameter is $\approx 250\,\mathrm{nm}$ [20]

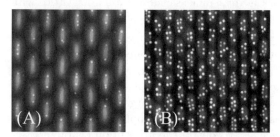

**Fig. 5.** (a) Atomic force micrograph of a lattice of mesas on which InAs islands were deposited using the strain engineered nucleation method. QDs appear as bright spots on top of the mesa. Note that the number of islands per mesa varies between one and three. The islands nucleate preferentially on the surface sites where the strain is maximum. (b) A double row of QDs is deposited on each mesa when the mesa top is larger. The mesa lattice parameter is 410 nm. Cross-section transmission electron microscopy of these devices reveals the existence of a thin wetting layer on the mesa top

Developing a processing that allows to reproducibly fabricate mesas with the desired dimensions and shapes is crucial to further progress in this field. The sub-surface stressor approach allows for the highest possible island density locally. Under optimal conditions and for a pyramidal shape mesa, one should be able to form lattices with one island per mesa.

Once a two-dimensional island lattice is formed on the surface, the growth of a three-dimensional island lattice using the strain coupling [21] between island layers is easily achieved. As shown in Fig. 6, the two-dimensional lattices of islands is replicated along the growth direction through the preferential nucleation of islands on top of each other along the growth direction if the interlayer spacing is smaller than $\approx 12\,\mathrm{nm}$.

The challenge for the future will be to reduce the size of the unit cell by developing a patterning process which is cheap and suitable for a rapid processing of large areas. Making this process compatible with the band gap engineering of the desired devices could also be required. This may be difficult

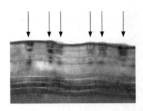

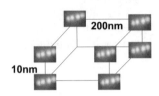

**Fig. 6.** Cross-section TEM through a three-dimensional quantum dot lattice. The InAs stacked quantum dots along the ⟨100⟩ growth direction are separated by a 10 nm thick GaAs layer. The quantum dots are detected through their strain fields. The *bottom* schematic is an idealized reconstruction of the three-dimensional quantum dot lattice with the dimensions of the unit cell indicated

in some instances since the InGaAs stressor layer behaves as a quantum well which may interfere with the intended functionality of the device.

The nucleation site engineering method may be a very useful approach for increasing the QD density and this may lead to large improvement in the gain characteristics of QD lasers [9]. The lateral coupling of QDs on mesa tops may also provide new avenues to control the absorption characteristics in QD infrared detectors [10].

## 4   Strain Effects and Growth Characteristics of Quantum Dots

Strain in the epitaxial layers controls not only the species surface diffusion and interdiffusion on the surface but also the kinetics of nucleation and growth of the islands. These effects are illustrated by experiments in which two electronically uncoupled QD layers have been deposited under identical or different growth conditions. These layers are close enough to each other to retain the strain alignment of the QDs from one layer to the other yet it is too large to allow electronic coupling of the energy levels between QDs of each layer. Their photoluminescence spectra (Fig. 7) provide a rapid method of analyzing their size and/or composition distribution. The distance (12 nm) between the two QD layers is sufficient to insure strain coupling and a stacking of the QDs in the 2 layers along the growth direction. This interlayer distance is too large to provide an electronic coupling between the stacked QDs. The growth conditions of the second layer are adjusted by using a partial coverage of the islands (PCI) with various amounts of GaAs prior to the final capping of the islands [22]. In Fig. 7 the PL spectra of island layers grown under identical conditions show two peaks indicative of differences in their size and/or composition.

According to the macro-PL spectra, the observed differences in the ground state energies (≈ 50 meV) can be reduced to zero by changing the growth

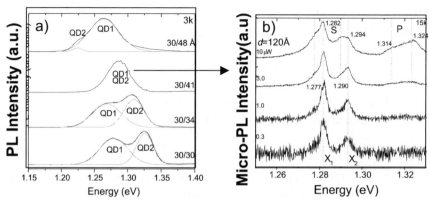

**Fig. 7.** (a) Photoluminescence spectra (excitation power $0.1\,\mu\text{W}$, $\lambda_{\text{exc}} = 633\,\text{nm}$, $T = 10\,\text{K}$) from samples with a high density of vertically stacked QDs ($d = 120\,\text{Å}$). The seed layer $QD_1$ in each sample has $30\,\text{Å}$ of GaAs partial capping, while the partial capping in the $2^{\text{nd}}$ $QD_2$ layer is varied as indicated on each spectrum. Gaussian curve fittings are applied to each curve to identify the two QD layer's ground state peaks. As the second $QD_2$ layer's PCI thickness is increased, the ground state energy of that QD decreases [23]. The numbers associated with each spectrum indicate the thickness (in $\text{Å}$) of the PCI layer for the first and second layer of QDs respectively. (b) Micro-photoluminescence spectra of a single QD pair from the sample indicated by the *arrow* in (a). The different micro-PL spectra correspond to different pump powers

conditions (PCI thickness of $\approx 4.1\,\text{nm}$, Fig. 7a) for the second QD layer [23]. Not surprisingly, the micro-PL spectra of this sample show that even with a single macro-PL peak, the energy difference between the ground state exciton luminescence for individual QD pairs is not zero and varies between 9 and $15\,\text{meV}$ depending on the QD pair. The energy separation for these lines cannot be associated with an electronic coupling between the QDs since the tunneling distance of $12\,\text{nm}$ is too large. Rather, a small change in the size or composition of the 2 QDs in a pair is responsible for the observed PL structure. From this type of experiment and studies of the QD luminescence in close proximity to a coherently strained quantum well, it has been shown that the nucleation and growth of QD layers in a stack of layers is strongly dependent on the presence of In floating or diffusing to the surface during growth [23].

We conclude this section by noting that much remains to be done to understand and control the nucleation, growth, positioning and ordering of self assembled quantum dots. QD systems other than the well-studied InGaAs/GaAs system should be explored. Surprisingly, little work has been done with QD materials systems with wider band gap and larger confinement energies. In QD systems such as AlInGaAs/AlGaAs [24] and GaN/AlGaN [12,25], the conduction and valence band offsets are larger than in

InGaAs/GaAs and the interband and inter-subband energy levels can be increased without further reducing the size of the QDs. A good control of these systems should permit the exploitation of QD devices which exhibit the full potential of the three dimensional confinement at room temperature.

# 5  Fundamental Electronic Properties of Self-Assembled Quantum Dots

The three-dimensional quantum confinement (3D) of carriers and excitons confers "atom like" properties to the quantum dots. However, this analogy should be carefully reconsidered since the contribution of the matrix (e.g., phonons) and the many-body effects should be taken into account. In this section we review some of the important physical QD properties.

## 5.1  The "Atom-Like" Shell Model for the Quantum Dots

One of the striking properties of the epitaxially deposited semiconductor QDs is their "atom like" electronic properties which originate from the three-dimensional carrier confinement inside the QDs. The expected discrete electronic energy level shell structure was first observed using capacitance (C-V) and infrared spectroscopy measurements on large ensemble of InAs QDs [26,27] embedded into a GaAs/AlGaAs MIS device structure. Figure 7 shows the schematics of the device and the associated conduction band diagram.

The capacitance versus voltage spectrum (Fig. 8) for a diode containing $\approx 10^6$ InAs QDs shows two series of peaks associated, respectively, with the s- and p-shell loading when electrons are tunneling from the back $n^+$ gate as a positive voltage is applied to the front one. The width of the peaks is due to the QD size dispersion and the large number of QDs ($\approx 10^6$) in the measured device. The Coulomb charging energy corresponding to loading of the second s electron peak into the s-shell is measured to be approximately 25 meV. The four peaks corresponding to the p-shell loading reflect the growth axis cylindrical symmetry of the QDs confining potential. The electron energy difference between the s- and p-shell levels is found to be $\Delta E_{e_{s-p}} \approx 50$ meV. A similar picture emerges for the hole shell structure of InAs QDs. However the energy level differences for s- and p-shell holes are smaller ($\Delta E_{h_{s-p}} \approx 10$ meV) than for electrons [27]. Modeling of the capacitance spectra based on the quantum tunneling of electrons from a back gate reservoir into the QDs ensemble has confirmed the C-V spectra interpretation [28]. Measurements and modeling of the intra-subband levels using FIR and capacitance spectroscopy have established that the confining potential is roughly parabolic [5,26]. For the InAs/GaAs QD system, the conduction and valence band offsets are $\approx 400–500$ meV and $\approx 70–100$ meV, respectively, and the number of electrons which can be confined to the QDs is controlled by their dimensions. Thus many-body effects due to exchange and Coulomb interactions are expected to dominate at low temperature.

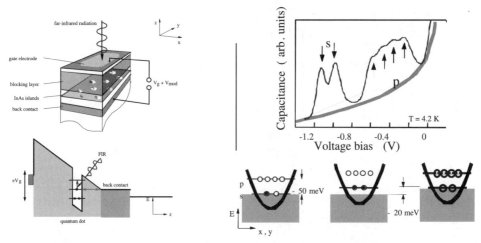

**Fig. 8.** Capacitance voltage spectra of an InAs QD ensemble ($10^6$ QDs) showing the loading of the $s$- and $p$-shell electrons. The electronic levels for the s and $p$-shell loading are indicated schematically with the measured values of the energy level differences. The schematic of the device structure and conduction band diagram along the growth direction are also shown [5]. In this example, up to 6 electrons can be confined to the QDs

## 5.2   Charged and Neutral Quantum Dots

Injection and trapping of charges can be done either through electrical [26] or optical injection [29,30] and provides the QDs with a wide range of physical properties which can then be used in practical QD devices. The associated many-body effects are important and readily observed using microprobe spectroscopy techniques that permit measurements of a single or a few QDs. Micro-photoluminescence (micro-PL) spectroscopy measurements have shown the formation of negatively or positively charged excitons or multi-exciton complexes [31,32,33,34]. We review briefly some of the micro-PL spectroscopy results for the InAs/GaAs QD system.

### 5.2.1   Photoluminescence Properties of Quantum Dots

Macro-PL spectra of QD ensembles as a function of increasing pump power reveal complex emission line structures which are difficult to interpret. This is mostly due to composition and size broadening effects and carrier exchange interactions and Coulomb interactions which can give rise to energetically closely spaced radiative recombination lines.

Micro-PL spectra [31,32] of single QDs on the other hand under low pump power conditions, can give rise to extremely narrow emission lines (FWHM ≈ 5 μeV) corresponding to a single $s$-shell exciton recombination [34].

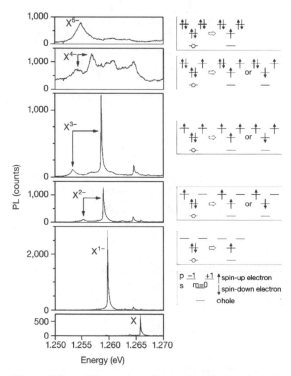

**Fig. 9.** Micro-PL spectra of a single QD as a function of the number of electrons in a single QD. The number of electrons is changed by applying a voltage bias while an exciton is generated in the QDs by resonant pumping of the wetting layer. The various charged states of the excitons are observed. The red shift of the luminescence associated with each additional exciton is accounted for by many-body effects and exchange interactions. Singlet and triplet states can be generated in the QDs after the exciton recombination [31]

These narrow spectral lines are consistent with the delta function density of states of the QDs.

To illustrate the importance of many-body interactions, we use as an example the micro-PL studies of QDs in which a controlled number of electrons is introduced using the device shown in Fig. 8. When excitons are generated by optical pumping at 850 nm with a solid state laser, a series of ultra sharp red shifted luminescence lines is observed as a function of the number of electrons present in the QD. Figure 9 shows the controlled formation of several excitonic charged states. The computed values of the red energy shift associated with each exciton charged state are accounted for theoretically when Coulomb and exchange interactions are taken into account [31].

The charged excitons $X^{2-}$, $X^{3-}$ and $X^{4-}$ give rise to bright exciton lines. This type of experiment allows the preparation of QDs with singlet or triplet electron spin states after exciton recombination and offers great opportunities for studies of spin coherence and spin dynamics in QDs [32].

Under non-resonant pumping and for increasing pump power, a multiplicity of narrow, red shifted lines is also observed. The 2X bi-exciton and charged exciton lines ($X^{1-}$, $X^{2-}$, $X^{3-}$) have been identified through a combination of both experiments and theory [31,32,33,34,35]. At the highest pump power blue shifted lines corresponding to the $p$-shell exciton recombination are observed. The charged exciton and bi-exciton energies have been computed and the differences in values reported by several group could very well be associated with the QD size and composition used in the experiments [36,37].

Studies of the recombination lifetimes for these various lines have clearly established the existence of cascaded relaxation processes associated with the recombination of the $p$-shell excitons followed by that of the $s$-shell exciton. The long recombination lifetimes measured ($\approx$ 1–5 ns) are associated with the efficient carrier confinement in the QDs. Intersubband electron relaxation times are in the 10–30 ps range.

Micro-PL measurements of the line broadening as a function of temperature show the important contribution of acoustic phonons on the relaxation processes of the QD excited states [38]. The presence of a superposition of QD excited states with a continuum of wetting layer states is responsible for the suppression of the expected phonon bottleneck in QDs [39]. At the highest temperature ($T > 100$ K), the luminescence intensity decays through thermal ionization of the carriers out of the QDs.

### 5.2.2 Upconversion Processes in Quantum Dots

Shell filling and exciton confinement give rise to efficient photon upconversion processes in the QDs [40,41]. As Fig. 10 indicates, upconversion of electrons and holes injected by optical pumping of a GaAs/InAs QD sample with below band gap photons ($E_{\mathrm{exc}} = 1.348$ eV) will generate higher energy excitons and emission of a GaAs ($D^0X$) at 1.428 eV. A study of the pump power dependence of this upconversion process supports a two-step photon absorption process which could be favored by the strong carrier localization in the quantum dots. Such processes could be important in the carrier relaxation and thermalization dynamics processes in QD lasers [42].

Efficient upconversion processes involving the recombination of QD bound states with states in the continuum of the wetting layer have also been observed in micro-PL, and micro-PL excitation experiments [41].

### 5.2.3 Excitons in Coupled Quantum Dots

The electronic coupling between two vertically stacked InAs QDs may offer a possible route for generating entangled states by forming a "QD molecule". However as already discussed in Sect. 3, the electronic coupling between two QDs using a simple growth approach cannot be achieved. The growth of two

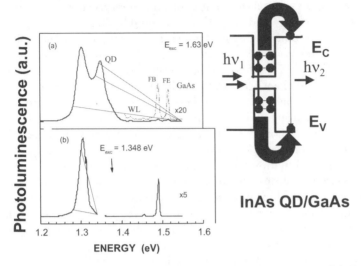

**Fig. 10. (a)** Photoluminescence spectra of a QD ensemble. **(b)** Excitation spectrum of a QD ensemble for an excitation energy of 1.348 eV. The schematic indicates a possible upconversion path corresponding to a two-step photon absorption process [40]

QDs with exactly the same size will be a very fortuitous event and a more practical approach to this problem is to imbed the QD pairs into a device structure which allows tuning of the QD electronic levels through an electric field [43]. This is achieved for example, by incorporating the QD pairs into an $n^+$-I-$n^+$ structure (Fig. 11a).

Figure 11b shows the micro-PL spectra of such a device. When the QDs forming a pair are grown under identical conditions, the voltage dependent micro-PL spectra of a single QD pair (labeled $QD_1$ and $QD_2$) with a spacing of 45 Å contain multiple peaks. These spectra are interpreted by modeling the carrier transport in the device as a function of the applied electric field [43]. Changes in this electric field allow transfer of an electron from the $QD_2$ "s" state to the adjacent $QD_1$ "f" state. From a comparison between the experimental spectra and the model (Fig. 11c), it follows that the main peaks in the extreme voltages spectra ($\pm 0.8$ V) are related to several configurations of neutral excitons in QD1: $1X_S$ (1.2599 eV), $2X_S$ (1.2577 eV), $3X_S$ (1.2484 eV, 1.2582 eV), $3X_P$ (1.291 eV). The spectrum at 0 V is dominated by configurations of negatively charged multi-excitons: $1X_S^-$ (1.2574 eV), $2X_S^-$ (1.2529 eV), $2X_P^-$ (1.2907 eV), $3X_S^-$ (1.2483 eV, 1.2503 eV, 1.2517 eV), $3X_P^-$ (1.2853 eV, 1.2928 eV). These micro-PL spectra were modeled by taking into consideration: a) a field dependent charge transfer with an acoustic phonon assisted tunneling and b) the many-body spectrum and exchange interactions in a quantum dot pair for different carrier configurations. The calculated spec-

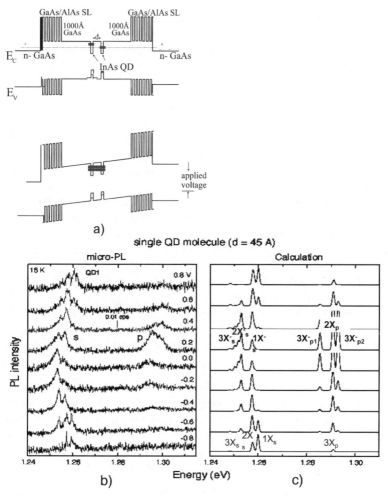

**Fig. 11.** (a) Device schematic. (b) Measured and computed (c) micro-PL spectra of QD pair separated by 45 Å as a function of the applied voltages. The computed spectra include a broadening parameter of 0.5 meV. The various charged states of the excitons are also indicated [43]

tra reproduce well the main experimental features when the field dependent carrier kinetics in the QD pairs and the many-body spectrum in the carrier population are simultaneously taken into account. However, in these experiments, there is no evidence of an electronic coupling or formation of a QD molecule. Similar experiments on QDs with nearly identical sizes, shapes and/or composition will have to be carried out to achieve electronic control of a QD molecule.

# 6    Quantum Dot Based Devices

Novel QD devices are making use of their unique physical properties and have been rapidly developing. Among the important QD characteristics which should impact the QD devices we find:

a) An ultra sharp density of states which confer to the QD "atom like properties". This analogy has been useful in applying some of the concepts of atomic quantum optics to the field of semiconductor QDs. Recent efforts to understand and control the spin dynamics and exciton dynamics [42,44,45,46,47] in QDs will certainly open up new avenues for producing and using entangled states for devices.
b) The three dimensional quantum confinement of carriers and excitons bringing into play important many-body effects and exchange interactions [35].
c) Carrier and exciton localization which reduce their interactions with the surrounding material and allow their manipulation using band structure-engineering techniques. Carrier localization in QDs makes possible their incorporation into extremely small devices since surface recombination effects will be greatly reduced [48].
d) The sequential character of the relaxation process for a QD filled with many excitons [44]. This characteristic allows the generation of single photons on command and will have a great impact in the field of quantum cryptography.

A wide variety of optoelectronic QD devices has already been investigated and spectacular progress has been made in the field of QD lasers [9] and QD infrared detector devices [10]. Since it is beyond the scope of this contribution to describe all the existing types of QD devices which have been made, we give two examples of devices which may find applications in the field of quantum computing and/or quantum cryptography. These are:

a) An exciton storage which uses carrier localization in QDs to write and read information using photons.
b) A quantum dot spin injection light emitting diode which shows the possibility of injecting spin polarized electrons or holes into QDs.

## 6.1    Information Storage using Quantum Dots

Trapping of carriers and excitons in a QD drastically reduces their recombination rate with ionized impurities or other defects in the surrounding material. This QD characteristic can be used for a memory device in which information is written and read by controlling charge storage in QDs. In the device example discussed here, light is used to write and read the information.

As shown schematically in Fig. 12, an exciton introduced in a narrow GaAs QW rapidly ($\approx 0.5\,\mathrm{ps}$) follows the electron relaxation to a lower energy X band minimum in the adjacent AlAs layer. The electron ends up in the

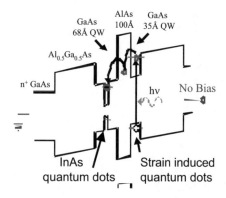

**Fig. 12.** The schematic band structure of the quantum dot memory device shows the dissociation process of the exciton during the write cycle of the quantum dot memory [8]

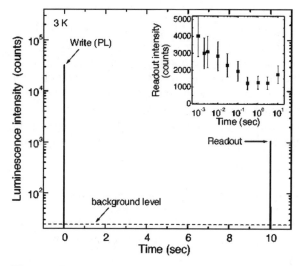

**Fig. 13.** Luminescence intensity from the peak emission line (1.25 eV) of the InAs QDs as a function of time. A 100 µs optical pulse excites the sample at $t = 0$, and a 10 µs bias pulse (3 V) is applied at $t = 10$ s. The *dashed line* represents the average background signal. The *inset* shows the readout integrated intensity as a function of delay time. The loss of signal as a function of storage time may be related to the presence of hole trapping centers in the AlAs layers [8]

adjacent InAs QD while the hole remains in the quantum well. The induced dipole produces an internal band bending (Fig. 12). The electron and hole are stored as a pair in the strain coupled, closely spaced, quantum dots. After storing, the exciton is reassembled by using an applied electric field to drive the hole into the quantum dot that contains the stored electron.

As shown in Fig. 13, very long storage times (well over 10 s) can be achieved with this device. These storage times are remarkably long when compared to the exciton lifetime in QDs ($\approx 1$ ns).

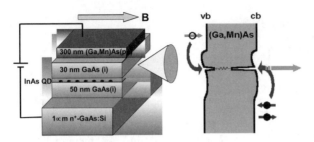

**Fig. 14.** Spin quantum dot light emitting diode layer structure and schematic band diagram. Polarized holes are injected from the GaMnAs layer into the QDs and the emitted circular polarization of the light is measured [53]

The carrier storage efficiency is preserved at temperatures up to $\approx 100\,\mathrm{K}$. Above this temperature, thermal ionization of carriers out of the QDs is occurring and destroys the light storing properties of the device. The recombination of the exciton is then read using standard photon detection techniques. The device uses the strain field associated with the InAs QDs to induce a GaAs QDs in a narrow GaAs quantum well adjacent to the InAs QDs [30]. However, a device with similar function could be engineered using a pair of strain coupled QDs of different materials or different sizes separated by a thin tunneling barrier.

## 6.2   A Spin Polarized Quantum Dot Light Emitting Diode

The recent demonstration of long spin coherence times in III–V compound semiconductors [49,50] and the possible use of spins for implementing quantum computing are important to the development of spin based optoelectronics. Quantum well (QW) based light-emitting diodes (LEDs) using magnetic semiconductors [51,52] have been used to demonstrate the possibility of coherent spin transfer across semiconductor heterojunctions. Polarized spin injection through the interface between the magnetic semiconductor and the QW will produce circularly polarized electroluminescence (EL).

Injecting polarized spins with high efficiency into QDs is attractive since the lateral quantum confinement of carriers broadens the selection rules for emitting circularly polarized light. Longer spin coherence times are expected since carriers are confined and interact less with their surrounding. Here we discuss a QD spin light emitting diode (SLED) which emits circularly polarized light with a magnetic field dependent polarization. The device is a hybrid structure composed of a GaMnAs ($p^+$) ferromagnetic semiconductor layer and GaAs (i)/InAs (i)/GaAs (n) semiconductor QD layers [53]. Under forward bias electrons and spin polarized holes are injected from the magnetized GaMnAs layer into the QDs while holes are injected from the GaAs (n) layer (Fig. 14).

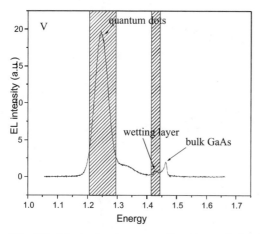

**Fig. 15.** Electroluminescence spectrum of the QD spin-LED at a forward bias of 1.7 V at 5 K. The quantum dot and emission wetting layer are indicated. The *shaded areas* correspond to the energy range over which the circular polarization is measured [53]

The electroluminescence (EL) circular polarization is analyzed as a function of the in-plane magnetization of the GaMnAs layer and temperature. The magnetic field is applied along the ⟨110⟩ easy magnetization axis and the emitted light is collected from the edge of the QDs SLED. Figure 15 shows the EL spectrum of the QDs at $T = 5$ K for a forward bias of 1.7 V.

The remanent circular polarization of the integrated EL at 5 K is shown in Fig. 16. The magnetic field dependence of the remanent circular polarization shows a clear hysteresis loop which coincides well with that of the GaMnAs measured using a SQUID magnetometer. On the other hand the polarization measurement of the photoluminescence (PL) of the same QD SLED does not show a measurable magnetic field dependence (Fig. 16). This result provides a clear indication that the injected spin polarized holes are responsible for the EL circular polarization.

The remanent circular polarization disappears above the Curie temperature ($T_c \approx 70$ K) and also gives a clear indication that the injection of spin polarized holes from the GaMnAs layer is responsible for the magnetic field dependence of the polarization. These experiments indicate that the amount of polarization of the EL is small (1–2%) and suggest that few spins make it to the QD layer without losing their polarization. Similar experiments carried out on quantum wells [52] also show a polarization dependence on the distance between the QW and the magnetic GaMnAs layer. A maximum polarization of 14% was reported in QWs. The small efficiency of the coherent spin transfer in this type of spin-LED is not yet understood.

Another promising avenue for injecting spins in QDs is the optical injection method. In a series of elegant experiments *Cortez* et al. [32] have used

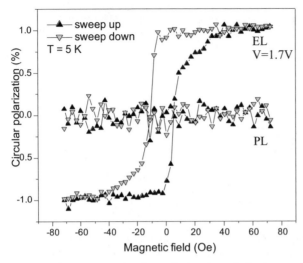

**Fig. 16.** Remanent circular polarization of the integrated EL for a QD SLED as a function of the in plane magnetic field at $T = 5\,\mathrm{K}$. The circular polarization of the photoluminescence (PL) for the same device is also shown as a function of magnetic field [53]

QDs which contain on the average a single electron. They demonstrate the writing of a spin in the QD by using circularly polarized light to spin polarize the resident electron in the QD. Reading of the spin as a function of a time delay with the polarization pulse is then carried out using a linearly polarized light pulse. Spin coherence times larger than 15 ns are observed at 5 K. These results are very promising and point out the potential applications of spin-polarized devices based on QDs to the field of spintronics and possibly quantum computation or quantum cryptography.

## 7    Conclusions

The growth of QDs is still an art and a science in progress. For the InAs/GaAs system discussed in this contribution, much remains to be done using crystal growth kinetics to achieve a better control of the nucleation, size, composition and shape of the QDs. This chapter presents one avenue for controlling the ordering and positioning of the QDs. However, there may be other fruitful ways to control nucleation and ordering by using surface reconstruction or vicinal surfaces. We believe that controlling QD positions and ordering will be important to the future application of QD devices. We have seen that the atomic "electron shell" model accounts well for the observed loading of electrons (or holes) in QDs but future use of this model will have to take into account the important coupling of the QDs to its crystal environment (acoustic phonons and wetting layer effects). The optical properties are dominated by many-body effects and the exchange interaction energies between carriers confined in the QDs. These will be important for understanding the exciton and spin dynamics in QDs. We have described here only a few devices which may have potential applications to the field of quantum computing and/or quantum cryptography. One of the remaining challenges to bring these devices into the realm of commercial products will be to develop a QD material system which minimizes or eliminates carrier thermal ionization out of the QDs at room temperature. A promising system is the GaN/AlN system where the band offsets for electrons and holes are much larger than for the InAs/GaAs system.

## Acknowledgements

The author wishes to thank the many students, post-doctoral researchers and colleagues who have contributed so much to this research. W. Schoenfeld, Y. Chee, B. Gerardot, J. Brown, I. Shtrichman, C. Metzner, G. Medeiros-Ribeiro, T. Lundstrom, J. Speck, D. Awschalom, A. Lorke, J. Kotthaus, K. Karrai and R. Warburton have all been active participants in this research which was supported through an ARO and a DARPA ONR grant #N 00014-99-1-1096 and an AFOSR grant #DAAD19-99-1-0372.

# References

1. D. Leonard, M. Krishnamurthy, C. M. Reaves, S. P. DenBaars, P. Petroff: Appl. Phys. Lett. **63**, 3203 (1993)
   J.-Y. Marzin et al.: Phys. Rev. Lett. **73**, 716 (1994)
   Q. Xie, A. Madhukar, P. Chen, N. P. Kobayashi: Phys. Rev. Lett. **75**, 2542 (1995)
2. G. Chen, T. H. Stievater, E. T. Batteh, Li Xiaoqin, D. G. Steel, D. Gammon, D. S. Katzer, D. Park, L. J. Sham: Phys. Rev. Lett. **88**, 117901 (2002)
3. W. Izumida, O. Sakai, S. Tarucha: Phys. Rev. Lett. **87**, 216803 (2001)
4. B. Lounis, H. A. Bechtel, D. Gerion, P. Alivisatos, W. E. Moerner: Chem. Phys. Lett. **329**, 399 (2000)
   L. J. Sham: Phys. Rev. Lett. **87**, 133603 (2001)
   S. A. Empedocles, R. Neuhauser, K. Shimizu, M. G. Bawendi: Adv. Mat. **11**, 1243 (1999)
5. P. M. Petroff, A. Lorke, A. Imamoglu: Phys. Today **54**, 5, 46 (2001)
6. D. Bimberg, M. Grundmann, N. N. Ledentsov: *Quantum Dot Heterostructures* (Wiley, Chichester 1998) and references therein
7. G. Abstreiter et al.: Jpn. J. Appl. Phys. **38**, 449 (1999)
   G. Yusa, H. Sakaki: Appl. Phys. Lett. **70**, 345 (1997)
8. T. Lundstrom, W. Schoenfeld, H. Lee, P. M. Petroff: Science **286**, 2312 (1999)
9. D. G. Deppe, D. L. Huffaker: Appl. Phys. Lett. **77**, 3325 (2000)
   P. Bhattacharya, S. Ghosh: Appl. Phys. Lett. **80**, 3482 (2002)
   N. N. Ledentsov, V. M. Ustinov, A. Yu Egorov, A. E. Zhukov, V. M. Maximov, I. G. Tabatadze, P. S. Kop'ev: Semicond. **28**, 832 (1994)

10. A. D. Stiff-Roberts, S. Chakrabarti, S. Pradhan, B. Kochman, P. Bhattacharya: Appl. Phys. Lett. **80**, 3265 (2002)
11. D. J. Eaglesham, M. Cerullo: Phys. Rev. Lett. **64**, 16, 1943 (1990)
12. J. Brown, C. Elsass, C. Poblenz, P. Petroff, J. Speck: Phys. Stat. Sol. **228**, 199 (2001)
    J. Simon, E. Martinez-Guerrero, C. Adelman, G. Mula, B. Daudin, G. Feuillet, H. Mariette, N. T. Pelekenos: Phys. Stat. Sol. **224**, 13 (2001)
    Y. Arakawa: IEEE J. Sel. Topics Quant. Electron. **8**, 823 (2002)
13. J. Marquez, L. Gedhaar, K. Jacobi: Appl. Phys. Lett. **78**, 2309 (2001)
14. D. M. Bruls, J. W. Vuyls, P. M. Koemrad, H. W. Salemink, J. H. Wolter, M. Hopkins, M. Skolnick, Fei Long, S. P. Gill: Appl. Phys. Lett. **81**, 1708 (2002)
    N. Liu, H. K. Lyeo, C. K. Shi, M. Oshiwa, T. Mano, N. Koguchi: Appl. Phys. Lett. **80**, 4345 (2002)
    K. J. Chao, C. K. Shi, D. W. Gotthold, B. G. Streetman: Phys. Rev. Lett **79**, 4822 (1997)
15. I. Kegel, T. H. Metzger, A. Lorke, J. Peisl, J. Stangl, G. Bauer, J. M. Garcia, P. M. Petroff: Phys. Rev. Lett. **85**, 1694 (2000)
    I. Kegel, T. H. Metzger, A. Lorke, J. Peisl, J. Stangl, G. Bauer, K. Nordlund, W. V. Schoenfeld, P. M. Petroff: Phys. Rev. B **63**, 0353181 (2001)
16. D. S. Mui, D. Leonard, L. A. Coldren, P. M. Petroff: Appl. Phys. Lett. **66**, 1620 (1995)
17. N. Tue, P. M. Petroff, H. Sakkaki, J. J. Merz: Phys. Rev. B **53**, 9618 (1996)
18. M. Ozdemir, A. Zangwill: J. Vac. Sci. Technol. A **10**, 684 (1992)
19. B. D. Gerardot, G. Subramanian, S. Minvielle, P. M. Petroff: J. Cryst. Growth **236**, 647 (2002)
20. H. Lee, J. A. Johnson, J. S. Speck, P. M. Petroff: J. Vac. Sci. Technol. B **18**, 2193 (2000)
    H. Lee, J. A. Johnson, J. S. Speck, P. M. Petroff: Appl. Phys. Lett. **78**, 105 (2001)
21. L. Goldstein, F. Glas, J. Marzin, M. Charasse, G. Le Roux: Appl. Phys. Lett. **47**, 1099 (1985)
22. J. M. Garcia, G. Medeiros-Ribeiro, K. Schmidt, T. Ngo, P. M. Petroff: Appl. Phys. Lett. **71**, 2014 (1997)
23. B. Gerardot, I. Shtrichman, D. Hebert, P. M. Petroff: J. Crystal Growth **236**, 4, 647 (2002)
24. R. Leon, P. M. Petroff, D. Leonard, S. Fafard: Science **267**, 5206 (1995)
25. N. Grandjean, N. Damilano, J. Massies: Proc. Int. Workshop Nitrides, Inst. Pure Appl. Phys., Tokyo, Japan (2000) p. 397
    J. Gleize, F. Demangeot, J. Frandon, M. A. Renucci, M. Kuball, B. Damilano, N. Grandjean, J. Massies: J. Appl. Phys. Lett. **79**, 686 (2001)
26. H. Drexler, D. Leonard, W. Hansen, J. H. Kotthaus, P. M. Petroff: Phys. Rev. Lett. **73**, 2252 (1994)
    R. Luyken, A. Lorke et al.: Appl. Phys. Lett. **74**, 2486 (1999)
27. G. Medeiros Ribeiro, D. Leonard, P. M. Petroff: Appl. Phys. Lett. **66**, 1767 (1995)
28. G. Medeiros-Ribeiro, F. G. Pikus, P. M. Petroff, A. L. Efros: Phys. Rev. B **55**, 1568 (1997)
29. K. F. Karlson, E. S. Moskalenko, P. O. Holtz, B. Moonemar: Appl. Phys. Lett. **78**, 2952 (2001)

30. W. V. Schoenfeld, C. Metzner, E. Letts, P. M. Petroff: Phys. Rev. B **63**, 205319 (2001)
31. R. J. Warburton, C. Schaflein, D. Haft, F. Bickel, A. Lorke, K. Karral, J. M. Garcia, W. Schoenfeld, P. M. Petroff: Nature **405**, 926 (2000)
32. S. Cortez, O. Krebs, S. Laurent, M. Senes, X. Marie, P. Voisin, R. Ferreira, G. Bastard, J. M. Gerard, T. Amand: Phys. Rev. Lett. **89**, 207401 (2002)
33. E. Dekel, D. Gershoni, E. Ehrenfreund, D. Spektor, J. M. Garcia, P. M. Petroff: Phys. Rev. Lett. **80**, 4991 (1998)
    L. Landin, M. S. Miller, M.-E. Pistol, C. E. Pryor, L. Samuelson: Science **280**, 262 (1998)
34. M. Bayer, P. Hawrylak, K. Hinzer, S. Fafard, M. Korkusinski, R. Wasilewski, O. Stern, A. Forchel: Science **291**, 451 (2001)
    M. Bayer, A. Forchel, P. Hawrylak, S. Fafard, G. Narvaez: Phys. Stat. Sol. B **224**, 331 (2001)
35. E. Dekel, D. Regelman, D. Gershoni, E. Ehrenfreund: Phys. Rev. B **62**, 11038 (2000)
    D. V. Regelman, E. Dekel, D. Gershoni, E. Ehrenfreund, A. J. Williamson, J. Shumway, A. Zunger, W. V. Schoenfeld, P. M. Petroff: Phys. Rev. B **64**, 16530 (2001)
36. G. W. Bryant: Phys. Rev. B **41**, 1243 (1989)
37. Ph. Lelong, G. Bastard: Solid State Commun. **98**, 819 (1996)
38. C. Kammerer, G. Cassabois, C. Voisin, C. De Lalande, P. Roussignol, A. Lemaitre, J. M. Gerard: Phys. Rev. B **65**, 33313 (2001)
39. H. Benisty, C. M. Satomayor, C. Weisbuch: Phys. Rev. B. **44**, 10945 (1991)
40. P. P. Paskov, P. O. Holtz, B. Monemar, J. M. Garcia, W. V. Schoenfeld, P. M. Petroff: Appl. Phys. Lett. **77**, 812 (2000)
41. C. Kammerer, G. Cassabois, C. Voisin, C. Delalande, Ph. Roussignol, J. M. Gerard: Phys. Rev. Lett. **87**, 207401 (2001)
42. P. Borri, W. Langbein, W. Schmeider, U. Woggon, R. Sellin, D. Ouyang, D. Bimberg: Phys. Rev. Lett. **87**, 15, 157401 (2001)
43. I. Shtrichman, C. Metzner, B. Gerardot, W. V. Schoenfeld, P. M. Petroff: Phys. Rev. B **65**, 081303(R) (2002)
44. D. V. Regelman, U. Mizrahi, D. Gershoni, E. Ehrenfreund, W. Schoenfeld, P. M. Petroff: Phys. Rev. Lett. **87**, 257401 (2001)
45. A. Zrenner, E. Beham, S. Stufler, F. Findeis, M. Bichler, G. Abstreiter: Nature **418**, 612 (2002)
46. A. Zrenner, F. Findeis, M. Baier, M. Bichler, G. Abstreiter, U. Hohenester, E. Molinari: Physica A **13**, 95 (2002)
47. R. J. Warburton, C. Schaflein, D. Haft, F. Bickel, A. Lorke, K. Karrai, J. M. Garcia, W. Schoenfeld, P. M. Petroff: Physica E **9**, 124 (2001)
48. A. Fiore, J. X. Chen, M. Illegems: Appl. Phys. Lett. **81**, 1752 (2002)
    J. K. Kim, T. A. Strand, R. L. Naone, L. A. Coldren: Conf. Proc. LEOS'98, IEEE Lasers and Electro-Optics Society 1998 Annual Meeting, Piscataway, NJ (IEEE Cat. No. 98CH36243, 1998) p. 111
49. D. Awschalom, J. Kikkawa: Physics Today **52**, 33 (1999)
    G. Schmidt. D. Hagele, M. Oestereich, W. Ruhle, N. Nestle, K. Eberl: Appl. Phys. Lett. **73**, 1580 (1998)
50. Y. Ohno, D. K. Young, B. Beschoten, F. Matsukura, H. Ohno, D. D. Awschalom: Nature **402**, 790 (1999)

51. D. K. Young, E. Johnston-Halperin, D. D. Awschalom, Y. Ohno, H. Ohno: Appl. Phys. Lett. **80**, 1598 (2002)
52. E. Johnston-Halperin, D. Lofgreen, R. K. Kawakami, D. K. Young, L. Coldren, A. C. Gossard, D. D. Awschalom: Phys. Rev. B **65**, 04130 (2002)
53. Y. Chye, M. White, E. Johnston-Halperin, B. D. Gerardot, D. D. Awschalom, P. M. Petroff: Phys. Rev. B **66**, 201302 (2002)

# Electronic Properties
# of Self-Assembled Quantum Dots

Pawel Hawrylak and Marek Korkusiński

Institute for Microstructural Sciences, National Research Council of Canada,
Ottawa, Ontario K1A 0R6, Canada
pawel.hawrylak@nrc.ca

**Abstract.** Theoretical investigation of electronic and optical properties of self-assembled quantum dots is reviewed. The single particle states of lens-shaped, disk-shaped, and vertically coupled dots are discussed. The spectrum of a single exciton, the exciton in the presence of many electrons, and many-exciton complexes are covered. The emission, absorption, far infrared spectroscopy, capacitance, photo-current spectroscopy and inelastic light scattering as a probe of the ground and excited states of quantum dots charged with electrons and excitons are discussed.

## 1  Introduction

We discuss here our theoretical work on self-assembled quantum dots (SADs), stimulated primarily by experimental programs at the Institute for Microstructural Sciences (IMS), National Research Council of Canada. The self-assembled dots grown at IMS are classified into three categories: lens-shaped dots, lens-shaped dots with height trimmed by the indium flush method, and vertical stacks of indium-flush dots. The next generation of structures, not discussed here, uses patterned substrates for positioning and alignment of individual dots. While many groups investigate quantum dots, the term "quantum dot" should only be applied to those structures where quantization of energy levels has been convincingly demonstrated. All the structures grown and studied at IMS show clear evidence of excited states, and their grouping into shells. The simplest way to demonstrate the excited states is in the interband emission spectrum under high excitation conditions. The spectrum shows up to five groups of peaks which split and restructure upon the application of a magnetic field. The actual spectrum, i.e., the number and spacing of peaks, depends on the lateral size and shape of the dot. This experimental evidence has been the cornerstone of our theoretical approach. We use simple one- and multi-band effective mass approaches to extract the dependence of single particle spectra on the shape and size of the quantum dot. We find convincing evidence that the lens-shaped quantum dots produce spectra which can be approximated by a spectrum of a pair of harmonic oscillators, with its characteristic spacing and degeneracies. Just as the plane waves were the single-particle states of choice in the jellium model of metals, we believe that the harmonic oscillator states are the single-particle states of

P. Michler (Ed.): Single Quantum Dots, Topics Appl. Phys. **90**, 25–91 (2003)

choice for self-assembled quantum dots filled with electrons and/or with excitons. Starting with the shell structure of 2D harmonic oscillators, we develop a theory of many particles, electrons and/or excitons, in self-assembled quantum dots. This theory covers a single exciton, many-excitons and a principle underlying their electronic structure, and Hund's rules and their impact on optical properties of dots charged with electrons. The goal here is the design of electronic and optical properties of many-particle systems. The design starts with single-particle levels but extends to many-particle states. This extension is crucial, as many nontrivial properties, from ferromagnetism, exciton condensation, to superconductivity are collective properties impossible to understand at the single-particle level. The effective mass approach misses on composition aspects and atomistic character of quantum dots in the same spirit as the jellium model does not differentiate sodium from aluminum except for density. Nevertheless, it is an interesting model in itself and serves as a guide to what is possible in the design of quantum systems. The model has been successfully confronted with experiment, and we refer the reader to relevant literature [1,2,3]. The chapter is intended to cover primarily our own work, and only a rudimentary attempt is made to cover very extensive literature on the subject.

## 2   Single-Particle States

This section reviews our understanding of single-particle levels, both of electrons and holes, in self-assembled quantum dots. The basic premise is that the dots are semiconductor structures and in semiconductors the concept of "electrons" in conduction band and "holes" in valence band as basic quasi-particles worked very well in the understanding of quantum wells and wires. We assume that when these quasi-particles are confined to a quantum dot by conduction/valence band offsets of the dot and barrier materials, they represent well the low-lying excited states, or quantized energy levels, of a quantum dot. The energy spectrum, with typical level spacing of tens of meV, much smaller then the gap of the dot material, is expected to be a sensitive function of the shape of a quantum dot, allowing for wave function engineering. In Fig. 1 we show a schematic energy spectrum which could be engineered by confining a particle with effective mass in different types of quasi-two-dimensional quantum structures: quantum disk, lens, ring, and a stack of quantum disks. In Fig. 2 we show transmission electron microscopy pictures of InAs quantum dots grown on GaAs at the Institute for Microstructural Sciences. These structures are designed to be in the form of lenses or disks, or stacks of them. In what follows we describe the calculation, in increasing degree of sophistication, of their energy spectra. These spectra turn out to be very well described by the schematic spectra of Fig. 1. In particular, as mentioned before, the spectrum of both electrons and holes in lens-shaped quantum dots, with its shell structure and degeneracies, turns out to be very

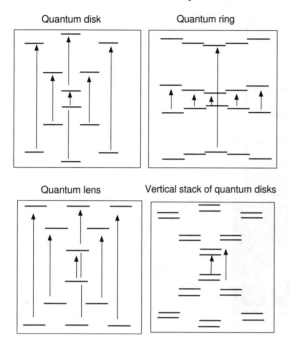

**Fig. 1.** Schematic energy levels of a quantum disk, lens, ring and a stack of disks

well approximated by that of two harmonic oscillators. We start with strain calculations to determine the details of the confining potential. We continue with the effective mass and $k \cdot p$ calculations which relate the shape to energy spectra. Since we might wonder why these approaches work so well, we end with a brief review of alternative tight-binding and "first principle" approaches.

## 2.1   Strain Effects

The strain in SADs is due to the mismatch between the lattice constants of the barrier and well materials. For example, the mismatch for InAs dots embedded in GaAs material is 6%, while for InAs in InP it is only $\sim 3\%$. Since the dot material has larger lattice constant, it is compressed, and upon partial relaxation, displaces the atoms of the surrounding barrier material from their equilibrium positions. Finding this displacement field, i.e., the vector field $d = r' - r$, with $r'$ $(r)$ being the displaced (unstrained) position of an atom, is the goal of the calculation. Once the displacements have been established, the strain tensor matrix elements are calculated as

$$\varepsilon_{ij} = \frac{1}{2} \left( \frac{\partial d_i}{\partial r_j} + \frac{\partial d_j}{\partial r_i} \right),$$   (1)

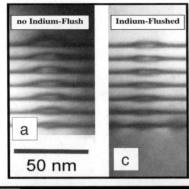

**Fig. 2.** TEM pictures of lens-shaped quantum dots (**a**),(**b**) and indium-flushed quantum disks (**c**) (courtesy of Dr. J. P. McCaffrey, IMS NRC Canada)

with $i$, $j = x$, $y$, $z$. These matrix elements enter the Bir–Pikus and the piezoelectric Hamiltonians.

The methods of calculation of the displacement field fall, in general, into two categories: the continuous elasticity models and the atomistic models.

The first group of methods is based on the assumption that the SAD and the barrier material are continuous, classical elastic media. Their elastic properties can be described either by a set of elastic constants or by the Poisson ratio and Young's modulus [4]. Thus these methods are natural extensions of classical theories of elasticity, based on Hooke's law. In this framework many detailed approaches have been developed, from the theory of inclusions [5,6] to finite element analysis [7].

Perhaps the most straightforward, although approximate, treatment of strain was put forward by *Downes* et al. [6] and *Davies* [8]. The key observation here is that the displacement field produced by an isotropic spherical inclusion of the dot material into the barrier material is identical to the electrostatic field produced by a spherical charge. If we now consider the quantum dot as a structure built out of infinitesimal spherical inclusions and assume the existence of a strain analogue of the superposition principle, we can formulate the isotropic displacement problem as a Poisson equation, in which the lattice mismatch plays the role of "charge density". To obtain the displacements we can now readily apply one of the widely available Poisson solvers. Here we shall only mention that in the case of disk-shaped dots with a natural choice of cylindrical coordinates we deal only with two nontrivial strain components: the in-plane strain along the radius, and the strain along the $z$ axis. Due to the isotropy the tangential strain component is identically zero, and the hydrostatic strain outside the dot vanishes in close proximity to the interfaces.

A more general method of calculation, not restricted to the isotropic case, requires a minimization of the total elastic energy of the system [4]:

$$
\begin{aligned}
E_{\text{elastic}} = \int d\boldsymbol{r} \frac{1}{2} C_{11}(\boldsymbol{r})(\varepsilon_{xx}^2 + \varepsilon_{yy}^2 + \varepsilon_{zz}^2) + C_{12}(\boldsymbol{r}) \left( \varepsilon_{xx}\varepsilon_{yy} + \varepsilon_{yy}\varepsilon_{zz} \right. \\
\left. + \varepsilon_{zz}\varepsilon_{xx} \right) + 2 C_{44}(\boldsymbol{r})(\varepsilon_{xy}^2 + \varepsilon_{yz}^2 + \varepsilon_{zx}^2) \\
- \alpha(\boldsymbol{r})(\varepsilon_{xx} + \varepsilon_{yy} + \varepsilon_{zz})\varepsilon_0 .
\end{aligned} \tag{2}
$$

The integration is carried out over the volume of the dot and the appropriately chosen part of the surrounding barrier. In the above formula, applicable for cubic crystals, the parameters $C_{ij}$, or the elastic moduli, are written explicitly as position-dependent, which permits us to model the elastic properties of the barrier and quantum dot materials, and also consider the appropriate type of interface – sharp or soft, depending on the system. The lattice mismatch is introduced by the parameter $\alpha$, whose value is $C_{11}+2C_{12}$ in the dot and zero in the barrier.

The implementation of this method [9,10,11,12] involves a discretization of the system's volume onto a grid of points and approximation of the strain tensor matrix elements in (1) by finite differences of displacements defined on these points. The elastic energy (2) is further minimized by finding the optimal field of displacements of each node.

As an example we present the values of strain tensor matrix elements in an InAs disk of radius 8 nm and height 2.5 nm embedded in the GaAs barrier material. The material parameters used in our calculations are similar to those used by *Pryor* et al. in [9]. Our results are presented in Fig. 3. We show the distribution of the strain tensor matrix elements $\varepsilon_{xx}$ and $\varepsilon_{zz}$ on the planes $XZ$ and $XY$ of the sample. Light-coloured areas exhibit negative strain (compression), and dark-coloured areas – positive strain (decompression). We see that the disk material is compressed along the $x$ direction ($\varepsilon_{xx}$ is negative), and decompressed in the growth direction ($\varepsilon_{zz}$ positive). Moreover, the strain effects extend far along the $z$ axis. This particular effect is used in layered growth of dots. This process involves a deposition and capping of one layer of dots, followed by the growth of another layer of dots on top of the existing structure. The propagating strain forces the dots of each subsequent layer to grow on top of the already existing buried dots.

The second group of strain calculations consists of atomistic models, which express the total elastic energy in the language of stretching and bending of interatomic bonds. The actual formula for the total elastic energy is usually taken in the following form [10]:

$$
E_{\text{elastic}} = \sum_{i \neq j} V_2(\boldsymbol{r}_i - \boldsymbol{r}_j) + \sum_{i \neq j \neq k} V_3 \left[ (\boldsymbol{r}_j - \boldsymbol{r}_i), (\boldsymbol{r}_k - \boldsymbol{r}_i), \theta_{ijk} \right], \tag{3}
$$

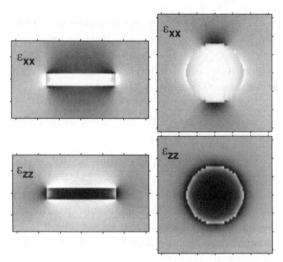

**Fig. 3.** Distribution of the strain tensor matrix elements $\varepsilon_{xx}$ and $\varepsilon_{zz}$ on the $XZ$ (*two left-hand panels*) and $XY$ (*two right-hand panels*) plane of the sample. The light (dark) colour denotes areas with compressive (tensile) strain

where $V_2$ is a two-body term describing the bond stretching, and $V_3$ is a three-body term describing the bond bending ($\theta_{ijk}$ is the angle created by atoms $i$, $j$ and $k$, subtended at the atom $i$). The summation extends over all atoms involved. Similarly to the discretization method in the continuous elasticity model, the above formula is calculated for a suitably chosen supercell, containing the quantum dot surrounded by a sufficient amount of barrier material. The choice of the potentials $V_2$ and $V_3$ depends on the degree of detail required. For strain calculations in low temperatures *Keating* for the diamond structure [13] and *Martin* for the zinc-blende crystals [14] assume the two-body and three-body terms as quartic functions of bond and angle distortions. This simple approach does not satisfactorily account for anharmonicity of the interatomic potentials [15], but is convenient in static calculations of the strain distribution. Its modified version has been successfully used to predict phonon spectra of diamond and zinc-blende semiconductors [16].

In order to describe such phenomena as melting of the crystal more realistic approximations are needed. *Stillinger* and *Weber* [17] in their study of melting of silicon crystals use a combination of power-law and exponential terms in $V_2$ and $V_3$ taken to allow for bond breaking. This pair of potentials is parametrized by seven material constants obtained by fitting of the results of the simulation to the experimental data – either general properties of bulk crystals, such as cohesive energy and density [18], or the lattice sums, melting point and structure of the liquid [17]. Application of the Stillinger–Weber potentials to Si/Ge quantum dots has also been reported [19].

*Tersoff* [20] proposed another pair of interatomic elastic potentials, of the form of Morse potentials, but with the local bond strength parameter. This model, similarly to the Stillinger–Weber approach, is also parametrized by seven empirical constants. It has been recently used to simulate the strain relaxation in InGaAs/GaAs quantum dots with nonuniform composition and illuminated such details of dot growth as variation of the indium distribution within the pyramidal island [21]. In this study the total number of atoms was of order $5 \times 10^5$.

As was mentioned before, these approaches may be used in two ways. The first implementation would be essentially equivalent to that of the discretized continuous elasticity: one minimizes the total elastic energy to find the optimal displacement field. The second implementation, consisting of a molecular dynamics study, allows one to compute the phonon spectra of any device small enough to be tractable.

### 2.1.1 Strain-Induced Corrections to the Dot Potential

As the unit cell of the crystal is deformed by strain, the bond lengths and angles change. This in turn influences the band structure. This is described within the $\mathbf{k} \cdot \mathbf{p}$ formalism by *Bir* and *Pikus* [22] for the diamond-type lattices. *Bahder* [23] extended their approach to the zinc-blende binary semiconductor alloys using the Löwdin perturbation theory and obtained an eight-band $\mathbf{k} \cdot \mathbf{p}$ Hamiltonian. This Hamiltonian can be used to calculate the local band edge profiles at each point of the grid used previously in the strain calculations [9]. If one is only interested in the band structure at the $\Gamma$ point of the Brillouin zone, the conduction band is not coupled to the valence band, and the strain-induced modification of the conduction band edge consists only of a shift proportional to the hydrostatic strain. The valence band, on the other hand, is much more complicated. All heavy and light hole band edges are shifted by a factor proportional to the hydrostatic strain. Further, the strain introduces a splitting between the heavy and light hole band edges proportional to the biaxial strain. Interestingly, the sign of the biaxial strain can change on the dot-barrier interface, which causes the light-hole and heavy-hole band edges to cross. This behaviour appears to take place for our model InAs/GaAs quantum disk at the top and bottom bases of the dot. The results of our calculations are shown in Fig. 4. Here we used the strain tensor matrix element distributions shown in Fig. 3. In general the qualitative behaviour of the band edges can be easily predicted even by inspection of these strain profiles. The strain tensor matrix elements $\varepsilon_{xx}$ and $\varepsilon_{zz}$ do not change their sign as we move through the interface along the disk radius, but they do change if we move along the $z$ axis. As a result, in the radial cross-section the heavy-hole band is always lower in energy than the light-hole band. On the other hand, at the top and bottom base of the disk we observe a band crossing.

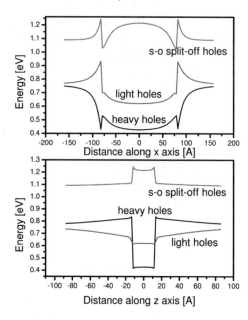

**Fig. 4.** Local band structure of the valence band in the disk-shaped quantum dot along the disk diameter (*top panel*) and along the rotational axis of the system (*bottom panel*). Note the reversal of the heavy- and light-hole band alignment at the interface at the top and bottom of the quantum dot

**Piezoelectric Potential.** The second strain-induced correction to the potential profile of the quantum dot is the piezoelectric potential. Its calculation starts with consideration of the piezoelectric charge density [15], which is induced by the shear strains. The piezoelectric potential can now be calculated as the solution of the Poisson equation, and turns out to be small, but longrange. Therefore care must be taken to solve this Poisson equation in a space larger than the computational supercell assumed for the strain calculation, and with Dirichlet boundary conditions [15]. In the InAs/GaAs systems the piezoelectric modulus is small and therefore the piezoelectric potential contributes little to the overall energy landscape of the system. Their importance lies in the fact that they lower the symmetry of the system – for example, for pyramidal dots the symmetry is lowered from $C_4$ to $C_2$ [9].

## 2.2    Effective Mass Adiabatic Models

This section describes the model calculation of the energy spectrum of a particle with effective mass confined to a quantum dot with disk and lens shape. The effective mass is treated as a phenomenological parameter.

### 2.2.1    Disk-Shaped Dots

We consider a disk-shaped quantum dot with height $H$ and radius $R$, positioned on a wetting layer of width $W$, as described in [12]. The system is shown schematically in Fig. 5 (bottom part). The dot height above the wetting layer is $H - W$. This system is surrounded by the barrier material. The

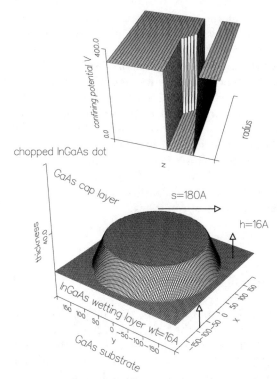

**Fig. 5.** *Bottom part*: A schematic picture of the quantum disk positioned on the wetting layer and capped with the barrier material. Numbers represent typical values of diameter and height characteristic for the InAs/GaAs structure. *Top part*: the effective vertical potential obtained from the analysis of vertical motion in the adiabatic approximation

conduction band offset between the barrier and dot materials is $V_0$. The material parameters enter through the effective Rydberg $\mathcal{R} = m_e e^4 / 2\epsilon^2 \hbar^2$ and the effective Bohr radius $a_{\mathrm{B}} = \epsilon \hbar^2 / m_e e^2$. The Hamiltonian of the disk-shaped quantum dot in polar coordinates is

$$\hat{H} = -\frac{1}{r^2} \left( r \frac{\partial}{\partial r} r \frac{\partial}{\partial r} + \frac{\partial^2}{\partial \theta^2} \right) - \frac{\partial^2}{\partial z^2} + V(r, z), \qquad (4)$$

where the potential $V(r, z) = -V_0$ inside the dot and the wetting layer, and $V = 0$ in the barrier. Since the dot is disk-shaped (the potential $V$ does not depend upon angle $\theta$), the angular motion can be readily decoupled, and the electron's angular momentum is a good quantum number.

Since the height of the dot is much smaller than its radius, the electron motion in the growth direction is strongly confined. Semiclassically, one could consider an electron being reflected off the bases of the disk with high frequency, whereas its planar motion is slow. This gives rise to the adiabatic approximation, where we assume that these two motions (planar and vertical) can be separated. Therefore we can write the approximate wave function of the electron as $\psi(r, \theta, z) = (1/\sqrt{2\pi})e^{im\theta} g_r^\nu(z) f_m^\nu(r)$. The exponential term in the above function describes the angular motion of the electron, with $m$ being the angular momentum. The function $g_r^\nu(z)$ describes the vertical motion. It is labelled by index $r$ because the potential in the $z$ direction seen by the electron depends upon the radial coordinate: it is different inside the dot, when $r < R$ (a wide quantum well), and different outside the dot (a narrow quantum well – the wetting layer). The solution of motion in the vertical direction will yield ground and excited states. These states are labelled by the subband index $\nu$.

For each angular momentum channel $m$ and each subband index $\nu$ the functions $f$ and $g$ satisfy the following set of equations:

$$\left[-\frac{\partial^2}{\partial z^2} + V(r, z)\right] g_r^\nu(z) = E_\nu(r) g_r^\nu(z), \tag{5}$$

$$\left[-\frac{1}{r^2}\left(r\frac{\partial}{\partial r} r \frac{\partial}{\partial r} - m^2\right) + E_\nu(r)\right] f_m^\nu(r) = E f_m^\nu(r). \tag{6}$$

Since the disk height is much smaller than its radius, the energy separation of subbands $\nu = 0$ and $\nu = 1$ is larger than the typical energy separation of the radial states. Therefore we shall confine our considerations to the lowest subband $\nu = 0$, whose energy $E_0(r)$ will be established by solving the first equation. $E_0(r)$ will be further substituted into the second equation as an effective potential for the radial motion.

As was mentioned before, for the vertical motion we distinguish two regions in our system. For $r < R$ we have a wide quantum well of width $H$ (the quantum dot on the wetting layer), and for $r > R$ we deal with a narrow quantum well with width $W$ (the wetting layer alone). To find the ground state energy $E_0(r)$ we employ the transfer matrix formalism. For a given energy $E$ we define the function $g_r^\nu(r)$ in three regions: (i) barrier below the structure, $z < z_1$, where $z_1$ is the position of the barrier-dot interface, $\kappa_1 = \sqrt{-E}$, $g_r^\nu(z) = A_1 \exp[\kappa_1(z - z_1)] + B_1 \exp[-\kappa_1(z - z_1)]$; (ii) the quantum well, $z_1 < z < z_2$, $k_2 = \sqrt{V_0 + E}$, $g_r^\nu(r) = A_2 \exp(-ik_2 z) + B_2 \exp(ik_2 z)$; (iii) barrier above the structure, $z > z_2$, $\kappa_3 = \sqrt{-E}$, $g_r^\nu(r) = A_3 \exp[-\kappa_3(z - z_2)] + B_3 \exp[\kappa_3(z - z_2)]$. The wave function $g_r^\nu(r)$ must be smooth and continuous at each interface. These conditions allow us to write the coefficients $A_1$, $B_1$ in terms of $A_3$, $B_3$:

$$A_1 = T_{11}(E)A_3 + T_{12}(E)B_3,$$
$$B_1 = T_{21}(E)A_3 + T_{22}(E)B_3,$$

with the transfer matrix defined as

$$T = L^{-1}(\kappa_1, z_1)U(k_2, z_1)U^{-1}(k_2, z_2)R(\kappa_3, z_2). \tag{7}$$

By virtue of the fact that $\kappa_1 = \kappa_3$ and setting $z_1 = 0$, the matrices $L$, $U$ and $R$ are defined as:

$$
L = \begin{bmatrix} 1 & 1 \\ \kappa & -\kappa \end{bmatrix}; \qquad
U = \begin{bmatrix} e^{-ikz} & e^{ikz} \\ -ike^{-ikz} & ike^{ikz} \end{bmatrix}; \qquad
R = \begin{bmatrix} 1 & 1 \\ -\kappa & \kappa \end{bmatrix}. \tag{8}
$$

It is convenient to carry out the multiplication $W = UU^{-1}$ to arrive at a modular expression for the transfer matrix:

$$
W = \begin{bmatrix} \cos(kh) & -\sin(kh)/k \\ k\sin(kh) & \cos(kh) \end{bmatrix}, \tag{9}
$$

where $h$ is the width of the quantum well. Now the transfer matrix can be written simply as $T = L^{-1}WR$. We find the allowed values of energy by imposing the boundary conditions $B_1 = 0$ and $B_3 = 0$ (i.e., absence of exponentially growing terms in the wave function inside the barrier). Since $B_1 = T_{21}(E)A_3 + T_{22}(E)B_3$, if $B_3 = 0$ we must have $T_{21}(E) = 0$. Hence the solution of the Schrödinger equation for the vertical motion has been reduced to finding zeros of the transfer matrix element $T_{21}$.

With this procedure we find the ground state energy $E_0$ of an electron in both radial regions $r < R$ and $r > R$. We obtain thereby the effective potential $E_0(r)$ for the radial motion, visualized in the top part of Fig. 5. This potential is constant in each radial region, with a step at $r = R$. To solve for the electronic energies $E$ in the radial direction we will also use the transfer matrix method. However, because of the cylindrical symmetry of the system we will build the wave function in terms of the Bessel functions. Inside the dot, i.e., for $r < R$, we have a propagating solution, so we set $k_1 = \sqrt{E - E_0^1}$ and form a function $f_m^0(r) = A_m^1 J_m(k_1 r) + B_m^1 Y_m(k_1 r)$. Outside the dot, i.e., for $r > R$, the solution is decaying, so we set $\kappa_2 = \sqrt{E_0^2 - E}$ and take $f_m^0(r) = A_m^2 K_m(\kappa_2 r) + B_m^2 I_m(\kappa_2 r)$. At the interface we match the wave function and its derivative to obtain the transfer matrix $T$. This matrix can now be written as $T = C^{-1}(R_1)D(R_1)$, where

$$
C(R_1) = \begin{bmatrix} J_m(k_1 R_1) & Y_m(k_1 R_1) \\ k_1 J_m'(k_1 R_1) & k_1 Y_m'(k_1 R_1) \end{bmatrix}, \tag{10}
$$

$$
D(R_1) = \begin{bmatrix} K_m(\kappa_2 R_1) & I_m(\kappa_2 R_1) \\ \kappa_2 K_m'(\kappa_2 R_1) & \kappa_2 I_m'(\kappa_2 R_1) \end{bmatrix}. \tag{11}
$$

The zeros of the transfer matrix element $T_{21}$ give the eigenenergies of the entire system. Results of the adiabatic calculation for a disk as a function of the disk radius $R$ are shown in Fig. 6. As can be seen, for the smallest radii $R \sim 50\,\text{Å}$ we only have one confined shell (the $s$-shell). Upon the increase of the disk radius first the doubly degenerate $p$-shell states (angular momenta $\pm 1$), and then the $d$-shell states emerge from the wetting layer continuum. Note the characteristic splitting of the $d$-shell. It consists of two low-energy states with angular momenta $\pm 2$ and one high-energy state with angular momentum 0.

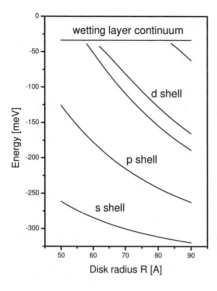

**Fig. 6.** Single-particle energies in the disk confinement for the confinement $V_0 = 600\,\text{meV}$ and the electronic effective mass $m^* = 0.053m_0$ as a function of the disk radius for typical radii of InAs/GaAs dots

### 2.2.2  Lens-Shaped Dots

The transfer-matrix approach described above has been applied to lens-shaped quantum dots [24], depicted schematically in Fig. 7. These dots are typically flat at the bottom, but their top surface has a shape of a spherical cap. The typical height-to-radius ratio is 0.25. In the transfer-matrix approach this surface is approximated as a series of concentric step-like rings, representing a shallow potential near the perimeter of the dot, and descending deeper as we move towards the dot centre. The consideration of the vertical motion consists of solving the Schrödinger equation in each ring separately. This yields the electronic ground states which constitute the effective radial potential $E_0(r)$, depicted in the top panel of Fig. 7. This potential has, of course, also a step-like structure, which enables us to use the transfer-matrix formalism in the radial problem too. However, in this case the formalism is more complicated. To explain it let us consider a simplest case of the potential being composed of two steps. In the innermost region, i.e., for $r < R_1$, the effective potential is constant and equal to $E_0^1$, in the intermediate region, i.e., for $R_1 < r < R_2$, the potential is $E_0^2 > E_0^1$, and finally in the outermost region, $r > R_2$, the potential is $E_0^3 > E_0^2$, which is the ground state energy in the wetting layer. In the first, innermost region we can only have propagating solutions, formulated using the Bessel functions $J_m$ and $Y_m$, as in the case of the disk. In the third, outermost region the bound state wave function must decay, so its wave function contains Bessel functions $K_m$ and $I_m$. However, in the intermediate region both propagating and decaying solutions are possible, and the wave function has to be defined separately in each case. Thus, for decaying solutions ($E_0^1 < E < E_0^2$) we define $\kappa_2 = \sqrt{E_0^2 - E}$ and take $f_m^2(r) = A_2 K_m(\kappa_2 r) + B_2 I_m(\kappa_2 r)$. For the

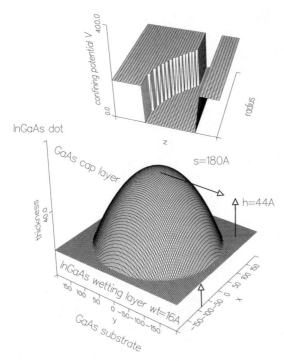

**Fig. 7.** *Bottom part*: A schematic picture of the lens-shaped dot. The values of height and radius correspond to the InAs/GaAs structure. *Top part*: The effective vertical potential obtained in the adiabatic approximation

propagating solution ($E_0^2 < E < E_0^3$) we define $k_2 = \sqrt{E - E_0^2}$ and take $f_m^2(r) = A_2 J_m(k_2 r) + B_2 Y_m(k_2 r)$. Now the transfer matrix can be written as $T = C^{-1}(R_1)S(R_1)S^{-1}(R_2)D(R_2)$, where the matrices $C$ and $D$ are defined analogously to (10), (11), and the matrix $S$ has to be defined separately for the decaying and propagating solutions:

$$S^{\mathrm{dec}}(R) = \begin{bmatrix} K_m(\kappa_2 R) & I_m(\kappa_2 R) \\ \kappa_2 K_m'(\kappa_2 R) & \kappa_2 I_m'(\kappa_2 R) \end{bmatrix}, \tag{12}$$

$$S^{\mathrm{prop}}(R) = \begin{bmatrix} J_m(k_2 R) & Y_m(k_2 R) \\ k_2 J_m'(k_2 R) & k_2 Y_m'(k_2 R) \end{bmatrix}. \tag{13}$$

Again, in order to establish the eigenenergies of the system, we look for the zeros of the transfer matrix element $T_{21}$. Results of this procedure are shown in Fig. 8. Similarly to the disk, in the case of the lens-shaped dot of radius small enough only the $s$-shell is bound. Upon the increase of the dot radius first the $p$, then the $d$-shell become bound. However here, unlike in the case of the disk-shaped dot, the $d$-shell consists of three almost degenerate states with angular momenta $0, \pm 2$, respectively. This degeneracy is a result of dynamical symmetry associated with the parabolic form of confinement. The parabolic

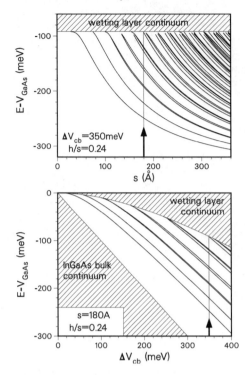

**Fig. 8.** Single-electron eigenenergies of an electron confined in a lens-shaped dot as a function of the dot radius $s$ (*top panel*) and the barrier-well conduction band offset (*bottom panel*). The ratio of height $h$ to radius $s$ is kept constant. (From [24])

confinement translates into a Hamiltonian of two coupled oscillators [25] with characteristic shell structure shown in Fig. 8.

## 2.3  $k \cdot p$ Models

The 8-band $k \cdot p$ models are most commonly used in calculating the energy spectra of InAs quantum dots [9,11,15,26]. The eight bands include two conduction bands (spin up and down), two heavy hole, two light hole, and two spin-split off bands. The 8-band model captures correctly the small wave vector (long-wavelength) behaviour of electron states in InAs based materials. The coupling of valence and conduction bands is needed due to the small bandgap of InAs. The parameters which enter the model are all extracted from the bulk and miss the atomistic character of interfaces between the dot and barrier material. Still, this model avoids the difficulty of defining the effective mass, and is the first step toward the understanding of the valence band. The model is too complicated for an approach other than numerical. The purpose here is to understand the confinement of holes, and this can be accomplished by considering the simplest 4-band model [27]. At the end we show the results of 8-band calculations for lens shaped quantum dots. These results support the notion of a harmonic oscillator spectrum for electrons, but reveal a somewhat more complicated picture for holes.

We consider a single hole on a quantum disk with radius $R$ and height $w$ in an external magnetic field applied in parallel to the rotational axis of the system (the $z$ axis). The Luttinger–Kohn $4 \times 4$ Hamiltonian in cylindrical coordinates reads

$$H_{\mathrm{L}} = \begin{pmatrix} P_+ & R & -S & 0 \\ R^* & P_- & 0 & S \\ -S^* & 0 & P_- & R \\ 0 & S^* & R^* & P_+ \end{pmatrix}, \tag{14}$$

where the basis of the above Hamiltonian consists of functions with total angular momentum $J = 3/2$ and $J_z = 3/2, -1/2, 1/2, -3/2$, respectively. The matrix elements of the Hamiltonian are

$$P_+ = \frac{\hbar^2}{2m_0} \left[ (\gamma_1 - 2\gamma_2)k_z^2 + (\gamma_1 + \gamma_2)k_\rho^2 \right], \tag{15}$$

$$P_- = \frac{\hbar^2}{2m_0} \left[ (\gamma_1 + 2\gamma_2)k_z^2 + (\gamma_1 - \gamma_2)k_\rho^2 \right], \tag{16}$$

$$R = \frac{\hbar^2}{2m_0} \left( -\sqrt{3} \right) \gamma_{23} k_-^2, \tag{17}$$

$$S = \frac{\hbar^2}{2m_0} \left( 2\sqrt{3} \right) \gamma_3 k_- k_z, \tag{18}$$

where $\boldsymbol{k} = -i\nabla - (e/c\hbar)\boldsymbol{A}$ $(e > 0)$, $k_- = k_x - ik_y$ and $k_\rho^2 = k_x^2 + k_y^2$. We use the axial approximation: $\gamma_{23} = (\gamma_2 + \gamma_3)/2$. The vector magnetic potential is taken in the symmetric gauge $\boldsymbol{A} = B/2(-y, x, 0)$.

The disk confinement is introduced as a potential $V(\rho, z)$ so that the total Hamiltonian is

$$H = H_{\mathrm{L}} + V(\rho, z). \tag{19}$$

However, in order to investigate the importance of the heavy hole–light hole mixing, we assume a simple confinement, i.e., a disk with infinite walls ($V(\rho, z) = 0$ inside, and infinity outside the disk), yielding simple wave functions.

In the absence of the mixing the wave functions for holes can be written as

$$\langle \boldsymbol{r} | m, n, \nu, m_j \rangle = \frac{\sqrt{2}}{R|J_{m+1}(k_n^m R)|} J_m(k_n^m \rho) \frac{1}{\sqrt{2}} \mathrm{e}^{im\phi} \xi^\nu(z) u_{m_j}(\boldsymbol{r}). \tag{20}$$

The above wave function consists of two parts: the periodic part of the Bloch function $u_{m_j}(\boldsymbol{r})$, for which $m_j = \pm 3/2$ for heavy holes and $m_j = \pm 1/2$ for light holes, and the envelope function. As the system possesses rotational symmetry, the angular momentum $m$ is a good quantum number, and the angular part of the envelope function is a simple exponent. The radial part of the envelope function is the Bessel function $J_m$ for the given angular momentum channel $m$, with $k_n^m$ being the hole wave vector defined in terms of the roots $\alpha_n^m$ of the Bessel function: $k_n^m = \alpha_n^m/R$, with $n$ being the radial quantum number. Finally the vertical part of the wave function $\xi^\nu(z)$ for a hard-wall disk can be expressed as $\xi^0(z) = \sqrt{2/w}\cos(\pi z/w)$ for the lowest subband, and $\xi^1(z) = \sqrt{2/w}\sin(2\pi z/w)$ for the second subband.

Upon the inclusion of mixing (terms $R$ and $S$ in the Hamiltonian $H_L$) the angular momentum of the envelope function $m$ and that of the Bloch function $m_j$ are no longer good quantum numbers. However, the eigenstates of the total Hamiltonian can be classified by the $z$ component of the total angular momentum: $L = m + m_j$. Moreover, the inversion symmetry of the system introduces yet another quantum number: the parity $\sigma =\downarrow$ or $\sigma =\uparrow$. Now the eigenstates of the Hamiltonian (14) take the form of spinors:

$$|L, N, \downarrow\rangle = \sqrt{\frac{2}{w}}\sum_n C_{n,m_j}^{L,N} \begin{pmatrix} f_n^{L-3/2}(\rho, \phi)\sin(2\pi z/w)|m_j = 3/2\rangle \\ f_n^{L+1/2}(\rho, \phi)\sin(2\pi z/w)|m_j = -1/2\rangle \\ f_n^{L-1/2}(\rho, \phi)\cos(\pi z/w)|m_j = 1/2\rangle \\ f_n^{L+3/2}(\rho, \phi)\cos(\pi z/w)|m_j = -3/2\rangle \end{pmatrix} \quad (21)$$

for parity $\sigma =\downarrow$, and

$$|L, N, \uparrow\rangle = \sqrt{\frac{2}{w}}\sum_n D_{n,m_j}^{L,N} \begin{pmatrix} f_n^{L-3/2}(\rho, \phi)\cos(\pi z/w)|m_j = 3/2\rangle \\ f_n^{L+1/2}(\rho, \phi)\cos(\pi z/w)|m_j = -1/2\rangle \\ f_n^{L-1/2}(\rho, \phi)\sin(2\pi z/w)|m_j = 1/2\rangle \\ f_n^{L+3/2}(\rho, \phi)\sin(2\pi z/w)|m_j = -3/2\rangle \end{pmatrix} \quad (22)$$

for parity $\sigma =\uparrow$. The quantum number $N$ enumerates the states with given total angular momentum and parity, and the function $f$ denotes the radial and angular part of the envelope function defined in (20).

We have calculated the eigenenergies of the $4 \times 4$ Luttinger–Kohn Hamiltonian for the $p$-type SiGe quantum disks [27]. In this system the holes are strongly confined, whereas the electrons exhibit only weak quantum confinement effects. This justifies our use of the $4 \times 4$ model, and substantiates the hard-wall disk approximation. The results of our calculations are shown in Fig. 9. At zero magnetic field the states with the same $L$ but opposite parity are degenerate (they form parity doublets). The energies of these states form a shell structure. The $s$-shell is composed of the states with $L = 0$. The $p$-shell consists of two parity doublets: states with $L = -1$ and $L = 1$, however these doublets do not have the same energy. Then there is a very broad $d$-shell, consisting of the parity doublets with $L = -2$ and $L = 2$ similar in energy, and the parity doublet with $L = 0$, whose energy is larger.

The parity degeneracy is removed in the magnetic field. Moreover, the energy spectrum evolves with the magnetic field into Landau levels. The lowest Landau level comprises states with parity $\sigma =\uparrow$ and has a slope corresponding to holes of large mass. The second Landau level is composed of the states with parity $\sigma =\downarrow$ and has a more light-hole character.

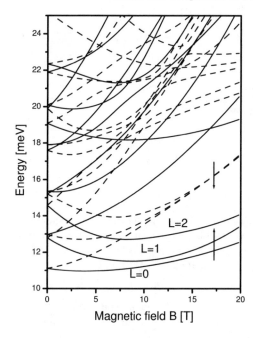

**Fig. 9.** Single-hole energy spectrum as a function of the magnetic field. *Solid lines* represent the parity $\sigma\ =\uparrow$ and *dashed lines* $\sigma\ =\downarrow$. (From [27])

The above considerations were carried out with an explicit assumption that the heavy- and light-hole band edges are degenerate. As described in the previous section, strain removes this degeneracy, so that the heavy hole band edge corresponds to a lower (less negative) energy than the light hole band edge. It means that as a result of the $4 \times 4$ Luttinger–Kohn approach the lowest hole states are expected to be predominantly of heavy-hole character. Moreover, in narrow-gap semiconductors the conduction–valence band mixing is large and it is necessary to consider an $8 \times 8$ Luttinger–Kohn model [9]. Calculation of the single-particle energies and eigenstates using the $8 \times 8$ model Hamiltonian proceeds along the following guidelines. First the distribution of the strain tensor matrix elements is calculated using the discretized continuous elasticity. The strain information is then used to calculate the piezoelectric charge density, which is input to a Poisson solver giving the piezoelectric potential. Then the Luttinger–Kohn and Bir–Pikus Hamiltonians together with the piezoelectric potential are constructed on the same discretization mesh as the one used for strain calculation, and diagonalized numerically [9,11]. Results of such a calculation for a model InAs/GaAs lens-shaped dot [28] with radius $R = 15.3$ nm and height $h = 2.3$ nm and perfectly sharp interface between InAs and GaAs (no intermixing) are shown schematically in Fig. 10. The gray area inside the structure in the right hand part of the figure represents the 98% probability isosurface for the lowest (*s*-type) state of predominantly electron character. The left-hand part of the figure shows the calculated energy levels: predominantly electron-like and pre-

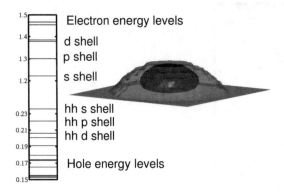

**Fig. 10.** *Right*: A schematic representation of the lens-shaped InAs/GaAs quantum dot used in the 8×8 Luttinger–Kohn calculation. The gray area inside the structure represents the 98% probability isosurface for the lowest electron state. *Left*: The calculated single-particle energies in the lens-shaped dot. (From [28])

dominantly hole-like. The electronic levels exhibit a clear shell-like structure with the equal energy spacing and proper degeneracies, with nondegenerate $s$ state, doubly degenerate $p$-shell, and three-fold degenerate $d$-shell. The hole levels, just like electronic levels, are almost equally spaced for this quantum dot. However, the degeneracies are not present. There is a splitting of the $p$-shell levels, and a larger splitting of the $d$-shell. The splitting of degenerate shells is due to the shear strain, which, as mentioned before, reduces the symmetry of the system. This symmetry-breaking effect is strong if the dot–barrier interface is sharp. Intermixing softens the interface and the amount of shear strain decreases. Thus the splitting of $p$ and $d$ hole shells is expected to decrease and the hole wave functions are expected to resemble those of a two-dimensional harmonic oscillator, justifying our approximations of energy levels of lens-shaped dots by the two-dimensional harmonic oscillators.

## 2.4   Tight-binding Models

The one-band and multi-band $\mathbf{k}\cdot\mathbf{p}$ methods described in previous sections do not include the atomistic structure of the quantum dot, including the effect of the interface between the dot and barrier materials. Yet, the results obtained from these effective mass approaches seem to work very well if one accepts the existence of very few effective parameters. This suggests that perhaps the details of atomic arrangement are washed out due to alloy fluctuations and only a small number of effective parameters, such as effective mass and band-offset, suffice to describe the ground and excited states. Obviously, one would like to know how to extract these microscopic parameters, including dielectric constant and confining potentials, from microscopic, i.e., atomistic structure of a quantum dot. A simulation of a structure containing an InAs self-assembled quantum dot embedded in GaAs material with a total size of $40\,\text{nm} \times 40\,\text{nm} \times 15\,\text{nm}$ would require a simulation of $\approx 1$ million atoms. With $sp^3$ orbitals per atom, i.e., 8 electrons, this means dealing with $\approx 10$ million electrons. One method, interpolating between first principle methods and $\boldsymbol{k} \cdot \boldsymbol{p}$ methods is the tight-binding approach. In the tight-binding method one selects most relevant orbitals $|a, i\rangle$ localized on the atom "$a$". The wave function of a system is expanded in the basis of these localized orbitals. The expansion coefficients and the energy spectrum are obtained by diagonalization of the tight binding Hamiltonian

$$H = \sum_{a,b,i,j} t_{i,j}^{a,b} c_{a,i}^{+} c_{b,j} \, , \tag{23}$$

where $c_{a,i}^{+}$ creates an electron on orbital "$i$" of atom "$a$". The energy spectrum is completely characterized by hopping matrix elements $t_{i,j}^{a,b}$. In a solid, one needs to know only a small number of the hopping matrix elements describing a unit cell. These elements control the energy spectrum, including gaps, bands, and the density of states. They are usually treated as phenomenological parameters fitted from an experiment. In a quantum dot one needs to modify these coefficients according to local strain, computed, e.g., by the valence force field (VFF) method. Full-fledged $sp^3d$ tight-binding models have been implemented by *Klimeck* and co-workers [31] and *Bryant* and co-workers [30]. A non-obvious extension of the tight-binding method to include optical transitions and excitonic effects was reported in [32]. This method requires a very sophisticated numerical apparatus and, in our opinion, the methodology is still under development. A simplified version of the tight binding model, the effective bond orbital model (EBOM), was developed by *Chang* and co-workers [29]. This approach uses one $s$-like anti-bonding and three $p$-like bonding effective orbitals. The effective orbitals, or rather tight-binding hopping matrix elements, are chosen in such a way as to reproduce the $\boldsymbol{k} \cdot \boldsymbol{p}$ band structure of constituent materials near the centre of the Brillouin zone. Thus the EBOM model is a real space version of the $\boldsymbol{k} \cdot \boldsymbol{p}$ model with proper effective masses, but also with elements of atomistic description built in. The electronic structure obtained using EBOM reported in [29] appears, not surprisingly, to be in good agreement with $\boldsymbol{k} \cdot \boldsymbol{p}$ calculations.

## 2.5   First Principle Methods

At the moment the first principle methods are limited to the empirical pseudopotential methods employed by *Zunger* and co-workers. They are nicely summarized in [34]. The "first principle" or "ab-initio" methods for "million atom" nanostructures are of course not solutions to the 10 million electron problem. Such a problem is beyond present capabilities. What these methods involve is a series of thoughtful approximations, starting with the replacement of the many-body problem by an effective one-body Kohn–Sham problem:

$$H\Psi_i(\boldsymbol{r}) = \left\{ -\frac{1}{2}\nabla^2 + V\left[\rho(\boldsymbol{r},\boldsymbol{r})\right] \right\} \Psi_i(\boldsymbol{r}) = E_i\Psi_i(\boldsymbol{r}) \,. \tag{24}$$

The effective potential in (24) is a self-consistent potential which is a functional of the charge density. Even when the self-consistent solution of the Kohn–Sham problem is reached, there is still an unresolved issue of being able to compute the response of the system to external perturbation. For a comparison of quantum-chemical and "solid state" methods applied to very small clusters, i.e., 6 atoms, the reader is referred to [33]. The extension and validity of solid-state methods, such as GW-Bethe-Salpeter equation and quantum Monte Carlo to a million atom cluster is obviously an issue.

Assuming that the self-consistent solution for a constituent material, e.g., InAs and independently GaAs, is reached, the Kohn–Sham self-consistent orbitals and effective self-consistent local density approximation (LDA) potentials are known. Since the gaps from LDA differ from the gaps obtained from experiment by a large margin ($\approx 30\%$), one replaces the LDA potential with an "empirical pseudopotential". This "empirical pseudopotential" aims at reproducing some properties of the constituent materials such as experimentally observed optical gaps and effective masses, similar in spirit to what is done in a tight-binding approach. Once the pseudopotentials for constituent atoms are developed, they are "glued" together at interfaces between the quantum dot and barrier atoms.

The overall computational procedure of Zunger and co-workers follows the following steps:

1. Assume the size, shape, and composition of the nanostructure.
2. Relax the atomic positions to minimize strain. If the dot is embedded in another material, the VFF model is used to relax all atomic positions.
3. Construct empirical pseudopotentials – "realistic" pseudopotentials that reproduce correct band gaps give (unlike LDA) correct effective masses and their anisotropies, and whose plane wave description can be accommodated with a small basis set (e.g., a cut-off of 5 Ry) are used.
4. Solve the single particle Schrödinger equation using one of the following methods: the folded spectrum method (FSM) or the linear combination of bulk bands (LCBB) method.
5. Calculate the dielectric screening function and the interparticle Coulomb and exchange integrals.
6. Using the single-particle levels and interparticle interactions, solve the problem of interacting electrons and holes.

This method is highly numerically involved, with results most likely dependent on the choice of pseudopotentials. To ascertain the predictive capability, let us recall that the "ab initio" LDA methods give excited states (optical gap) of bulk materials with accuracy of at most several percent. It is difficult to expect that this method is capable of quantitatively predicting excited states of quantum dots with accuracy of a few meV. Such accuracy appears to be required in the understanding of the effect of geometry on the second and third shell of quantum dots. However, the "first principle empirical pseudopotential" method is very useful in exploring material-related trends in electronic and optical properties of quantum dots. For details of the success of the method and relevant references the reader should consult [34] and references therein.

The calculations described in this section all assume the shape and composition of the quantum dot. In principle, one would like to be able to predict the shape and composition based on first principle calculations, calculate their optical properties, and correlate, e.g., the PL and absorption spectrum with growth conditions. While this program, to our knowledge, has not been realized, an effort toward the understanding of size, shape, and composition of quantum dots from growth condition has been undertaken by *Scheffler* and co-workers [35].

In summary, much work remains to be done toward the development of reliable microscopic theoretical understanding of the relationship between the substrate, constituent materials and growth condition and optical and electronic properties of self-assembled quantum dots. In the end, however, we must obtain a small number of excited states which can hopefully be understood based on a simple empirical model such as, e.g., the harmonic oscillator model.

## 2.6   Coupled Dots

Previous sections discussed different methods of calculating single-particle energy levels in a quantum dot. Different shapes were considered as means of tailoring the energy levels. A natural extension of this engineering of wave functions and density of states is to couple different dots.

In this section we will apply the effective mass adiabatic method to a vertically coupled dot system [12]. The method is applicable to an arbitrary number of dots, but for simplicity only two dots are discussed. The system is shown schematically in Fig. 11a. It consists of two disk-shaped quantum dots of height $H$ and radii $R_1$ and $R_2$, respectively ($R_1 < R_2$). Both disks are positioned on their respective wetting layers, and the wetting layer distance is $D$. Since we already understand the single-particle spectrum of a single disk, we want to understand how the single-particle energies evolve in this coupled system as we change the distance $D$. As described before for a single dot, our adiabatic calculation starts with considering the vertical motion of the electron. In our system we deal with three vertical regions with different potential profiles, as denoted with vertical lines in Fig. 11a. In the first (inner) region we have a symmetric double-well potential, each well comprising the combined thickness of the disk and the wetting layer. In the second region the bottom part consists only of the narrow wetting layer, whereas in the top part we still have the broad quantum well due to the disk. Finally in the third (outer) region we deal with two narrow quantum wells due to the presence of the wetting layers. We must now solve for the vertical motion in each of these three regions separately. We can use the transfer matrix formalism described for a single disk, but now we must modify it to account for the barrier between the wells. The expression for the transfer matrix in this case takes the form:

$$T = L^{-1}W_1B_1W_2R, \tag{25}$$

where the matrices $L$, $R$ and $W(k,h)$ have been defined before (8), (9). The matrix $B_1$ describing the middle barrier is also derived from the continuity conditions for the wave function and its derivative at respective interfaces and takes the form:

$$B(\kappa, d) = \begin{bmatrix} \cosh(\kappa d) & -\sinh(\kappa d)/\kappa \\ -\kappa\sinh(\kappa d) & \cosh(\kappa d) \end{bmatrix}, \tag{26}$$

where $\kappa = \sqrt{-E}$ $(E < 0)$ and $d$ is the total thickness of the barrier. This time, however, we are interested in the ground and first excited eigenstates in this double-well problem for each region. To this end we look for the first two zeros of the matrix element $T_{21}(E)$. These energies and the corresponding eigenfunctions are shown in Fig. 11b (left panel). Note that in the innermost region the ground state wave function is symmetric, and the first excited state wave function is antisymmetric. In the outermost region, when we deal only with the wetting layer, only the ground, symmetric solution is bound. These energy levels constitute the radial potential, separately for the symmetric and antisymmetric case. This potential, as seen in Fig. 11b (right panel), is a two-step quantum well, discussed before in the case of the lens-shaped single quantum dot. The ground and excited states of this potential are the sought eigenenergies of the system.

As asserted in the description of the adiabatic approximation for a single disk, the procedure needs two input parameters: the conduction band offset between the barrier and dot material, and the effective mass. Both of these parameters are influenced by strain. However, the band offset can be assessed by calculating the strain distribution in the sample and using the Bir–Pikus formalism to calculate the local band structure, as described in previous sections. In our case the disks have radii $R_1 = 8.5$ nm, $R_2 = 8$ nm, height $H = 2$ nm and the wetting layer has thickness 0.54 nm. Simulation of strain and local band structure in this InAs/GaAs system gives the conduction band offset $V_0 = 600$ meV [12]. The effective mass of the confined carrier cannot be evaluated so easily and we treat it as a parameter permitting us to fit our result either to experiment or to a more microscopic calculation. We performed a fitting to an $8 \times 8$ ($\mathbf{k} \cdot \mathbf{p}$) calculation carried out by *Pryor* [36] and obtained a value $m^* = 0.053m_0$. The comparison of our calculation with this effective mass and Pryor's is shown in the inset of Fig. 12. Once the

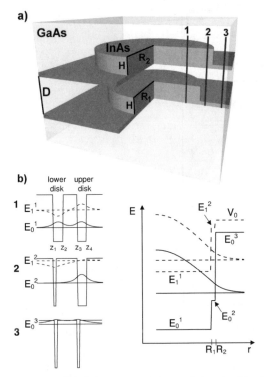

**Fig. 11.** (a) Schematic picture of the double-dot system. The top disk radius is larger than that of the bottom one. (b) *Left panel*: Potential profiles for the vertical motion in three radial regions of the system, together with the ground (symmetric – *solid lines*) and first excited (antisymmetric – *dashed lines*) electronic states. *Right panel*: The effective potential profile for the radial problem. (From [12])

parameters have been established, we calculated all bound energy levels in our double dot system as a function of the wetting layer separation (quantum dot layer distance) $D$. These results are shown in Fig. 12. We find that the spectrum is composed of pairs of states, one of symmetric, and one of antisymmetric character. At a very large distance $D$ between the dots these energy levels are quasi-degenerate and form shells, of which $s$, $p$ and $d$ are bound (are below the wetting layer continuum). We recover the characteristic degeneracies of the $p$-shell, as well as degeneracies and splitting of the $d$-shell, observed previously for a single disk. Thus we find a structure and behaviour of the energy level similar to that in a diatomic molecule.

As the quantum dot layer separation becomes small, the tunnelling between disks causes the symmetric-antisymmetric pairs of states to split. The splitting causes a crossing of the symmetric $p$-shell and antisymmetric $s$-shell states at $D = 4.5$ nm.

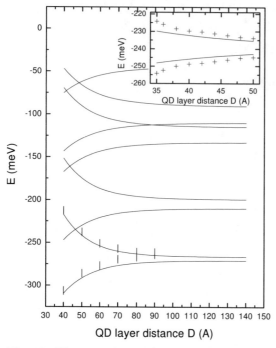

**Fig. 12.** Electronic levels of a double-dot system for $V_0 = 600\,\text{meV}$ and $m^* = 0.053m_0$ as a function of the quantum dot layer separation $D$ calculated using the adiabatic approximation. *Inset* shows the result of fitting of two lowest energy levels to the result of *Pryor* [36]. *Bars* show the results of an exact diagonalization calculation. (From [12])

The adiabatic approximation relies on the assumption that the vertical and lateral motion of the charge carrier can be separated due to the disparity of the vertical and radial dimensions of the structure. To test this approximation we performed an exact diagonalization study of the Hamiltonian (4) with $V(r, z)$ being our double-dot potential, and the units of length and energy are the effective Bohr radius and Rydberg, respectively. We discretize this continuous Hamiltonian on a grid of points, similarly as it is done in the $k \cdot p$ model, and diagonalize the resulting large sparse Hamiltonian matrix using the recursive Green's function technique. The result of the calculation is shown in Fig. 12 as vertical bars. We find a very good agreement between our approximation and the exact diagonalization results. The disparity between data at large quantum dot layer separation $D$ is due to the discretization errors: the thickness of the wetting layer is too small compared to the overall dimension of the computational grid to be represented by a sufficient amount of grid points.

# 3   Excitons in Quantum Dots

In this section we determine the ground and excited states of an exciton in zero-dimensional systems [37,38], its magnetic field dependence, coupling with phonons, electrons (trions) and a second exciton (biexcitons). We consider its interaction with phonons and how it is measured in Raman scattering, and discuss its coupling to photons when a self-assembled quantum dot (SAD) is inserted into an optical cavity as a source of single photons.

## 3.1   Excitons in Jacobi Coordinates

The variation of the exciton density of states of a single self-assembled quantum dot [1,3] with size has been measured by photoluminescence excitation spectroscopy (PLE) [39,40,41,42,43]. New features in the absorption spectra, previously hidden in spectra from large ensembles of dots due to inhomogeneous broadening [44], were observed. These features can be understood by constructing a theoretical model of an exciton in a "coherent exciton" basis consistent with Jacobi coordinates [43,45]. This basis captures important symmetries of the interacting electron–hole pair. The trivial variation of the single-particle shell structure is translated into a nontrivial modification of the interacting electron–hole system.

As discussed in Sect. 2, the lens-shaped self-assembled quantum dots are excellent models of zero-dimensional (0D) systems. The single particle states $|n, m\rangle$ can be well approximated by those of a pair of harmonic oscillators with quantum numbers $m$ and $n$ (the Fock–Darwin spectrum). These states form a finite number of degenerate shells, with $S = m + n$ constant within each shell. Each shell has degeneracy $g_S = S + 1$ and energy $E(S) = \Omega \cdot S$, where $\Omega = \Omega_{e(h)}$ for electrons and holes, respectively. The angular momentum of electrons is $l^e = n_e - m_e$, the angular momentum of holes is opposite to that of electrons due to opposite charge. The electrons and holes interact with each other via the Coulomb interaction, forming excitons. The exciton Hamiltonian can be written in any basis of electron–hole states [46]. We focus here on the construction of a basis which captures all the symmetries of the problem and allows us not only to compute but also to understand the exciton spectrum and its dependence on the number of electronic shells. The exciton states are superpositions of products of electron and hole states $|i\rangle|j\rangle$ (where $|i\rangle = |n_e, m_e\rangle$, and $|j\rangle = |n_h, m_h\rangle$). Coulomb scattering conserves the total angular momentum $R = l_i - l_j$ of an electron–hole pair, so all states can be classified by their total angular momentum. The absorption spectrum $A(\omega)$ of a photon with frequency $\omega$ due to transitions from the initial vacuum state $|0\rangle$ of a SAD to all final states $|f\rangle$ is given by Fermi's golden rule: $A(\omega) = \sum_f |\langle f|\mathcal{P}^\dagger|0\rangle|^2 \delta[\omega - (\mathcal{E}_f - \mathcal{E}_0)]$. Here $\mathcal{E}_f$ and $\mathcal{E}_0$ are the energies of the final and initial states of the absorption process, respectively. Initial and final states are coupled through the interband polarization operator $\mathcal{P}^\dagger|0\rangle = \sum_j |j\rangle|j\rangle$. $\mathcal{P}^\dagger$ creates electron–hole pairs on identical orbitals $j$, i.e., with zero total angular momentum of the pair. Let us now rename the single-particle orbitals in terms of their shell index $S$ and angular momentum $l$ as $|S, l\rangle$. We can now generate and classify the electron–hole pairs with zero total angular momentum. These states fall into two classes. On each shell $S$ there are what appear to be optically active pair states of the form $|S, l\rangle|S, -l\rangle$. Their number equals the degeneracy of each shell, $S + 1$, and the energy of each pair is $(\Omega_e + \Omega_h) \cdot S = t \cdot S$. Pairs on different shells have different energy. The density of states corresponding to these pairs in a dot with three shells is shown in Fig. 13a. While these pairs correspond to vertical transitions, photons do not create individual pairs but rather a linear superposition of them $\mathcal{P}^\dagger |0\rangle = \sum_S [\sum_l |S, l\rangle|S, -l\rangle]$. Hence only one "coherent" state $|\mathcal{S}\rangle = \sum_l |S, l\rangle|S, -l\rangle$ from each $S + 1$-fold degenerate shell couples to photons. The remaining $S$ states are dark. It is therefore convenient to create a coherent exciton representation which explicitly accounts for this symmetry. This is done by a transformation into Jacobi-like coordinates where one of the orthogonal basis states is the "coherent" state $|\mathcal{S}\rangle$. In Jacobi coordinates one can immediately decide which states are coupled to the coherent state by Coulomb interactions and become optically active. In fact, for symmetric interactions $V_{ee} = V_{hh} = -V_{eh}$ the coherent states are exact eigenstates of the interacting exciton Hamiltonian restricted to a given shell. Hence even in the presence of interactions each shell contains exactly one optically active configuration. This symmetry and procedure is analogous to Kohn's theorem for intraband transitions, and can be loosely described as "Kohn's theorem for excitons" [47].

The next step is the search for other possible two-particle configurations. The most important configurations, if they exist, are those with energy resonant with the coherent exciton states. At first glance the only candidates are pairs of states from different shells, e.g., $|S, l\rangle|S', -l\rangle$. The energy of these states, $\Omega_e \cdot S + \Omega_h \cdot S'$, is different from the energy of coherent exciton states. It is therefore difficult to expect these states to acquire a significant oscillator strength apart from accidental, parameter dependent, degeneracies. However, when we construct a superposition of these electron–hole states from different shells in the form

$$|S, p\rangle = \frac{1}{\sqrt{2}}(|S - p, l\rangle|S + p, -l\rangle \pm |S + p, l\rangle|S - p, -l\rangle),\qquad(27)$$

we find that their energy,

$$E(S, p) = \frac{1}{2}\left[\Omega_e \cdot (S - p) + \Omega_h \cdot (S + p)\right] + \left[\Omega_e \cdot (S + p) + \Omega_h \cdot (S - p)\right],(28)$$

equals the energy $(\Omega_e + \Omega_h) \cdot S$ of the coherent exciton state on the shell $S$. What is even more important, this conclusion is valid for any ratio of the electron and hole kinetic energy. These configurations have not been constructed from Jacobi coordinates and therefore Coulomb interactions mix them with coherent exciton states. The mixing makes the configurations with the $+$ sign

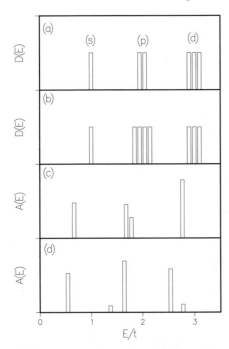

**Fig. 13.** (a) Density of states of optical transitions and (b) density of states of an exciton in the coherent exciton representation showing degeneracies of excitonic shells. (c), (d) Calculated absorption spectra showing the splitting of the $p$-like transition for different sets of quantum dot parameters. (From [43])

optically active. These are precisely the nontrivial states which strongly modify the absorption spectrum. Now we can immediately construct degenerate shells of electron–hole states with degeneracies depending on the number of confined single particle levels $(n, m)$. For two shells, $(s, p)$, we have degeneracies $(1, 2)$, for three shells, $(s, p, d)$, we have $(1, 4, 3)$, while for five shells we have $(1, 4, 9, 12, 5)$. Hence the degeneracies of excitonic shells depend on the number of confined single particle levels. It is best to illustrate these constructions on a simple example of a dot with two and three shells.

For a dot with three lowest shells, $S = 0, 1, 2$, all possible configurations are shown in Fig. 14. There are 6 optically active configurations: (a)–(f), or $(|00\rangle|00\rangle)$, $(|11\rangle|1-1\rangle, |1-1\rangle|11\rangle)$, $(|22\rangle|2-2\rangle, |2-2\rangle|22\rangle, |20\rangle|20\rangle)$. In addition there are two configurations: (g) and (h), originating from different shells: $|00\rangle|20\rangle$, $|20\rangle|00\rangle$. These optically inactive states involve an electron (hole) in the zero angular momentum state $|00\rangle$ of the $s$-shell and a hole (electron) in the zero angular momentum state $|20\rangle$ of the $d$-shell. They are degenerate with the coherent exciton states of the $p$-shell. Hence the exciton is composed of 8 states with zero total angular momentum. The next step is the transformation into coherent exciton states (Jacobi coordinates). The $s$-shell is unchanged, $|A\rangle = |a\rangle$. The $p$-shell contains two configurations: $|B\rangle = \frac{1}{\sqrt{2}}(|b\rangle + |c\rangle)$ and $|C\rangle = \frac{1}{\sqrt{2}}(|b\rangle - |c\rangle)$. The $d$-shell is replaced with $|D\rangle = \frac{1}{\sqrt{3}}(|d\rangle + |e\rangle + |f\rangle)$, $|E\rangle = \frac{1}{\sqrt{2}}(|d\rangle - |e\rangle)$, and $|F\rangle = \frac{1}{\sqrt{6}}(|d\rangle + |e\rangle - 2|f\rangle)$. The two configurations, (g) and (h), are replaced with $|H\rangle = \frac{1}{\sqrt{2}}(|g\rangle + |h\rangle)$

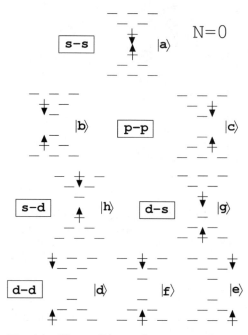

**Fig. 14.** All possible configurations of an electron–hole pair with total angular momentum $R = 0$ on three lowest shells in the conduction and valence band

and $|G\rangle = \frac{1}{\sqrt{2}}(|g\rangle - |h\rangle)$, respectively. The density of states of an electron–hole pair in Jacobi coordinates is shown in Fig. 13b. The key result is that there are now four degenerate states, B, C, G, and H, on the $p$-shell. The Jacobi coordinates allow us then to separate the optically active states from the dark states in the interacting system. By expanding the interband polarization operator in the new basis: $\mathcal{P}^\dagger|0\rangle = |A\rangle + \sqrt{2}|B\rangle + \sqrt{3}|D\rangle$, we see that only three coherent states, $|A\rangle$, $|B\rangle$, and $|D\rangle$ are optically active. Clearly the states $|C\rangle$, $|E\rangle$, and $|G\rangle$, which are the antisymmetric superpositions of pair states, cannot couple to the remaining states. Therefore there are five active states: state $|A\rangle$ from the $s$ shell, states $|B\rangle$ and $|H\rangle$ from the $p$-shell, and states $|D\rangle$, $|F\rangle$ from the $d$-shell. Hence the spectrum of a SAD with three shells will consist of five peaks: one derived from the $s$-shell, two from the $p$-shell, and two from the $d$-shell. The calculated absorption spectra $A(\omega)$ showing the splitting of the $p$-shell transition are shown in Fig. 13c,d. In Fig. 13c the frequencies $\Omega_e = \Omega_h$ and the masses are the same; in Fig. 13d, $\Omega_e = 2\Omega_h$ and the hole mass is twice that of the electron. In addition, a small splitting of the $d$-shell energy levels was introduced to simulate the energy levels of the lens shaped self-assembled quantum dots. In the absence of interactions the transition energies, shown in Fig. 13a, are equally spaced in units of electron–hole energy $t$. The inclusion of the electron–hole scattering changes the absorption spectrum. The attractive electron–hole interaction renormalizes transition energies and leads to additional structures in the spectrum. Here the most important effect is the splitting of the $p$-shell absorption, shown in Fig. 13c,d. The oscillator strength of these transitions depends on system parameters: it is almost equally distributed between the $p$–$p$-like transition $|B\rangle$ and the $s$–$d$-like transition $|H\rangle$ in Fig. 13c in contrast to the situation in Fig. 13d. The absorption into the $d$-shell is also split into two lines, one strong and one weak, and the oscillator strength of these lines depends on dot parameters. It is important to note that the splitting of the $p$-shell absorption is an excitonic feature due to mixing of configurations. These configurations, and the splitting, become possible in dots with at least three electronic shells and are absent in dots with only two shells. The splitting is not related to the splitting of single particle levels due to dot asymmetries. To test these predictions experimentally we chose identical $In_{0.60}Ga_{0.40}As$ SADs with different sizes: one dot with only two shells and one with three shells. The experimental results, discussed in the Chapter by *Bayer*, agreed very well with our theoretical predictions.

### 3.2   Magneto-Excitons

The calculation of magneto-excitons in quantum dots with the Fock–Darwin (FD) spectrum has been described in detail in [48]. Here we present the results of a calculation by *Wojs* et al. [24] for the InGaAs sample studied in [49].

   In Fig. 15 we show the evolution of the absorption spectrum with increasing magnetic field $B$. The top frame corresponds to a noninteracting electron–hole pair, and the bottom frame to the interacting magneto-exciton. The areas of black dots are proportional to the intensities of discrete peaks, the energies of which (measured from the valence band-conduction band gap) are given on the vertical axis.

   For the noninteracting electron–hole pair all peaks are of equal strength and their evolution in $B$ repeats the FD pattern of each particle. The FD orbitals $|nm\rangle$, on which the pair is created, have been indicated for a few peaks. This picture illustrates the destruction and restoration of the dynamical symmetries of the FD spectrum by the magnetic field. Since for certain values of $B$ (0, 12, 19, 25 T) the levels cross, intensities of neighbouring peaks add up around these values of $B$, and, e.g., for $B = 0$ the Coulomb interaction mixes many electron–hole states and leads to a richer energy spectrum, shown in the bottom frame of Fig. 15. In particular, the two absorption peaks of the $p$-shell discussed in previous section are clearly visible. The $p$-shell absorption does not split into two peaks as in the noninteracting FD picture, with one line increasing, and the second line decreasing in energy. Instead, the two peaks show weak magnetic field dependence and exchange of oscillator strength. Despite these subtle features, the magneto-exciton spectrum resembles quite well the FD spectrum of a noninteracting electron–hole pair, in particular the restoration of symmetries and the reappearance of gaps in the spectrum

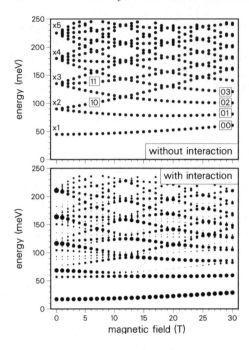

**Fig. 15.** Evolution of absorption spectrum of a SAD magneto-exciton with the magnetic field. Areas of dots are proportional to intensities of transitions. *Upper* and *lower frames* present graphs for noninteracting and interacting systems, respectively. In the *upper frame* the degeneracies of overlapping peaks at $B = 0$ have been indicated. (From [24])

around 13 T. This is because in such small quantum dots the characteristic confinement energy: $\omega_{e,0} + \omega_{h,0}$ ($\sim 45\,\text{meV}$) exceeds the characteristic Coulomb energy ($\sim 25\,\text{meV}$, slightly $B$-dependent). The main effect of including the interaction on the eigenenergies is therefore the parallel red-shift of the entire spectrum. It is worthwhile pointing out that the magneto-absorption spectrum of a single quantum dot has not been measured yet.

### 3.3  Exciton in a Double Quantum Dot Molecule

Let us now consider an exciton localized on a double quantum dot molecule. Our goal is to demonstrate that this exciton state is an entangled state. Let us first assume that our disks are far apart, so no tunnelling occurs, and that they are small enough to hold only one single-particle state each. We denote the orbital on the lower dot by $|0\rangle$, with corresponding energy $E_s^0$, and the orbital on the upper dot by $|1\rangle$, with corresponding energy $E_s^1$, and assume $E_s^0 = E_s^1 = E_s$. Equivalently we can consider two states of an isospin: isospin down corresponds to the state $|0\rangle$, and isospin up corresponds to the state $|1\rangle$. These two orbitals span the basis of our single-particle problem. Inclusion of tunnelling leads to the appearance of the off-diagonal matrix elements $\langle 0|H|1\rangle = \langle 1|H|0\rangle^* = -t$, and we assume $t$ real. Thus our Hamiltonian with tunnelling, written in the basis of localized orbitals, has the form:

$$H = \begin{bmatrix} E_s & -t \\ -t & E_s \end{bmatrix}, \tag{29}$$

with eigenenergies $E_\pm = E_s \mp t$ and corresponding eigenvectors $|\pm\rangle = (1/\sqrt{2}) \cdot (|0\rangle \pm |1\rangle)$. In the isospin language the tunnelling will then lead to a rotation of the isospin. Note that this rotation, leading to the appearance of symmetric and antisymmetric states, does not depend upon the magnitude of the tunnelling matrix element $t$. The splitting between the ground and first excited states, $2t$, depends on the quantum dot layer distance $D$, therefore we recover the behaviour of the symmetric and antisymmetric energy levels of the $s$-shell shown in Fig. 12. This splitting plays the role of the "Zeeman energy" for the isospin, and is equivalent to the Zeeman energy of a single spin in a magnetic field.

Let us now consider an electron and a hole localized on the quantum molecule [50]. The Hamiltonian of the system takes the form:

$$H = \sum_i E_i^e c_i^+ c_i + \sum_\sigma E_i^h h_i^+ h_i - \sum_{ijkl} \langle ij|V|kl\rangle c_i^+ h_j^+ h_k c_l \,, \tag{30}$$

where $c_i$ ($h_i$) is the electron (hole) annihilation operator in the basis of localized orbitals, and $\langle ij|V|kl\rangle$ denotes the two-body Coulomb interaction matrix element. Now the Hilbert space of this problem is spanned by products of localized orbitals for each particle, and we deal with the following basis set: $\{|0\rangle_e|0\rangle_h, |0\rangle_e|1\rangle_h, |1\rangle_e|0\rangle_h, |1\rangle_e|1\rangle_h\}$. In the presence of tunnelling, but in the absence of the Coulomb interaction both particles behave independently. We will then deal with a rotation of two independent isospins to the symmetric and antisymmetric states $|\pm\rangle_e$ and $|\pm\rangle_h$. The basis of our Hilbert space is now rotated and assumes the form $\{|+\rangle_e|+\rangle_h, |+\rangle_e|-\rangle_h, |-\rangle_e|+\rangle_h, |-\rangle_e|-\rangle_h\}$. The rotations of individual isospins have been performed independently, so the basis of the two-particle Hilbert space is built as a simple tensor product of the single-particle bases. Therefore no quantum entanglement is encountered at this point. The states $|+\rangle_e|+\rangle_h$ and $|-\rangle_e|-\rangle_h$ can be created optically, as in each case the symmetry of the electron and hole states is the same. The two remaining states – of mixed symmetry – are dark.

Now we include the Coulomb interaction and build and diagonalize the full Hamiltonian matrix in our basis. To this end we assume that both disks forming the quantum molecule are identical, and that the wetting layer is absent. This permits us to approximate the electron and hole wave functions in the form $\langle r|+\rangle = N_+ J_0(k_0^\rho \rho) \sum_l A_l^+ \cos(k_l^z z)$, where $N_+$ is the normalizing factor, $J_0$ is the Bessel function of 0-th order, depending on the radial coordinate only, the angular momentum is zero (the orbital is of $s$-type), and the vertical behaviour, including the symmetry properties, is modelled by a series of cosines with appropriately chosen amplitudes $A_l^+$. The analogous form is taken for parity $(-)$. Now the Coulomb interaction does not couple the subspace of optically active and dark states, and the Hamiltonian describing bright excitons can be simply written as

$$H_X = \begin{bmatrix} E_{++} + V_{++} & V_{+-} \\ V_{+-} & E_{--} + V_{--} \end{bmatrix} \,. \tag{31}$$

a)

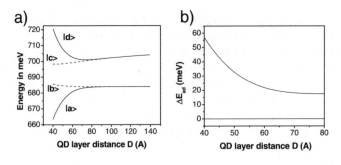

b)

c)

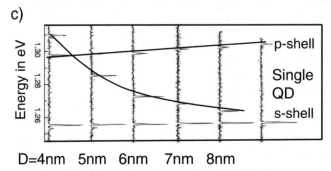

D=4nm   5nm   6nm   7nm   8nm

**Fig. 16.** (a) Exciton energies (*dark* and *bright*) calculated using the two-isospin model of an electron–hole pair. States $|a\rangle$ and $|d\rangle$ are optically active. (b) Splitting between the energy levels of the optically active states as a function of the quantum dot layer distance. (c) Photoluminescence spectra of quantum dot molecules with various interdot distances, recorded at low excitation power and temperature $\sim 60\,\mathrm{K}$. The PL spectrum of a single dot is shown as a reference. (From [50])

The evolution of the resulting energy levels of the exciton (dark and bright) is shown in Fig. 16a. When the dots are far apart, the energy states form two groups split by the Coulomb interaction. As the dots are shifted closer together, the tunnelling increases the splitting between the bright states. The energy of dark states remains approximately constant, as the energy shifts of the electron and hole tunnelling are of opposite signs and nearly cancel out.

As can be easily seen from diagonalizing the Hamiltonian (31), the wave functions corresponding to the bright excitons with interactions can be written as:

$$|a\rangle = \alpha_1 [|0\rangle_e |0\rangle_h + |1\rangle_e |1\rangle_h] + \beta_1 [|0\rangle_e |1\rangle_h + |1\rangle_e |0\rangle_h], \tag{32}$$

$$|d\rangle = \alpha_2 [|0\rangle_e |0\rangle_h + |1\rangle_e |1\rangle_h] + \beta_2 [|0\rangle_e |1\rangle_h + |1\rangle_e |0\rangle_h]. \tag{33}$$

These states cannot be written as tensor products of single particle (individual isospin) states. Tunnelling and Coulomb interaction then lead to an entanglement of isospins.

The validity of our simple approximation was tested in a recent photolu-minescence experiment performed on quantum dot molecules with different interdot distances [50]. The single dot molecules were isolated in mesas to suppress the inhomogeneous broadening. The experiment was done at low excitation powers, but at the temperature $\sim 60\,\mathrm{K}$, making it possible to observe transitions from ground and excited states. The recorded PL spectra are shown in Fig. 16c. As the measurements were done on different systems, we refrain from comparing the absolute values of energy corresponding to the PL peaks. Instead we aligned all traces according to the lowest-energy PL line. A single quantum dot trace is also shown as a reference. We see that the splitting induced by tunnelling and interactions is about $10\,\mathrm{meV}$ at interdot separation $8\,\mathrm{nm}$, and increases to about $60\,\mathrm{meV}$ when the dots are only $4\,\mathrm{nm}$ apart. The behaviour of the PL peaks belonging to the $s$-like states follows qualitatively that predicted theoretically. In the experimental spectra we also see a third PL line, representing the $p$-shell emission. This line is not seen in our model, as we take into account the $s$-shell orbitals only. For more details concerning this experiment we refer the reader to the Chapter by *Bayer* in this volume.

## 3.4   Exciton, Biexciton, and Charged Exciton

The excitonic emission spectrum of a quantum dot is in the same spectral range as that of a biexciton or a charged exciton. This can be easily under-stood since the exciton is a charged neutral complex and it interacts weakly with other excitons or charges. This implies that shifts in energy levels of these complexes are a result of both deviations from charge neutrality, i.e., how different holes are from electrons, and of mixing of configurations. The mixing of configurations is in turn a sensitive function of parameters such as the ratio of electron to valence hole mass, the confining potential for either carrier, and the number of confined shells. To investigate these effects, we performed numerical calculations of an exciton, biexciton and charged $X-$ and $X+$ complexes localized on a single quantum dot [51]. The results of these calculations are presented in Fig. 17. We assumed here the electronic energy level spacing to be of order of $\Omega_0^e = 50\,\mathrm{meV}$, while the hole energy spacing was taken to be $\Omega_0^h = 20\,\mathrm{meV}$, so that the kinetic energy difference between the $p$-shell and the $s$-shell exciton is $70\,\mathrm{meV}$. The Coulomb inter-actions strongly renormalize the excitonic energies, but this renormalization depends on the type of complex considered. This follows from the fact that the size of the basis of the Hilbert space is different in each case: for five confined shells we deal with 29 states for a single exciton, 1276 states for the biexciton, and 186 states for each charged complex. In Fig. 17 we show the re-sults of our calculations for three different ratios of the electron to hole mass. The middle panel, with $m_e/m_h = 0.4$, corresponds to the "symmetric" inter-actions, i.e., the $e-e$, $h-h$, and $e-h$ interactions are almost identical. In this case we predict the emission line attributed to the biexciton recombination to

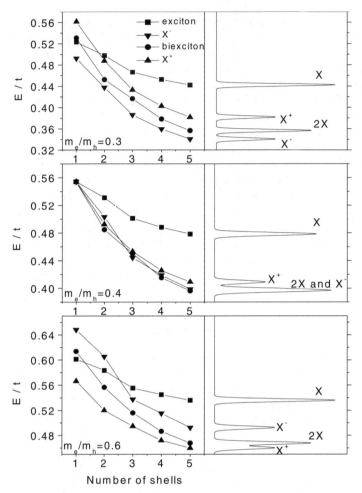

**Fig. 17.** Calculated exciton, biexciton and negatively and positively charged exciton emission energies as a function of the number of shells confined in the dot and the ratio of electron to hole mass. The *right-hand side* shows the possible PL spectrum for five shells (the oscillator strength for each transition is assigned arbitrarily). (From [51])

be almost degenerate with the $X-$ line. The emission from the other charged complex appears between the single exciton and biexciton peaks. For the ratio $m_e/m_h = 0.3$ and five confined shells the $X-$ line is found below, and the $X+$ line – above the biexciton line. For the mass ratio $m_e/m_h = 0.6$ the position of the lines corresponding to charged complexes is reversed. Hence without knowing the structure of excited states it is very difficult to assign the origin of PL lines in quantum dots, perhaps with the exception of biexcitons due to different excitation power dependence.

## 3.5   Exciton–Phonon Interaction and Raman Scattering

Inelastic light scattering is a useful probe of the interaction of the interband polarization of quantum dots with phonons and free carriers in semiconductor nanostructures [52,53]. We describe here briefly the interaction of excitons with acoustical phonons and its contribution to the Raman scattering cross-section of a self-assembled quantum dot [38]. Due to the large barrier, the quantum dots showed very clear absorption resonance related to the second excited 2D subband. This created unique conditions for inelastic light scattering experiments, as the photoluminescence signal was separated from the Raman spectrum.

The mean lateral size of the $Ga_{0.8}In_{0.2}As$ quantum disks was 80 nm, and their thickness was $\approx$ 5–6 nm [54,55]. The disparity between the thickness and diameter is here even larger than that encountered previously in the adiabatic calculations. Thus the electronic energy levels and wave functions can be well separated into those pertaining to the vertical and lateral motions. For the vertical energy levels, a very large conduction band discontinuity between the indium-rich disk and the $Ga_{0.5}Al_{0.5}As$ barrier, of $\approx$ 650 meV, leads to two confined subbands, with calculated energy spacing $\Delta E_s \approx$ 280 meV. The lateral confining potential, $V(r)$, is influenced by the decreasing disk thickness and amount of strain as we move from the centre to the edge of the SAD. Hence inside the disk this effect is described by a slowly varying parabolic potential, $m_e \omega_0^2 r^2 / 2$ with a characteristic frequency $\omega_0$, and at the interface we add a hard-wall component $V_0 \Theta(r - R)$, where $V_0$ is the discontinuity of the potential at the edge of the disk. The effective mass of strained $In_{0.2}Ga_{0.8}As$ is $m_e \approx 0.067$, and the dielectric constant $\epsilon \approx 12.5$. These parameters enter through the effective Rydberg and Bohr radius as the energy and length units, respectively. By virtue of these approximations the disk energy levels and wave functions are again of the FD type, with the effects of the hard-wall potential included perturbatively. The states $|smn\rangle$ and energies $E_{smn}$ are then those of a pair of anharmonic oscillators with energies $E_{smn} = E_s + \Omega_-(n + 1/2) + \Omega_+(m + 1/2) + \gamma_{m,n}(m - n)^2$, attached to the subband $s$, where $\Omega_\pm = (\Omega \pm \omega_c)/2$, $\Omega^2 = \omega_0^2 + 4\omega_c^2$, $\omega_c = eB/cm_e$ is the cyclotron energy, and $\gamma_{m,n}$ contains the effect of the hard wall potential on states with increasing angular momentum $R = m - n$. The valence band holes are treated in the effective mass approximation as a positively charged particle with angular momentum $R^h_{mn} = n - m$, opposite to the electron, and FD energies $E^h_{mn} = E^h_s + \Omega^h_+(n + 1/2) + \Omega^h_-(m + 1/2) + \gamma^h_{m,n}(m - n)^2$ (neglecting the semiconductor gap).

As described earlier, we form the correlated electron–hole pair states $|\nu\rangle$ as linear combinations of electron–hole pair states $|k\rangle = c_i^+ h_j^+ |0\rangle = |m_e, n_e, m_h, n_h\rangle$ with zero total angular momentum $R = m_e - n_e - m_h + n_h$: $|\nu\rangle = \sum_k A_k^\nu |k\rangle$. The electron–hole pair amplitudes $\boldsymbol{A}^\nu$ are the solutions of the Hamiltonian $H_0$, i.e., $\sum_{k'} H_0^{k,k'} A_{k'}^\nu = E_\nu A_k^\nu$.

We define the photon creation operators $a_\lambda^+$ for photons with polarization $e_\lambda$ and energy $\omega_\lambda$. The electron–photon Hamiltonian $H_{e-\text{phot}}$ corresponds to the process of creating electron–hole pairs by annihilating photons, and conversely:

$$H_{e-\text{phot}} = \sum_{i,j,\lambda} \mu_i^\lambda c_i^+ h_i^+ a_\lambda + \mu_i^\lambda h_i c_i a_\lambda^+ . \tag{34}$$

The optical selection rules are hidden in dipole matrix elements. We have $\mu_i^\lambda = \mu^\lambda \delta_{m_e, m_h} \delta_{n_e, n_h}$, i.e., only vertical transitions with zero angular momentum are allowed. The number of excitonic states is larger then the number of optically allowed states.

The interaction of electrons and holes with phonons involves scattering of either electrons or holes by phonons:

$$H_{e-\text{phon}} = \sum_{i,i',q} [M_{i,i'}^e(q) c_{i'}^+ c_i + M_{i,i'}^h(q) h_{i'}^+ h_i](b_q + b_{-q}^+) , \tag{35}$$

where $b_q^+$ creates phonons with wave vector $q$ and frequency $\omega_q$, and $M_{i,i'}^e(q)$ are scattering matrix elements which depend on the type of phonons involved. The exciton–phonon Hamiltonian follows from (35). Even though a scattering event changes the state of an electron or a hole, the final scattered state may remain an optically active exciton state. Such states play an important role in light scattering spectra where a photon is virtually absorbed, creates an exciton, an exciton undergoes phonon scattering, and re-emits a photon. Only states which remain excitonic after phonon scattering may re-emit light.

The inelastic light scattering measures the probability that a photon with frequency $\omega_i$ and polarization $e_i$ decays into a scattered photon with frequency $\omega_s$ and polarization $e_s$ and a phonon. The probability of the resonant one-phonon scattering is given by Fermi's golden rule [53]:

$$\Gamma = \sum_q | \sum_{\nu,\nu'} (\mu A^\nu) \cdot \frac{\{A^\nu \cdot [M^e(q) + M^h(q)] \cdot A^{\nu'}\}}{(E_\nu - \omega_s)(E_{\nu'} - \omega_i)} \cdot (A^{\nu'} \mu)|^2$$
$$\times \delta(\omega_i - \omega_s - \omega_q) . \tag{36}$$

The scattering rate involves a virtual absorption of a photon, a creation of an exciton which emits a phonon, and subsequently, a photon. Only states which remain optically active after the phonon emission contribute to the cross-section.

We are primarily interested in acoustical phonon emission. The final expression for the scattering rate can be reduced to a single integral over a dimensionless variable $Q = q\ell/2^{1/2}$, where $\ell = \sqrt{1/m_e 2\omega_0}$ is the effective length in Bohr radii:

$$\Gamma = 2\Gamma_0 \int_0^\infty dQ Q \frac{\Omega^2 f_{11}^2(Q)}{\sqrt{\Omega^2 - Q^2}}$$

$$\times \left| \sum_{\nu,\nu'} (\boldsymbol{\mu} \boldsymbol{A}^{\nu}) \cdot \frac{\left\{ \boldsymbol{A}^{\nu} \cdot \left[ M^e(\boldsymbol{q}) + M^h(\boldsymbol{q}) \right] \cdot \boldsymbol{A}^{\nu'} \right\}}{(E_{\nu} - \omega_s)(E_{\nu'} - \omega_i)} \cdot (\boldsymbol{A}^{\nu'} \boldsymbol{\mu}) \right|^2 . \qquad (37)$$

Here $\Omega = (\omega_i - \omega_s)\ell/(2^{1/2}c_s)$ is the frequency shift measured in the energy of the phonon with wave vector inversely proportional to the length scale of the dot (spacing of states). The form factor $f_{11}(Q)$ is the Fourier transform of the excited subband wave function, and $\Gamma_0 = (D^e)^2 \hbar \sqrt{2}/(4\pi^2 \rho c_s^2)$ where $D^{e(h)}$ is the electron(hole) deformation potential, $\rho$ is the density, and $c_s$ is the velocity of sound.

The dependence of the transition rate on the Raman frequency shift $\omega_i - \omega_s$ for incident energy close to the excitonic resonance of the first excited subband was analyzed in [38] and compared with experiment. These were preliminary experiments, and more experiments on self-assembled InAs quantum dots are needed to understand the interaction of excitons with phonons.

## 3.6   Exciton and Biexciton in an Optical Cavity

Electron–electron interactions and the coupling of the quantum dot to confined cavity photon modes, discussed extensively by *Gerard* in this volume, are important ingredients of a single photon source based on a semiconductor quantum dot [56]. In this proposal a quantum dot is strongly excited by a short intense pulse and populated with electron–hole pairs. These pairs recombine by emitting photons and reduce their numbers. The last electron–hole pair emission produces a photon with a definite energy, and it is this last exciton which produces a single photon. This scheme rests on the fact that the emission from a dot with exciton number different than one is always sufficiently different from the energy to emit a single exciton. This difference arises only due to electron–electron interactions. Hence one needs to understand the effects of interactions on electronic states of a quantum dot filled with $N$ excitons and a corresponding emission spectrum. This is discussed in detail in the following section. Once a single photon is emitted it has to be efficiently collected. This is accomplished by placing the quantum dot in a cavity. But this implies that electron–hole pairs and photons in a cavity are interacting and forming polaritons. Hence one needs to understand the coupling of electronic excitations with photons, and the renormalization effects associated with it.

The simplest but relevant model consists of a quantum dot interacting with two confined photon modes corresponding to the photon polarization $(+)$ and $(-)$ [95]. If we limit ourselves to the lowest number of excitations in the dot, we must consider four states: (1) the vacuum state $|0\rangle$ with energy $E_0$; (2) the $X+$ state $|+\rangle$ with energy $E_+$, created from the vacuum state by absorption of one $\sigma^+$ photon; (3) the $X-$ state $|-\rangle$ with energy $E_-$, created from the vacuum state by absorption of one $\sigma^-$ photon; (4) the biexciton state $|+-\rangle$, with energy $E_{+-}$. The energy levels and coupling to photons are shown in Fig. 18a. The relevant Hamiltonian of the coupled exciton–photon system can be written as:

$$
\begin{aligned}
H = {} & E_0|0\rangle\langle 0| + E_+|+\rangle\langle +| + E_-|-\rangle\langle -| + E_{+-}|+-\rangle\langle +-| \\
& + \omega_+ a_+^+ a_+ + \omega_- a_-^+ a_- + \gamma_+ \left\{ P_+^+ a_+ + P_+ a_+^+ \right\} + \gamma_- \left\{ P_-^+ a_- + P_- a_-^+ \right\} \\
& + V_+(t) \left\{ a_+ + a_+^+ \right\} + V_-(t) \left\{ a_- + a_-^+ \right\} .
\end{aligned}
\tag{38}
$$

Here $\omega_+$, $\omega_-$ are photon energies, $\gamma_+$, $\gamma_-$ are the dipole matrix elements which control coupling of photons and excitons, and $V_+(t)$, $V_-(t)$ describe the external light field. In a coupled system the external laser field creates/annihilates cavity photons while cavity photons in turn create/annhilate excitons. The dynamics of the coupled system will be described elsewhere; here we describe briefly the polariton spectrum. In the absence of the driving field we can construct polariton states of the coupled exciton/cavity system as simple products of excitonic and photonic states: $|n_+ n_-\rangle|X\rangle$, where $X = 0, +, -, +-$, and $n_+$, $n_-$ are nonnegative integers. The states can be grouped according to energy shells as shown in the left panel of Fig. 18. Once the interactions are switched on, the states from degenerate shells couple in a characteristic fashion described by arrows in Fig. 18. The result is the removal of degeneracy and a renormalized spectrum, as shown in the right panel. The exact eigenstates are next used to study the interaction of coupled system with the external field by solving the time-dependent problem. Further inclusion of higher exciton population and inelastic relaxation processes is needed to simulate the operation of the single photon gun, and this will be reported in the future.

## 4    Multi-Exciton Complexes

In this section we discuss the ground-state properties and emission spectra of quantum dots filled with electrons and holes, an important ingredient of a quantum dot laser and a quantum-dot-based single photon source. For this reason the electronic properties of electrons and holes in quantum dots have been investigated theoretically by a number of groups [40,41,57,58,59,60,61] [62,63,64]. For a review of experimental results in this domain the reader is referred to the Chapter by *Bayer*. Here we focus primarily on model Hamiltonians but nontrivial physics. Whether this problem is nontrivial is best decided by exploiting the analogy between the quantum Hall (QHE) and quantum dot physics. Both the magnetic field and the confining potential turn the continuous spectrum of a quasi-two-dimensional electron into a discrete spectrum of two harmonic oscillators. The most important aspect of these spectra is the presence of degeneracies. In a dot with a parabolic confining potential in zero magnetic field the dynamical symmetries are responsible for the formation of shells of degenerate states with different angular momenta. For free electrons in a magnetic field, the degeneracies of Landau levels have

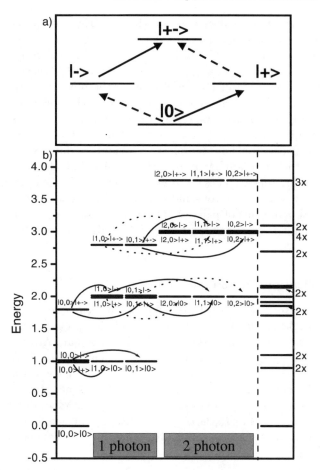

**Fig. 18.** (a) Energy levels and coupling scheme of quantum dot/cavity system. (b) Exciton–phonon energy levels forming almost degenerate shells (*left panel*) and their splitting due to the exciton–photon interaction (*right panel*). (From [95])

strong implications for the ground state of interacting electrons. When only electrons are present in a partially filled, spin polarized lowest Landau level, very complicated and not completely understood ground states of the fractional quantum Hall effect exist [65]. Adding to spin polarized electrons an equal number of spin polarized particles with opposite charges, i.e., holes, changes the ground state drastically. Electrons and holes form a condensate of magneto-excitons due to "hidden symmetries". This many-body ground state is exactly known for any fractional filling [66]. The experimental manifestation of this condensate should be the emission spectrum which does not depend on the filling fraction. Unfortunately, this has not been observed yet.

We have exploited these QHE analogies in studying electrons, and electrons and holes on degenerate shells of a quantum dot. For electrons, exact diagonalization techniques established "generalized Hund's rules" for quantum dots (see the Chapter by *Schweitzer* et al.), in analogy to atoms. However, we were not aware of the existence in nature of atoms composed of particles of opposite charges, and hence of a guiding principle, such as Hund's rules, in their electronic properties. Here we review the results of calculations showing that there are rules, similar to "hidden symmetries", which govern the electronic structure of electron–hole complexes in quantum dots. Some of these symmetries are the ones familiar from QHE physics, but also new symmetries, related to spin, appear. These new symmetries involve not just an electron and a hole but pairs of electrons and pairs of holes as one finds in a biexciton. The unusual behaviour of pairs of carriers can also be found in the Hubbard models of high-$T_c$ materials, and we find it a rather unexpected and exciting development.

## 4.1    Exact Diagonalization Technique

The exact diagonalization technique (EDT) is a numerical technique useful in studying strongly correlated electron systems. The starting point in this method is the single particle spectrum and interacting Hamiltonian. At zero magnetic field the electron energies $E_{mn} = \Omega_0(n + m + 1)$, eigenstates $|mn\rangle$ and angular momenta $L = m - n$ are those of two harmonic oscillators with energy $\Omega_0$ and electronic shells [1,25]. The valence band hole is again treated in the effective mass approximation as a positively charged particle with angular momentum $L = n-m$, opposite to the electron. With a composite index $j = [m, n, \sigma]$ the Hamiltonian of the interacting charge-neutral electron–hole system may be written in compact form as:

$$H = \sum_i E_i^e c_i^+ c_i + \sum_i E_i^h h_i^+ h_i - \sum_{ijkl} \langle ij|V_{eh}|kl\rangle c_i^+ h_j^+ h_k c_l$$
$$+ \frac{1}{2}\sum_{ijkl} \langle ij|V_{ee}|kl\rangle c_i^+ c_j^+ c_k c_l + \frac{1}{2}\sum_{ijkl} \langle ij|V_{hh}|kl\rangle h_i^+ h_j^+ h_k h_l . \tag{39}$$

The operators $c_i^+$ ($c_i$), $h_i^+$ ($h_i$) create (annihilate) the electron or valence band hole in the state $|i\rangle$ with the single-particle energy $E_i$. The two-body Coulomb matrix elements are $\langle ij|V|kl\rangle$ for electron–electron ($e$–$e$), hole–hole ($h$–$h$) and electron–hole ($e$–$h$) scattering, respectively [48,67]. We shall discuss the case of symmetric $e$–$e$, $h$–$h$, and $e$–$h$ interactions. To elucidate the physics, we shall often refer to the matrix elements by specifying which type of carrier, which shell, and whether direct or exchange scattering is involved. For example, $V_{ee}^{pp,x}$ denotes electron–electron exchange scattering involving two electrons on a $p$-shell. It is important to note that the $p$-shell electron–hole scattering matrix elements $V_{eh}^{pp}$ are equal to equivalent $e$–$e$ exchange matrix elements $V_{ee}^{pp,x}$, $\langle 10, 10|V_{eh}|01, 01\rangle = \langle 10, 01|V_{ee}|10, 01\rangle$. The $p$ to $s$-shell electron–hole scattering matrix elements $V_{eh}^{ps}$ are equal to the equivalent $e$–$e$ exchange matrix elements $V_{ee}^{ps,x}$, $\langle 10, 10|V_{eh}|00, 00\rangle = \langle 10, 00|V_{ee}|10, 00\rangle$.

The eigenstates $|\nu\rangle$ of the electron–hole system with $N$ excitons are expanded in products of the electron and hole configurations $|\nu\rangle = (\prod_{i=1}^{N} c_{j_i}^{+}) \times (\prod_{i=1}^{N} h_{k_i}^{+})|0\rangle$. A total of 30 single-particle states (5 shells) of the dot, including spin, was used by *Wojs* and *Hawrylak* [49,61]. The configurations were labelled by total angular momentum $L_{\text{tot}}$ and the $z$-th component of total spin $S_z^{\text{tot}}$. We concentrate here on the optically active subspace of $L_{\text{tot}} = 0$ and $S_z^{\text{tot}} = 0$. An example of typical electronic configurations for zero magnetic field and $N = 7$ spin polarized excitons ($N = 13$ excitons when spin is included) is illustrated in the inset to Fig. 19a. The energy levels are equally spaced, and their degeneracy increases with energy. The shells $s$, $p$, $d$ are completely filled and a single exciton is present on the shells $f$ and $g$. Arrows indicate some relevant configurations with increasing kinetic energies. Due to the large confinement, the lowest kinetic energy configurations are an excellent approximation in the case of filled shells. When electrons and holes partially fill up a degenerate shell the states and energies are completely determined by their mutual interactions, and need to be treated numerically. The calculations for up to $N = 6$ excitons were carried out exactly and a combination of numerical diagonalization in a partially filled shell and the Hartree–Fock approximation extended calculations up to $N = 20$ excitons. In Fig. 19a we show the calculated addition spectrum $\mu - \mu_0$ of excitons added into independent shells of the SAD. Here $\mu = E_G(N) - E_G(N - 1)$ is the chemical potential of the interacting system and $\mu_0$ is the chemical potential of the noninteracting system. The shell energies are separated by $\Omega_+^e = \Omega_-^e = 30\,\text{meV}$ for electrons and $\Omega_+^h = \Omega_-^h = 15\,\text{meV}$ for holes. The addition energy measures the interaction energy of an added exciton with excitons already present in the dot. The lack of dependence of $\mu$ on the number of particles indicates that the electron–hole pairs in each independent shell do not interact. This is to be contrasted with a renormalization of energy for one exciton which is very large, i.e., comparable to the kinetic energy quantization. This large renormalization agrees with our intuition which suggests that in quantum dots electrons and holes are brought closely together and should interact strongly.

Excitons begin to interact weakly when scattering to empty shells with higher energy is allowed. In Fig. 19b we show the results of calculations of addition energy including scattering to higher shells. The addition energy now begins to depend on the number of excitons present on each shell. It also forms a characteristic doublet structure which separates into odd and even numbers of excitons present, suggesting a condensation of biexcitons.

Up to now we analyzed the situation of excitons on isolated shells. Now we consider the effect of filled shells. The electrons and holes in a partially filled shell interact and exchange with particles in filled shells. Hence we can think of quasi-electrons and quasi-holes forming excitons. The energy of

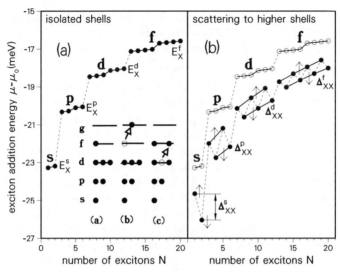

**Fig. 19.** Exciton addition spectrum in a quantum dot: (**a**) addition spectrum of excitons into isolated individual shells ($s$, $p$, $d$ and $f$), (**b**) addition spectrum including scattering to higher shells. *Empty circles* correspond to the isolated shells. *Inset*: illustration of different configurations in the many electron–hole system. (From [61])

quasi-particles is renormalized in what appears as a bang-gap renormalization, as shown in Fig. 20.

In the inset to Fig. 20 we summarize the evolution of the addition spectrum, including the kinetic energy, with the number of excitons $N$ in the dot. The plateaus in the addition spectrum imply that the energy to add/subtract an exciton to/from a partially filled shell does not depend on the filling of this shell. This result, i.e. plateaus in the exciton addition spectrum, were also found in a sophisticated quantum chemistry calculation by *Brasken* et al. [58].

## 4.2 Hidden Symmetries

We now turn to reconcile the notion of strong interactions in quantum dots with the picture of seemingly noninteracting excitons. This lack of interaction can be traced to symmetries in the Hamiltonian. We start by constructing the relevant operators and symmetries. The interband optical processes in a quantum dot are described by the set of interband polarization operators $(P_\sigma^+, P_\sigma^-, P_z)$, which form an algebra of angular momentum. $P_\sigma^+$ ($P_\sigma^-$) creates (annihilates) electron–hole pairs: $P_\sigma^+ = \sum_i c_{i\sigma}^+ h_{i,-\sigma}^+$ ($P_+^- = \sum_i h_{i,-\sigma} c_{i\sigma}$) by annihilating (creating) photons with definite circular polarization [61]. The operator $P_z = (N_\sigma^e + N_\sigma^h - N_{\text{tot}})/2$ measures the population inversion. Possible symmetries are associated with commutation properties of polarization

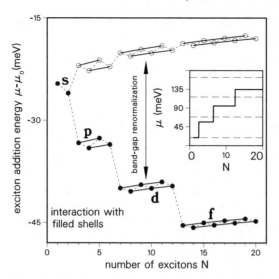

**Fig. 20.** Exciton addition spectrum including interaction with filled shells. *Inset* shows the effect of filling of the dot with excitons on the addition/subtraction spectrum. (From [61])

operator $P^+$ with the Hamiltonian [61]. This commutator,

$$[H, P^+] = \sum_i (E_i^e + E_i^h) c_i^+ h_i^+ - \sum_{ijk} \langle ij|v_{eh}|kk\rangle c_i^+ h_j^+$$

$$+ \frac{1}{2} \sum_{ijkl} (\langle ij|v_{ee}|kl\rangle - \langle ik|v_{eh}|jl\rangle)(c_i^+ h_i^+ c_j^+ c_k - c_i^+ h_k^+ c_j^+ c_l)$$

$$+ \frac{1}{2} \sum_{ijkl} (\langle ij|v_{hh}|kl\rangle - \langle ik|v_{eh}|jl\rangle)(c_l^+ h_i^+ h_j^+ h_k - c_k^+ h_i^+ h_j^+ h_l) \qquad (40)$$

involves a family of two- and four-particle operators. However, if we neglect the spin degrees of freedom, and consider only states of a degenerate shell the degeneracy of single particle levels combined with the symmetry of $(ee)$, $(hh)$, and $(eh)$ interactions cause a remarkable cancellation of the four particle contribution and lead to a very simple result: $[H, P^+] = E_X P^+$, where $E_X$ is the exciton energy for a given shell. This commutation relation enables the construction of exact eigenstates $|N\rangle = (P^+)^N |0\rangle$ of the Hamiltonian by a multiple application of $P^+$ on vacuum. The energy of these states depends linearly on the number of excitons. Hence the energy of addition/subtraction of excitons from these states does not depend on the number of excitons $N$. This is the essence of the "hidden symmetry", a quantum-dot analogue of the QHE. The main difference between a quantum dot and a QHE system is due to spin. Spin starts playing a role when we have more than one carrier of the same type. This has been anticipated by constructing another relevant operator, $Q^+$, creating singlet biexcitons [61]:

$$Q^+ = \frac{1}{2}\sum_{i,j}(c_{i\downarrow}^+ c_{j\uparrow}^+ + c_{j\downarrow}^+ c_{i\uparrow}^+)(h_{i\uparrow}^+ h_{j\downarrow}^+ + h_{j\uparrow}^+ h_{i\downarrow}^+). \tag{41}$$

The singlet operator involves creation of pairs of electrons and pairs of valence holes, a much more complex object. Rather surprisingly, this biexciton operator was shown to satisfy a similar commutation relation to the polarization operator: $[H, Q^+] = E_{XX}Q^+$, with $E_{XX} = 2E_X$. The application of $Q^+$ to the vacuum generates a coherent state of singlet biexcitons, and the energy of this biexciton state is twice the energy of a single exciton. The multiplicative state $(P^+)^2|0\rangle$ and the biexciton state $Q^+|0\rangle$ are degenerate. However, any small perturbation lowers the singlet–singlet state. At this stage we have constructed a family of relevant operators for the construction of exact eigenstates of many electrons and holes on degenerate shells. We shall now illustrate the role of these operators by explicit construction of many-exciton states and diagonalization of the Hamiltonian.

The first case where the role of "hidden symmetries" comes to play is the ground state of the four exciton complex [64]. We start with the triplet–triplet configuration, $|t\rangle = (c_{10\downarrow}^+ h_{10\uparrow}^+)(c_{01\downarrow}^+ h_{01\uparrow}^+)|XX\rangle$, where $|XX\rangle$ is the fully occupied $s$-shell state. By inspection, the triplet state is a multiplicative state generated by $P^+$, i.e. $|t\rangle = (P^+)^2|XX\rangle$. We have written this state explicitly as a product of two electron–hole pairs, each with energy $E_\mathrm{p}$. Two electrons and two holes have parallel spins and lower their respective energy by exchange. The ground state energy, measured from the energy of the filled $s$-shell, is $E_{4Xt} = (2[2T - (V_{ee}^{sp,x} + V_{hh}^{sp,x}) - V_{eh}^{pp,d}] - (V_{ee}^{pp,x} + V_{hh}^{pp,x})$, where $T$ is the total kinetic energy of the electron–hole pair. Because the exchange interaction and electron–hole scattering interactions are equal and attractive, $V_{ee}^{pp,x} = V_{eh}^{pp}$, the energy of the two additional excitons in the $p$-shell of the four exciton complex is exactly twice the energy of a single exciton in a three exciton complex, $E_{4Xt} = 2E_{3X}$. This is so because the lowering of energy due to mixing of electron–hole configurations of the three exciton complex is exactly equal to the exchange energy in a single four exciton configuration.

When both electrons and holes are in singlet configurations, the total number of possible configurations increases. While there was only one triplet–triplet configuration, there are three possible singlet–singlet configurations:

a)  $|a\rangle = (c_{10\uparrow}^+ c_{10\downarrow}^+)(h_{10\downarrow}^+ h_{10\uparrow}^+)|XX\rangle$,

b)  $|b\rangle = \frac{1}{\sqrt{2}}(c_{10\uparrow}^+ c_{01\downarrow}^+ + c_{01\uparrow}^+ c_{10\downarrow}^+)\frac{1}{\sqrt{2}}(h_{10\downarrow}^+ h_{01\uparrow}^+ + h_{01\downarrow}^+ h_{10\uparrow}^+)|XX\rangle$,

c)  $|c\rangle = (c_{01\uparrow}^+ c_{01\downarrow}^+)(h_{01\downarrow}^+ h_{01\uparrow}^+)|XX\rangle$.

The electron–hole scattering can move an electron–hole pair from configuration (b) to either (a) or (c) and mix configurations. The final lowest eigenvalue and eigenstate is [64] $E_{4Xs}(1) = 2E_p - 2V_{eh}^{pp}$ and $|1\rangle = \frac{1}{\sqrt{3}}(|a\rangle + |b\rangle + |c\rangle)$. We see that $E_{4Xs}(1) = 2(E_{3X})$, i.e., the energy of the lowest singlet–singlet state is twice the energy of the single exciton on the $p$-shell. The singlet–singlet and triplet–triplet configurations are degenerate. Moreover, by direct inspection, the state generated by the biexciton operator $Q^+|XX\rangle$ is identical to the singlet–singlet ground state $|1\rangle \approx |a\rangle + |b\rangle + |c\rangle$. In this state all charge-neutral biexciton configurations are occupied with equal probability. Hence we could have written the ground state of the interacting system explicitly without resorting to numerical work.

Now let us check if the same principle applies to higher shells [96]. Let us consider the next shell, the $d$-shell. It consists of two orbitals: $|2,0\rangle$ and $|0,2\rangle$ with angular momentum $\pm 2$ and one orbital $|1,1\rangle$ with angular momentum 0. Let us put two excitons on the $d$-shell in a singlet–singlet configuration and zero total angular momentum. We find eight possible configurations shown in Fig. 21. Configurations (1)–(3) are completely analogous to those of the $p$ shell. For example, the configuration (2) reads $|2\rangle = \frac{1}{\sqrt{2}}(c_{20\uparrow}^+ c_{021\downarrow}^+ + c_{02\uparrow}^+ c_{20\downarrow}^+)\frac{1}{\sqrt{2}}(h_{20\downarrow}^+ h_{02\uparrow}^+ + h_{02\downarrow}^+ h_{20\uparrow}^+)|0\rangle$. Configurations (4)–(6) involve singlets including the additional orbital $|1,1\rangle$. By inspection, configurations (1)–(6) are included in the definition of the operator $Q^+|0\rangle \approx \sum_{i=1,6}|i\rangle$, which is just a sum of these six configurations with equal weight. Note that configurations (1)–(6) involve only orbitals occupied by both electron and a hole so there is local neutrality on each orbital. In a sense, holes shadow electrons as they move from orbital to orbital. The additional configurations (7)–(8) do not satisfy this condition as they involve, e.g., an electron singlet on orbital $|1,1\rangle$ and a hole singlet on orbitals $|2,0\rangle$ and $|0,2\rangle$. Note that the configurations (7)–(8) are not included in the state generated by operator $Q^+$. We have constructed an explicit $8 \times 8$ Hamiltonian for configurations (1)–(8) using symmetric interactions (interaction among electrons, holes, and electrons–holes is identical) and diagonalized it. The energy spectrum, measured from configuration (1), is shown in Fig. 21 without and with the mixing of configurations. We see that without mixing there are three low energy states, corresponding to singlets occupying each of the three orbitals, i.e., configurations (1), (3), and (5). The configuration (5) has the lowest energy as it corresponds to a zero angular momentum state which maximizes (slightly) the electron–hole attraction. The higher-energy band at $\sim 0.2$ energy units is shifted by the repulsive exchange energy (singlet configurations) and corresponds to configurations (2), (4) and (5). Finally, the highest energy band corresponds to the additional configurations (7),(8) which can be understood as due to lack of local charge neutrality. The correlated spectrum, i.e., the one calculated including the mixing of configurations, is much broader in energy, with one well isolated ground state. The amplitudes $A_i$ of the ground-state wave function $|GS\rangle \approx \sum_{i=1,6} A_i|i\rangle$ are shown in Fig. 21. We see that these coefficients are almost exactly equal to $1/\sqrt{6} = 0.408$ for the first six configurations, and almost zero for the last two configurations. There

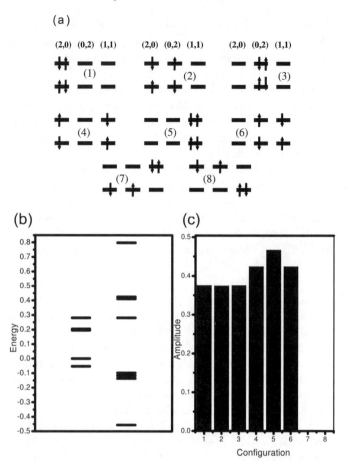

**Fig. 21.** (a) Two-exciton configurations on the $d$-shell. (b) Energy spectrum in units of the $s$-shell Coulomb attraction for a system without (*left*) and with (*right*) configuration mixing. (c) Amplitude of all the configurations in the ground state. (From [96])

is a small enhancement of the amplitude corresponding to configuration (5) as the $|1,1\rangle$ orbital leads to slightly different Coulomb matrix elements. The overlap of the ground state and the state created by the operator $Q^+$ equals 0.996. When we force the orbital $|1,1\rangle$ to give the same Coulomb matrix elements as orbitals $|2,0\rangle$, $|0,2\rangle$, the state $Q^+|0\rangle$ is the exact eigenstate of the interacting system. When pairs of holes shadow pairs of electrons, they cannot be scattered in or out of this coherent state, and the state remains a ground state. By the same analysis we can construct the second and third multiplicative state: $(Q^+)^2|0\rangle$ and $(Q^+)^3|0\rangle$ as ground states of the four and six excitons on the $d$-shell. This simple analysis nicely illustrates the power of "hidden symmetries".

## 4.3   Excited States and Emission Spectra

Hidden symmetries tell us that the energy needed to remove an exciton from a partially filled $p$-shell does not depend on the filling of the shell. Hence the emission spectrum of the $p$-shell does not depend on the population of this shell. In order to distinguish spectra corresponding to different numbers of excitons we need to investigate the removal of excitons from a filled $s$-shell as a function of the filling of the $p$-shell. These processes leave the final-state exciton droplet in an excited state, and hidden symmetry no longer applies. The calculated emission spectra [64] for a typical ratio of the Coulomb to kinetic energy $V_0/t = 0.5$ are shown in Fig. 22. The $1X$ and $2X$ recombination spectrum corresponds to a single recombination line. The $2X$ emission is at a slightly lower energy and its amplitude is twice that of the single exciton because there are two final exciton states, with two different spin orientations. In the emission spectra from the ground state of the three-exciton complex the final $2X$ states can be either triplets or singlets. Therefore one expects to see one emission line for the emission from the $p$-shell and two emission lines from the $s$-shell. However, only the emission to the triplet biexciton state is visible. The emission energy is almost identical to that of a single exciton. The emission to the singlet–singlet final state is quenched by the quantum interference effect. The remaining possibility is the emission to singlet/triplet and triplet/singlet final states (not shown).

The emission from the $4X$ complex to the excited states of the $3X$ complex consists of two spectra originating from almost degenerate singlet–singlet and triplet–triplet initial $4X$ states. The recombination spectra consist of three groups of states: (a) the recombination from the $p$-shell which is degenerate with the recombination from the $3X$ complex, (b) the recombination to the final $3X$ triplet–triplet state, which is close in energy to the recombination from a single exciton, and (c) the lower energy band of the $3X$ excited states.

The emission from the $5X$ complex to the singlet–singlet and triplet–triplet $4X$ final states shows a very strong emission from the $p$-shell and almost no emission from the $s$-shell. The weak emission from the $s$-shell is caused by a similar interference effect to that for the $3X$ complex. The matrix element for the recombination from the $5X$ ground state to the ground and one excited singlet–singlet $4X$ states is given by $|\langle (4X, ss), f|P^-|5X\rangle|^2 = |\sqrt{3/2}A_0^f + A_+^f/2|^2$, where coefficients $A^f$ are eigenvectors of the four exciton singlet–singlet Hamiltonian. These coefficients have identical sign for the emission to the ground state but opposite sign for the emission to the excited state. The numerical coefficients multiplying $A^f$ lead to almost complete cancellation of the matrix element for the $s$-shell recombination. Hence the recombination to the ground state is enhanced and to the excited state, with a missing exciton in the $s$-shell, is reduced. There is also emission to the singlet–triplet $4X$ final states, but it is not included in this figure.

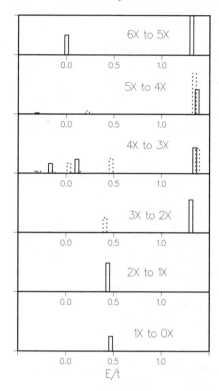

**Fig. 22.** Emission spectra from $N$-exciton ground states to $(N-1)$-exciton ground and excited states for the ratio of interaction to kinetic energies $V_0/t = 0.5$. Only same total spin configurations for electrons and holes in the final state are shown. (From [64])

The final result, Fig. 22, shows that in the $5X$ complex the emission from the $s$-shell is suppressed in comparison with the lower density quantum dot, i.e, with either 4 or 3 excitons, a rather counterintuitive result.

The emission from the $s$-shell is recovered when our dot is completely filled with $6X$.

The emission from the $s$-shell corresponds now to the removal of a $s$-shell exciton, without mixing with other configurations. The energy of this quasi-exciton is renormalized by the exchange interaction of the $s$-shell electron and a hole with electrons and holes on the $p$-shell. This lowers the energy of the emission band by the exchange self-energy of the exciton $2(V_{ee}^{sp,x} + V_{hh}^{sp,x})$.

To summarize, we discussed here the electronic properties of artificial atoms composed of a fixed number of electrons and holes confined to a parabolic quantum dot. We showed that the emission spectra show strong variation with the number of excitons in the dot. The variations are due to filling different electronic shells and for partial filling, due to the condensation of electrons and holes into coherent states due to relevant symmetries. We identified the relevant "hidden symmetries" and showed that they replace Hund's rules in two-component systems. These calculations were used as a fingerprint of excitonic artificial atoms observed in the "single dot spectroscopy" experiments described in *Bayer* et al. [68] and by *Bayer* in this volume.

# 5    Charged Quantum Dots

In this section we discuss the electronic properties of quantum dots charged with electrons by either modulation-doping or tunnelling. We start by reviewing the ground state properties and follow up with an extensive discussion of the expected emission, absorption, far infrared and photo-current spectra as a function of the number of electrons. We discuss the generalized Hund's rules for parabolic dots which were theoretically established early on [69,70]. The Hund's rules were verified in vertical quantum dots [71], but not yet in self-assembled dots. For this reason we focus on their manifestation in optical experiments. We discuss the band-gap renormalization, the shake-up, and the ground state emission as a function of the number of electrons. We emphazise a nontrivial effect which takes place in the recombination process, namely the breakdown of the optically created hole in the electronic state at the bottom of the filled band and a splitting of the emission line. Similar electron-number-sensitive spectra are obtained in the absorption and intraband spectroscopy.

The electronic properties of modulation doped quantum dots are determined by the number and structure of bound single particle levels, the form of the Coulomb interaction among carriers, and many-particle configurations which depend strongly on the number of carriers. The single particle levels of electrons and holes in smooth confining potentials are described by the coupled harmonic oscillator spectrum with characteristic degeneracies and shell spacing $\omega_0$. The electronic structure depends on the ratio of the kinetic energy quantization $\omega_0$ to the interaction energy $V_0 \sim \sqrt{\omega_0}$. For large confining energy the electronic quantum dots satisfy Hund's rules [69,70], while the excitonic quantum dots are controlled by "hidden symmetries" [61,64]. For small confining energy many different correlated phases are expected [2,25,72,73] [74,75]. Quantum dots containing free carriers are interesting examples of strongly correlated electron and electron–hole systems. For confined electrons one might expect that their electronic structure, and in particular electronic correlations, are a function of the confining potential. However, this is not the case. In smooth confining potentials, the centre of mass motion does not depend on the confining potential [25] and the electron–electron interaction contribution to the energy, and hence correlations, is the same as in the translationally invariant system. Hence quantum dots offer a unique insight into correlated electronic states, both those already existing in extended quasi-two-dimensional systems, and those particular to finite systems. This is especially true in a strong magnetic field where these correlated states are analogues of incompressible liquids encountered in the FQHE [73,74,75], or spin excitations encountered in quantum Hall ferromagnets [75,76,77]. The second obvious property of large quantum dots is that they are finite electronic systems and therefore have edges. In strong magnetic fields, edge excitations of quantum Hall droplets are examples of chiral Fermi and Luttinger liquids [78,79]. Optical properties of edge excitations of quantum Hall droplets were discussed elsewhere [46]. Here we focus on small self-assembled quantum dots.

## 5.1   Ground State of Charged Dots

Our discussion of ground-state properties follows *Wojs* and *Hawrylak* [69,70]. The single-particle spectrum is that of two harmonic oscillators. The many-body Hamiltonian of a charged quantum dot is given as:

$$H = \sum_{nm\sigma} (E_{nm} + g\mu_B B\sigma)\, c_{nm\sigma}^\dagger c_{nm\sigma}$$

$$+ \frac{1}{2} \sum_{\substack{n_1 m_1 \sigma_1\ n_2 m_2 \sigma_2 \\ n_1' m_1'\quad n_2' m_2'}} \langle n_1' m_1', n_2' m_2' | V_{ee} | n_2 m_2, n_1 m_1 \rangle$$

$$\times c_{m_1' n_1' \sigma_1}^\dagger c_{m_2' n_2' \sigma_2}^\dagger c_{m_2 n_2 \sigma_2} c_{m_1 n_1 \sigma_1}\,, \tag{42}$$

where $c_{nm\sigma}^\dagger$ ($c_{nm\sigma}$) are the operators creating (annihilating) an electron with the spin $\sigma$ in the harmonic oscillator (FD) state $|nm\rangle$, $V_{ee}$ is the electron–electron Coulomb repulsion and $g\mu_B B\sigma$ is the Zeeman energy. The two-body Coulomb matrix elements $\langle n_1' m_1', n_2' m_2' | V_{ee} | n_2 m_2, n_1 m_1 \rangle$ in the FD basis are given in [67]. The Coulomb interaction conserves the total angular momentum $R$ and the $z$-th component of the total spin $S$ hence the diagonalization of the Hamiltonian (42) has been performed separately for each $(R, S_z)$ subspace. In addition to these quantum numbers, there are other symmetries. One is associated with total spin $S^2$ which commutes with the Hamiltonian. All eigenstates of the Hamiltonian can be classified by total spin. There is however another symmetry related to the centre of mass (CM) motion. The CM operator generates an algebra similar to the algebra generated by the "hidden symmetry" operators and allows for the additional classification of states. The total-spin resolved and CM-resolved calculations in second quantization will be reported soon. An explicit spin and CM resolved calculation was reported in [25].

In what follows we use a simple configuration–interaction approach. The basis of the few-electron Hilbert space was constructed from the properly antisymmetrized products of single-electron FD states (including spin $\sigma$): $|n_1 m_1 \sigma_1, n_2 m_2 \sigma_2, \ldots, n_N m_N \sigma_N\rangle \equiv \hat{A}\, \Pi_{i=1}^N |n_i m_i\rangle |\sigma_i\rangle$. The calculations were carried out for InGaAs quantum dot with shell spacing of $\approx 30$ meV. For each value of magnetic field the lowest four FD states were kept and the mixing between confined FD states and the wetting layer continuum was neglected. Figure 23 shows the ground-state (GS) energies of the interacting $N$-electron systems as a function of magnetic field. The curves corresponding to different $N$ have been vertically shifted so that all of them can be shown in a single frame. The zero-field GS energies have been marked on the vertical axis and the arrow in the top-left corner gives the vertical scale common to all curves. The pairs of numbers in brackets (total angular momentum, total

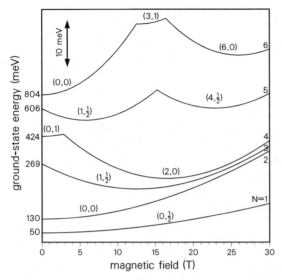

**Fig. 23.** Ground state energies of the many-electron systems as a function of the magnetic field. The *curves* have been vertically shifted to fit in the frame – the energies in the zero field are indicated on the *vertical axis* and the scale common for all curves is given in the *top-left corner*. Configurations of the states $(R, S_z)$ have also been shown. (From [69])

spin) $= (R, S_z)$ indicate the evolution of GS configurations with number of electrons.

The GSs of $N = 1$, 2 and 3 electrons are well approximated by the lowest kinetic energy noninteracting states: $|00 \downarrow\rangle$, $|00 \downarrow, 00 \uparrow\rangle$, and $|00 \downarrow, 00 \uparrow, 01 \downarrow\rangle$, respectively. The first magnetic-field-induced transition in the GS takes place for $N = 4$ electrons, i.e., two electrons in a partially filled $p$-shell. For low magnetic fields the two electrons on the half-filled second shell take advantage of the degeneracy of the two FD orbitals ($|01\rangle$ and $|10\rangle$) and maximize their total spin $S_z = 1$ to lower their energy by an exchange-interaction term $\langle 01, 10|V_{ee}|01, 10\rangle$. This is Hund's rule for self-assembled dots. In non-zero fields the degeneracy of the second shell is removed and the kinetic energy of the maximum spin configuration increases. At $B \approx 2.8$ T the gain in exchange energy is overtaken by an increase of kinetic energy and the system reverts to a spin singlet $S_z = 0$ lower kinetic energy configuration.

The transition in the $N = 5$ electron system at 15 T corresponds to changing of the FD orbital by the outer electron (from $|10\rangle$ into $|02\rangle$) in the presence of the rigid four-electron core. It occurs earlier than the FD levels actually cross since the new many-body state has a lower interaction energy. In the case of six electrons the situation is more complex as there are two GS transitions. These transitions correspond to the singlet–triplet–singlet transitions. The singlet–triplet transition is associated with the magnetic-field-induced

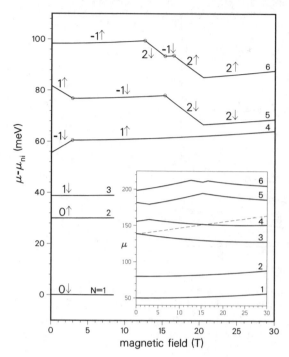

**Fig. 24.** Charging energies (difference in the chemical potential due to interaction) of the SAD as a function of the magnetic field. The angular momenta and spins of orbitals into which the electron is added are indicated. *Inset* shows the chemical potential. (From [69])

degeneracy of two orbitals $|10\rangle$ and $|02\rangle$. There is a Hund's rule associated with this degenerate shell, with the triplet ground state corresponding to two electrons occupying the degenerate $|10\rangle$ and $|02\rangle$ orbitals. Past the triplet transition the six electrons occupy, with spin up and down, the successive levels $|m,0\rangle$ of the lowest Landau level and form a small, filling factor 2, quantum Hall droplet [80].

In Fig. 24 we present the chemical potential of the SAD for up to six electrons, extracted from the data shown in Fig. 23. The figure shows the charging energy, i.e., the difference between the chemical potentials of the interacting and noninteracting systems. The transitions in the four, five and six electron GSs have been marked with open circles. The change of slope for $N = 5$ and 6 at 20.5 T is due to the crossing of FD levels. This crossing results in the singlet–triplet–singlet transition in the $N = 6$ electron droplet.

The inset in Fig. 24 shows the chemical potential $\mu$, a directly measured quantity. The dashed line shows the energy of the bottom of the wetting layer continuum. In the absence of a positive charge layer nearby, we find that only $N = 3$–4 electrons can be injected to the dot with energy below the wetting layer. This number increases with the size of the dot and inclusion of positive charges. Taking the dot diameter of 250 Å yields the inter-shell separation $\omega_0$ measured by *Drexler* et al. [81] ($\approx$ 41 meV). For this size one finds that the dot can be charged with up to $N = 5$ electrons.

## 5.2   Emission

The emission from charged self-assembled quantum dots has been investigated experimentally by a number of groups [82,83,84] and theoretically by *Wojs* et al. [70]. The purpose of these investigations was to establish whether one can determine the number of electrons $N$ in the dot from the emission spectrum. This is a valid question since the recombination process involves the recombination of valence holes from the $|0,0\rangle$ topmost valence level with the lowest electron level $|0,0\rangle$ of the dot. This process leaves an empty electron state at the bottom of the occupied band. In the absence of electron–electron interactions the emission energy does not depend on the number $N$ of electrons in the dot, and hence PL is an ideal candidate to measure the effect of electron–electron interactions.

To describe the initial state we consider a system of $N$ electrons and a single valence-band hole [46,48,61]. The final state corresponds to a system of $N - 1$ electrons and a photon. Using composite indices $i$ to describe single-particle states $|i\rangle$, the initial-state many-particle Hamiltonian can be written as:

$$H = \sum_i \varepsilon_i^e \, c_i^\dagger c_i + \sum_i \varepsilon_i^h \, h_i^\dagger h_i + \frac{1}{2} \sum_{ijkl} \langle ij|V_{ee}|kl\rangle \, c_i^\dagger c_j^\dagger c_k c_l$$

$$+ \sum_{ijkl} \langle ij|V_{eh}|kl\rangle \, c_i^\dagger h_j^\dagger h_k c_l \,. \qquad (43)$$

The operators $c_i^\dagger$ and $c_i$ ($h_i^\dagger$ and $h_i$) create and annihilate an electron (a valence hole) in the single-particle state $|i\rangle$. The first and the second terms in (43) describe the kinetic energy of electrons and a valence hole in the dot, and the third and fourth terms describe the electron–electron and electron– valence hole scattering. Here $\langle ij|V|kl\rangle$ are the two-body matrix elements of the electron–electron (*ee*) [67] and electron–valence hole (*eh*) [48] Coulomb interactions, respectively.

The PL intensity $E(\omega)$ as a function of the photon frequency $\omega$ is given by Fermi's golden rule:

$$E(\omega) = \sum_f |\langle \nu_f|\mathcal{P}|\nu_i\rangle|^2 \delta(\mathcal{E}^i - \mathcal{E}^f - \omega) \,, \qquad (44)$$

where $|\nu_i\rangle$ and $|\nu_f\rangle$ are the initial and final states of the system, with corresponding energies $\mathcal{E}^i$ and $\mathcal{E}^f$ (we take $\hbar = 1$). The interband polarization operator $\mathcal{P} = \sum_{kl} \langle k|l\rangle \, c_k h_l$ removes an electron–valence hole pair from the ground state $|\nu_i\rangle$ of the photo-excited system. The selection rules are hidden in the electron–valence hole overlap matrix element $\langle k|l\rangle$. Since there

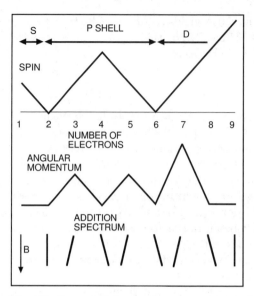

**Fig. 25.** Ground state angular momentum and total spin as a function of the number of electrons in a quantum dot. The occupation of successive orbitals leads to the magnetic field behaviour of the addition spectrum as shown at the *bottom*. (From [70])

is only a single valence hole we can define a purely electronic initial state $\left|\nu_i^k\right\rangle = \sum_l \langle k|l\rangle h_l |\nu_i\rangle$. The operator $\mathcal{P}$ creates a hole in the initial state: $\mathcal{P}|\nu_i\rangle = \sum_k c_k \left|\nu_i^k\right\rangle$, and the PL spectrum can be viewed as the spectral function of this hole.

In self-assembled dots with large kinetic energy quantization $\omega_0$ the electronic configurations are filled according to their increasing kinetic energy. The only difficulty arises for partially filled degenerate shells which are filled according to Hund's rules, i.e., electrons tend to maximize their total spin. Hund's rules for quantum dots are summarized in Fig. 25, where total spin and total angular momentum of the calculated ground state of a quantum dot as a function of $N$ are shown. We see that closed shells have zero spin and partially filled shells have maximum spin. The rules of occupying single particle orbitals impact the magnetic behaviour of the energy of adding an electron. Below the total angular momentum behaviour of the ground state we show also a qualitative dependence of the addition energy with the magnetic field. This characteristic dependence of addition energy has been observed in vertical tunnelling quantum dots [71]. Far infrared (FIR) spectroscopy of charged quantum dots [81,85] has also shown features specific to a number of particles in the dot [69].

Let us qualitatively describe the results of our investigation of emission from charged quantum dots by considering in detail the emission from the

$N = 4$ electron dot. The $N = 4$ electron dot has a filled $s$-shell and a partially filled $p$-shell. The two $p$-electrons are in a triplet state. In the initial state we have the 4-electron complex and a valence hole with two possible spin orientations, as shown in Fig. 26a,b. Specific spin orientation of the valence hole selects an electron with a specific spin for the recombination process, and hence leaves different spin configurations of an electron left in the partially filled $s$-shell. For initial spin configuration Fig. 26b there is only one final state with all electrons with parallel spins. Since we removed an electron with spin opposite to all other electrons, the energy of recombination for this spin configuration is exactly the same as the energy of a single exciton $E(1)$. The same is true for $N = \{1, \ldots, 4\}$. For the spin configuration shown in Fig. 26a the final configuration involves a spin down electron in the $s$-shell and two electrons with parallel spins in the $p$-shell. This configuration belongs to a group of 4 degenerate configurations. The most important group A, B, C is obtained by simultaneously flipping the spin of the $p$-shell electron and the $s$-shell electron. Alternatively, we can think of the spin-down electron circulating on a plaquette of three sites [25]. The three degenerate states strongly mix, but in the end only two of them have a finite overlap with the optically created configuration A. Due to the resonant configuration mixing the optically created configuration ceases to exist and breaks into two configurations with energies $E^1 = E(A) + 0.46E_0$ and $E^2 = E(A) - 0.27E_0$ and transition probabilities of 0.63 and 0.37, respectively [94]. The splitting of the two peaks is very large, i.e., $\approx 70\% E_0$. $E_0$ equals the binding energy of the $s$-shell electron–hole pair.

In addition to the spin-related resonant configuration mixing there are resonant Auger processes, illustrated in Fig. 26. This process involves one electron jumping down and annihilating the optically created hole in the $s$-shell, and a second electron jumping up to the unoccupied $d$ level. This process involves final state electrons in $s$ and $d$-shells and contributes much less into the breakdown of the optically created hole. In addition to resonant configurations, we also have the excited final state configurations coupled to the optically created configuration. These shake-up configurations involve an electron excited by twice the electron kinetic energy $2T = 2\omega_0$ and the emission line shifted down in energy by the same amount. In general, electron–electron interactions tend to lower the energy of emitted photons, an effect called band-gap renormalization. In a quantum dot a new effect, an emission at higher energy, appears when the $d$-shell becomes populated. The initial and final configurations for $N = 9$ are shown in Fig. 27. The three electrons on the $d$-shell are needed to populate the $|1, 1\rangle$ state with angular momentum zero. The configuration with the optically created hole is mixed by Coulomb interactions with a configuration with hole relaxed to the zero angular momentum state on the $d$-shell. This is clearly a lower energy configuration and emission from this configuration will appear at higher energy, shifted approximately by $2T$ from the main emission line. In Fig. 28 we summarize the

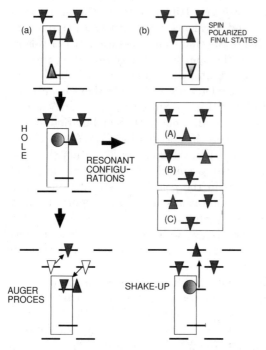

**Fig. 26a,b.** Initial and final configurations in the emission from a 4-electron quantum dot. (From [94])

effect of electrons on the emission spectrum of a quantum dot [94]. The effect of free carriers is to broaden and shift the emission spectrum to lower energy. The main emission spectrum is accompanied by a shake-up band below the main emission line, at energy ≈ twice the kinetic energy quantization for electrons. When the zero angular momentum channel of higher shells is populated, the photo-created hole in the electronic system can relax to the Fermi level and an emission at higher energies (approximately at the kinetic energy of an even number of electrons) should be observable. A preliminary observation of this high energy band has been reported in [82]. Perhaps the most startling and interesting phenomenon is associated with partially filled shells. Due to a large number of spin degrees of freedom of a partially filled shell, the simple picture of a photo-created hole at the bottom of the occupied band breaks down. This quasi-particle breakdown manifests in the splitting of the emission line for a partially filled shell and a return to a single emission line for closed shells (broadened by Auger processes). Hence the width of the recombination spectrum should oscillate as a function of the filling factor of successive shells. The physical picture of invoking spin degrees of freedom of partially filled shells of degenerate levels is identical to the picture invoked

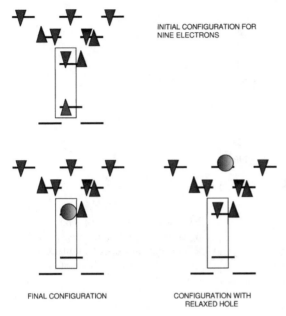

INITIAL CONFIGURATION FOR
NINE ELECTRONS

FINAL CONFIGURATION          CONFIGURATION WITH
                              RELAXED HOLE

**Fig. 27.** Initial and final configurations in the emission from a 9-electron quantum dot. (From [94])

to explain anomalies and oscillations in the emission line from the 2DEG at odd and even filling factors [86].

## 5.3 Absorption

In absorption, an electron–hole pair is added to the existing $N$ electron system. The photo-injected electron cannot be distinguished from electrons already present, and the final state Hamiltonian describes $N + 1$ electrons and one valence hole. The final state Hamiltonian is the same as the initial state Hamiltonian in the emission.

The final state Hamiltonian describes $N + 1$ electrons interacting with a single valence hole. The system before optical excitation contained $N$ electrons. We wish to know the properties of this "initial state" $N$ electron system. Let us assume that we know exact eigenstates of the initial $|u\rangle_N$ and final $|v_{N+1}\rangle$ electronic systems. The final state droplet is negatively charged and attracts a valence hole. To account for this interaction we must form many-particle Wannier states $|f\rangle = \sum_{i,v} A^f_{i,v} h^+_i |v_{N+1}\rangle$ as a sum of products of many-electron and valence hole states with amplitudes $A^f_{i,v}$ determined by interactions. This linear combination means that the hole mixes

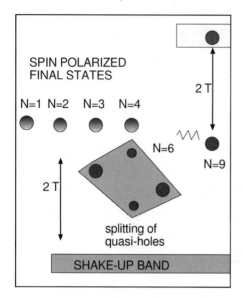

**Fig. 28.** Summary of characteristic features in emission spectrum from a modulation doped quantum dot as a function of the number of electrons. (From [94])

many-electron states and increases the number of degrees of freedom. The absorption spectrum, given by

$$A(\omega) = \sum_f |\langle f|P^+|u_N\rangle|^2 \delta\{\omega - (E_f(N+1) - E_u(N))\} \tag{45}$$

describes the addition of an electron–valence hole pair. The transition probability matrix element $\langle f|P^+|u_N\rangle = \sum_{i,v} A_{i,v}^f \langle v_{N+1}|c_i^+|u_N\rangle$ differs from the matrix element associated with just adding an electron $\langle v_{N+1}|c_i^+|u_N\rangle$ into a single particle state $i$. The interference effects, contained in squares of the transition matrix element $|\langle f|P^+|u_N\rangle|^2$, lead to nontrivial effects, such as the Fermi edge singularity (FES) [87], even in the absence of $e$–$e$ interactions. Little is known about the effect of electron–electron interactions on the FES. This aspect has been addressed in [45]. In related work, *Warburton* et al. [44] measured absorption spectra of ensembles of charged quantum dots as a function of the number of electrons $N$ in the dot. His experiments nicely demonstrated the Pauli exclusion principle, i.e., that once electronic shells are fully occupied the absorption process into these shells is blocked. Simple mean-field calculations of a few-electron and single hole configuration reasonably well described the energetic shifts of the inhomogeneously broadened absorption spectrum as a function of the number of carriers [88]. However, a detailed study [45] showed that an exciton in the presence of electrons ceases to be a well defined quasi-particle, leading to much more complex spectra. This is well illustrated by the absorption of the $N = 4$ electron dot.

The lowest kinetic energy state with energy $t = 6\Omega_e$, and the first excited states, having an excess of kinetic energy equal to $\Delta = 2\Omega_e$, of the $N = 4$

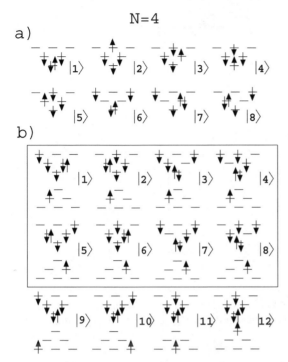

**Fig. 29.** (a) Electronic configurations for $N = 4$ electrons in the lowest kinetic energy state compatible with Hund's rule and low-lying excitations. (b) Low-lying states for $N + 1 = 5$ electrons plus a valence hole. (From [45])

electron dot, are shown in Fig. 29a. Excited $N = 4$ electronic states separate into two groups: i) four configurations 2–5 with singly occupied orbitals, and ii) three configurations 6–8 with doubly occupied orbitals. Configurations in group i) differ only by the position of the spin up electron. Alternatively, we can think of these states as generated by the hopping of the spin up electron on a four orbital plaquette.

The Hamiltonian of $N = 4$ interacting electrons was diagonalized within the Hilbert space shown in Fig. 29 for different ratios of Coulomb to kinetic energy $V_0/\Omega_e$ [45]. The ground state for $V_0/\Omega_e = 0.5$ was very well approximated by the lowest kinetic energy configuration $|1\rangle$ with $|A_1^G|^2 = 0.985$. Increasing the strength of Coulomb interactions to $V_0/\Omega_e = 1$ gives $|A_1^G|^2 = 0.895$. Configurations $|2\rangle$, $|7\rangle$ and $|8\rangle$ have a small contribution to the ground-state: $|A_2^G|^2 = 0.011$, $|A_7^G|^2 = 0.042$, and $|A_8^G|^2 = 0.042$. Increasing the value of Coulomb interactions penalizes doubly occupied configurations and favors singly occupied configurations. The quadruplet configuration composed of configurations $|2\rangle$–$|5\rangle$ becomes the lowest energy configuration for $V_0 > 2.5\Omega_e$.

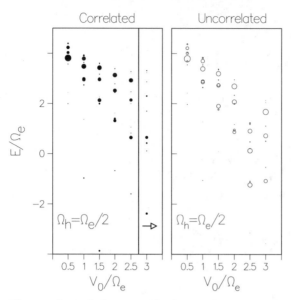

**Fig. 30.** Optical absorption for $N = 4$ electrons in the SAD with $\sigma_-$ polarized light. *Left*: correlated initial state. *Right*: ground state given by a single configuration. $\sigma_-$ polarization. (From [45])

We now turn to the absorption spectrum. In Fig. 29b we show all final state configurations built from configurations shown in Fig. 29a by the addition of an electron–hole pair. These configurations can be viewed as excitations of the ground state in the presence of the photo-excited electron–hole pair. They are allowed because the valence hole recoils and absorbs energy and momentum. In Fig. 29 the state $|1\rangle$ can be obtained from the optically created configuration $|9\rangle$, exciting a spin-up electron from state $|00\rangle$ to state $|10\rangle$. This new configuration corresponds to an electron–hole pair excitation of the ground state coupled with the relaxation of the valence hole. All other states can be regarded as one electron–hole pair excitations: $|3\rangle$ being an excitation of state $|9\rangle$; $|2\rangle$, $|4\rangle$, and $|6\rangle$ are excited states of configuration $|11\rangle$; $|5\rangle$, $|7\rangle$, and $|8\rangle$ are excited states of $|10\rangle$.

Figure 30 shows the optical absorption spectrum obtained for the correlated ground state (left panel) for different values of the ratio $V_0/\Omega_e$. For comparison, the optical absorption neglecting correlations is also shown in the right panel. The absorption spectrum for a correlated ground state shows a band of peaks corresponding to initial and final states in Fig. 29. New optical transitions appear due to the interference effects in the optical matrix element. These transitions slightly broaden the main peaks. Up to $V_0/\Omega_e \leq 2.5$ the effects of correlations are small. The ground state is dominated by the configuration $|1\rangle$ consistent with Hund's rules and neglecting correlations in the ground state does not lead to significant changes in the absorption spectrum.

For $V_0/\Omega_e > 2.5$ the ground state is no longer described by Hund's rules and is a quadruplet. The absorption spectrum changes drastically at this point to reflect the changes in the ground state.

For a systematic discussion of absorption spectra as a function of the number of electrons in the dot see [45].

## 5.4    Far-Infrared Spectroscopy

*Drexler* et al. [81] and *Fricke* et al. [85] reported FIR absorption measurements of SADs in a magnetic field. The dots were charged with up to $N = 6$ electrons filling the $s$ and $p$ electronic shells. The infra-red spectroscopy was used to study the electronic excitations of the dots as a function of the number of electrons and the magnetic field.

The excitations of SAD reflect the electronic structure and the number of electrons in the dot. For an infinite parabolic confinement only the centre of mass excitations with frequencies $\Omega_+$ and $\Omega_-$ (generalized Kohn's theorem) [1] can be measured in FIR. In SAD, a finite number of confined FD levels leads to additional transitions in the IR spectrum related to the magnetic-field-induced changes in the GS, e.g. spin triplet to spin singlet transition discussed above.

The FIR absorption for $N$ electrons can be conveniently expressed in terms of the FD creation/annihilation operators $a$ and $b$ [67,69,89]:

$$A(\omega) \propto \sum_f |\langle f| \sum_{j=1}^N (a_j + a_j^\dagger + b_j + b_j^\dagger)|i\rangle|^2 \delta(E_f - E_i - \omega), \qquad (46)$$

where $|i\rangle$ is the initial (ground) state and the summation is over all bound final states $|f\rangle$. The infrared radiation connects only the states with the same $S_z^{tot}$ and total angular momenta different by $\pm 1$. In Fig. 31a we show the magnetic field evolutions of the IR spectra calculated for the SAD with $N = 4$ electrons. The area of each circle is proportional to the intensity $A(\omega)$. The solid lines show the transition energies $\Omega_\pm$ of the noninteracting system and a vertical line marks the spin transition in the GS. The GS and two excited single particle configurations for $B \geq 2.8\,\mathrm{T}$ are shown in Fig. 31b. The two excited configurations responsible for the splitting of the transition for $N = 4$, 5 are coupled through Coulomb interactions. Experiments by *Fricke* et al. [85] indeed showed the splitting predicted here which illustrates the desired sensitivity of the intraband optical transitions to the number of electrons $N$.

## 5.5    Photocurrent Spectroscopy

Photocurrent (PC) spectroscopy [90,91,92,93] is an attractive alternative to interband, capacitance, and FIR spectroscopies of modulation doped self-

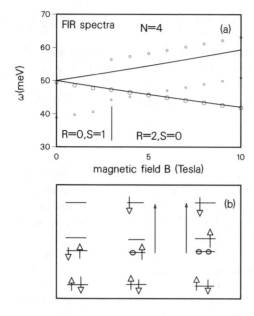

**Fig. 31.** (a) Far-infrared spectra of SAD with $N = 4$ electrons; (b) ground state and excited single particle configurations with *arrows* indicating transitions

assembled quantum dots. In PC spectroscopy an absorption of infrared photons results in transitions from populated bound states in the quantum dot to scattering states in the GaAs barrier region. The photo-excited electrons are collected by electrodes and current is measured as a function of photon frequency and polarization. The main advantage of PC over interband spectroscopy is that it involves only conduction electrons and leaves out all the complications of the valence band. PC spectroscopy is related to FIR spectroscopy, which however couples primarily to centre-of-mass excitation of bound electrons. Understanding PC spectroscopy is important for quantum dot-based photodetectors. In [93] theoretical foundations for PC spectroscopy were outlined, and preliminary experimental results presented. These results show the sensitivity of the photocurrent to both the frequency and polarization of incident radiation. More experiments and theoretical work is in progress.

# References

1. L. Jacak, P. Hawrylak, A. Wojs: *Quantum Dots* (Springer, Berlin, Heidelberg 1998)
   M. Grundmann, N. Ledentsov, D. Bimberg: *Quantum Dot Heterostructures* (Wiley, New York 1998)
   P. M. Petroff, S. P. Denbaars: Superlatt. Microstruct. **15**, 15 (1994)
   M. Kastner: Physics Today **46**, 24 (1993)
   T. Chakraborty: Comments Cond. Matt. Phys. **16**, 35 (1992)

S. Fafard, Z. R. Wasilewski, C. N. Allen, D. Picard, P. G. Piva, J. P. McCaffrey: Superlatt. Microstruct. **25**, 87 (1999)

2. R. C. Ashoori: Nature **379**, 413 (1996)
3. P. Hawrylak, S. Fafard, Z. Wasilewski: Cond. Matt. News **7**, 16 (1999)
4. L. D. Landau, E. M. Lifshitz: *Theory of Elasticity* (Butterworth-Heinemann, Oxford 1995)
5. J. R. Downes, D. A. Faux, E. P. O'Reilly: Mater. Sci. Eng. B **35**, 357 (1995)
   J. R. Downes, D. A. Faux: J. Appl. Phys. **77**, 2444 (1995)
   D. A. Faux, J. R. Downes, E. P. O'Reilly: J. Appl. Phys. **80**, 2515 (1996)
   D. A. Faux, J. R. Downes, E. P. O'Reilly: J. Appl. Phys. **82**, 3754 (1997)
   A. D. Andreev, J. R. Downes, D. A. Faux, E. P. O'Reilly: J. Appl. Phys. **86**, 297 (1999)
6. J. R. Downes, D. A. Faux, E. P. O'Reilly: J. Appl. Phys. **81**, 6700 (1997)
7. M. Grundmann, O. Stier, D. Bimberg: Phys. Rev. B **52**, 11969 (1995)
8. J. H. Davies: J. Appl. Phys. **84**, 1358 (1998)
9. C. Pryor, M. E. Pistol, L. Samuelson: Phys. Rev. B **56**, 10404 (1997)
   C. Pryor: Phys. Rev. B **57**, 7190 (1998)
10. C. Pryor, J. Kim, L. W. Wang, A. J. Williamson, A. Zunger: J. Appl. Phys. **83**, 2548 (1998)
11. M. Tadić, F. M. Peeters, K. L. Janssens, M. Korkusiński, P. Hawrylak: J. Appl. Phys. **92**, 5819 (2002)
12. M. Korkusiński, P. Hawrylak: Phys. Rev. B **63**, 195311 (2001)
13. P. N. Keating: Phys. Rev. **145**, 637 (1966)
14. R. M. Martin: Phys. Rev. B **1**, 4005 (1970)
15. O. Stier, M. Grundmann, D. Bimberg: Phys. Rev. B **59**, 5688 (1999)
16. E. O. Kane: Phys. Rev. B **31**, 7865 (1985)
17. F. H. Stillinger, T. A. Weber: Phys. Rev. B **31**, 5262 (1985)
18. K. Ding, H. C. Andersen: Phys. Rev. B **34**, 6987 (1986)
19. W. Yu, A. Madhukar: Phys. Rev. Lett. **79**, 905 (1997)
    Phys. Rev. Lett. **79**, 4939 (1997)
20. J. Tersoff: Phys. Rev. Lett. **56**, 632 (1986)
    Phys. Rev. B **39**, 5566 (1989)
21. M. A. Migliorato, A. G. Cullis, M. Fearn, J. H. Jefferson: Phys. Rev. B **65**, 115316 (2002)
22. G. E. Pikus, G. L. Bir: Sov. Phys. Solid State **1**, 1502 (1960)
    G. L. Bir, G. E. Pikus: *Symmetry and Strain-Induced Effects in Semiconductors* (Wiley, New York 1974)
23. T. B. Bahder: Phys. Rev. B **41**, 11992 (1990)
    Phys. Rev. B **45**, 1629 (1992)
24. A. Wojs, P. Hawrylak, S. Fafard, L. Jacak: Phys. Rev. B **54**, 5604 (1996)
25. P. Hawrylak: Phys. Rev. Lett. **71**, 3347 (1993)
26. O. Stier, M. Grundmann, D. Bimberg: Phys. Rev. B **58**, 9955 (1998)
    A. Schliwa, O. Stier, R. Heitz, M. Grundmann, D. Bimberg: Phys. Stat. Sol. B **224**, 2232 (2001)
    W. Sheng, J. P. Leburton: Phys. Rev. Lett. **88**, 167401 (2002)
27. L. C. G. Rego, P. Hawrylak, J. A. Brum, A. Wojs: Phys. Rev. B **55**, 15694 (1997)
28. W. Sheng (unpublished)
    M. Korkusiński. W. Sheng, P. Hawrylak: Phys. Stat. Sol. B **238**, 246 (2003)
29. S. J. Sun, Y.-C. Chang: Phys. Rev. B **62**, 13631 (2000)

88     Pawel Hawrylak and Marek Korkusiński

30. G. W. Bryant, W. Jaskolski: Physica E **11**, 72 (2001)
31. G. Klimeck, F. Oyfuso, T. B. Boykin, R. C. Bowen, P. von Allmen: Computer Model. Engin. Sci. **3**, 601 (2002)
32. S. Lee, L. Jönsson, J. W. Wilkins, G. W. Bryant, G. Klimeck: Phys. Rev. B **62**, 13631 (2001)
33. J. C. Grossman, M. Rohlfing, L. Mitas, S. G. Louie, M. L. Cohen: Phys. Rev. Lett. **86**, 472 (2001)
34. A. Canning, L. W. Wang, A. Williamson, A. Zunger: J. Comp. Phys. **160**, 29 (2000)
    A. J. Williamson, A. Zunger: Phys. Rev. B **59**, 15819 (1999)
    Phys. Rev. B **58**, 6724 (1998)
    A. J. Williamson, A. Zunger, A. Canning: Phys. Rev. B **57**, R4253 (1998)
    A. Franceschetti, A. Zunger: Europhys. Lett. **50**, 243 (200)
35. N. Moll, M. Scheffler, E. Pehlke: Phys. Rev. B **58**, 4566 (1998)
    P. Kratzer, C. G. Morgan, M. Scheffler: Phys. Rev. B **59**, 15246 (1999)
    L. G. Wang, P. Kratzer, N. Moll, M. Scheffler: Phys. Rev. B **62**, 1897 (2000)
36. C. Pryor: Phys. Rev. Lett. **80**, 3579 (1998)
37. G. W. Bryant: Phys. Rev. B **37**, 8763 (1988)
    Y. Z. Hu, H. Giessen, N. Peyghambarian, S. W. Koch: Phys. Rev. B **53**, 4814 (1996)
    V. Halonen, T. Chakraborty, P. Pietiläinen: Phys. Rev. B **45**, 5980 (1992)
    W. Que: Phys. Rev. B **45**, 11036 (1992)
    U. Bockelmann: Phys. Rev. B **48**, 17637 (1993)
    S. V. Nair, T. Takagahara: Phys. Rev. B **55**, 5153 (1997)
    S. V. Nair, T. Takagahara: Phys. Rev. B **56**, 12652 (1997)
    A. Franceschetti, A. Zunger: Phys. Rev. Lett. **78**, 915 (1997)
38. P. Hawrylak, M. Potemski, D. J. Lockwood, H. J. Labbe, H. Kamada, H. Weman, J. Temmyo, T. Tamamura: Physica E **2**, 652 (1998)
39. K. Brunner, U. Bockelmann, G. Abstreiter, M. Walther, G. Böhm, G. Tränkle, G. Weimann: Phys. Rev. Lett. **69**, 3216 (1992)
    A. Zrenner, L. V. Butov, M. Hagn, G. Abstreiter, G. Böhm, G. Weimann: Phys. Rev. Lett. **72**, 3382 (1994)
    D. Gammon, E. S. Snow, B. V. Shanabrook, D. S. Katzer, D. Park: Science **273**, 87 (1996)
    D. Gammon, S. W. Brown, E. S. Snow, T. A. Kennedy, D. S. Katzer, D. Park: Science **277**, 85 (1997)
    D. Gammon, E. S. Snow, B. V. Shanabrook, D. S. Katzer, D. Park: Phys. Rev. Lett. **76**, 3005 (1996)
    M. Bayer, A. Kuther, A. Forchel, A. Gorbunov, V. B. Timofeev, F. Schäfer, J. P. Reithmaier, T. L. Reinecke, S. N. Walck: Phys. Rev. Lett. **82**, 1748 (1999)
    A. Kuther, M. Bayer, A. Forchel, A. Gorbunov, V. B. Timofeev, F. Schäfer, J. P. Reithmaier: Phys. Rev. B **58**, 7508 (1998)
40. L. Landin, M. S. Miller, M.-E. Pistol, C. E. Pryor, L. Samuelson: Science **280**, 262 (1998)
41. E. Dekel, D. Gershoni, E. Ehrenfreund, D. Spektor, J. M. Garcia, P. M. Petroff: Phys. Rev. Lett. **80**, 4991 (1998)
42. A. Zrenner, M. Markmann, E. Beham, F. Findeis, G. Böhm, G. Abstreiter: J. Electron. Mater. **28**, 542 (1999)
43. P. Hawrylak, G. Narvaez, M. Bayer, O. Stern, A. Forchel: Phys. Rev. Lett. **85**, 389 (2000)

44. R. J. Warburton, C. S. Dürr, K. Karrai, J. P. Kotthaus, G. Medeiros-Ribeiro, P. M. Petroff: Phys. Rev. Lett. **79**, 5282 (1997)
45. G. A. Narvaez, P. Hawrylak: Phys. Rev. B **61**, 13753 (2000)
46. P. Hawrylak, A. Wojs, J. A. Brum: Phys. Rev. B **54**, 11397 (1996)
47. R. Rinaldi, P. V. Giugno, R. Cingolani, H. Lipsanen, M. Sopanen, J. Tulkki, J. Ahopelto: Phys. Rev. Lett. **77**, 342 (1996)
48. A. Wojs, P. Hawrylak: Phys. Rev. B **51**, 10880 (1995)
49. S. Raymond, P. Hawrylak, C. Gould, S. Fafard, A. Sachrajda, M. Potemski, A. Wojs, S. Charbonneau, D. Leonard, P. M. Petroff, J. L. Merz: Sol. Stat. Commun. **101**, 883 (1997)
50. M. Bayer, P. Hawrylak, K. Hinzer, S. Fafard, M. Korkusiński, Z. R. Wasilewski, O. Stern, A. Forchel: Science **291**, 451 (2001)
    M. Korkusiński, P. Hawrylak, M. Bayer, G. Ortner, A. Forchel, S. Fafard, Z. R. Wasilewski: Physica E **13**, 610 (2002)
51. K. Hinzer, P. Hawrylak, M. Korkusiński, S. Fafard, M. Bayer, O. Stern, A. Gorbunov, A. Forchel: Phys. Rev. B **63**, 075314 (2001)
52. D. J. Lockwood, P. Hawrylak, P. D. Wang, C. M. Sotomayor-Torres, A. Pinczuk, B. S. Dennis: Phys. Rev. Lett. **77**, 354 (1996)
53. R. M. Martin: Phys. Rev. B **4**, 3676 (1971)
    M. Cardona, G. Güntherodt (Eds.): *Light Scattering in Solids VI*, Topics Appl. Phys. **68** (Springer, Berlin, Heidelberg 1991)
54. R. Notzel, J. Temmyo, T. Tamamura: Nature **369**, 131 (1994)
55. H. Weman, H. Kamada, M. Potemski, J. Temmyo, R. Notzel, T. Tamamura: Solid State Electron. **40**, 379 (1996)
56. J. M. Gérard, B. Gayral: J. Lightwave Technol. **17**, 2089 (1999)
    P. Michler, A. Kiraz, C. Becher, W. V. Schoenfeld, P. M. Petroff, Lidong Zhang, E. Hu, A. Imamoglu: Science **290**, 2282 (2000)
    C. Santori, M. Pelton, G. Solomon, Y. Dale, Y. Yamamoto: Phys. Rev. Lett. **86**, 1502 (2001)
    E. Moreau, I. Robert, L. Manin, V. Thierry-Mieg, J. M. Gerard, I. Abram: Phys. Rev. Lett. **87**, 183601 (2001)
57. A. J. Williamson, A. Franceschetti, A. Zunger: Europhys. Lett. **53**, 59 (2001)
    J. Shumway, A. Franceschetti, A. Zunger: Phys. Rev. B **63**, 155316 (2001)
58. M. Brasken, M. Lindberg, D. Sundholm, J. Olsen: Phys. Rev. B **61**, 7652 (2000)
59. U. Hohenester, F. Rossi, E. Molinari: Solid State Commun. **111**, 187 (1999)
60. A. Barenco, M. A. Dupertuis: Phys. Rev. B **52**, 2766 (1995)
61. A. Wojs, P. Hawrylak: Sol. Stat. Commun. **100**, 487 (1996)
    P. Hawrylak, A. Wojs: Semicond. Sci. Technol. **11**, 1516 (1996)
62. M. Bayer, T. Gutbrod, A. Forchel, V. D. Kulakovskii, A. Gorbunov, M. Michel, R. Steffen, K. H. Wang: Phys. Rev. B **58**, 4740 (1998)
63. A. Zrenner, M. Markmann, A. Paassen, A. L. Efros, M. Bichler, W. Wegscheider, G. Böhm, G. Abstreiter: Physica B **256**, 300 (1998)
64. P. Hawrylak: Phys. Rev. B **60**, 5597 (1999)
65. T. Chakraborty P. Pietiläinen: *The Quantum Hall Effects* (Springer, Berlin, Heidelberg 1995)
66. I. V. Lerner, Yu. E. Lozovik: Zh. Eksp. Teor. Fiz. **80**, 1488 (1981) [Sov. Phys. JETP **53**, 763 (1981)]
    A. B. Dzyubenko, Yu. E. Lozovik: Sov. Phys. Solid State **25**, 874 (1983)
    D. Paquet, T. M. Rice, K. Ueda: Phys. Rev. B **32**, 5208 (1985)

A. H. MacDonald, E. H. Rezayi: Phys. Rev. B **42**, 3224 (1990)

Yu. A. Bychkov, E. I. Rashba: Phys. Rev. B **44**, 6212 (1991)

67. P. Hawrylak: Sol. Stat. Commun. **88**, 475 (1993)
68. M. Bayer, O. Stern, P. Hawrylak, S. Fafard, A. Forchel: Nature **405**, 923 (2000)
69. A. Wojs, P. Hawrylak: Phys. Rev. B **53**, 10841 (1996)
70. A. Wojs, P. Hawrylak: Phys. Rev. B **55**, 13066 (1997)
71. S. Tarucha, D. G. Austing, T. Honda, R. J. van der Hage, L. P. Kouvehoven: Phys. Rev. Lett **77**, 3613 (1996)
72. R. C. Ashoori, H. L. Störmer, J. S. Weiner, L. N. Pfeiffer, K. W. Baldwin, K. W. West: Phys. Rev. Lett. **71**, 613 (1993)
73. P. A. Maksym, T. Chakraborty: Phys. Rev. Lett. **65**, 108 (1990)

    P. A. Maksym, T. Chakraborty: Phys. Rev. B **45**, 1947 (1992)
74. A. H. MacDonald, S. R. Eric Yang, M. D. Johnson: Aust. J. Phys. **46**, 345 (1993)
75. A. Wojs, P. Hawrylak: Phys. Rev. B **56**, 13227 (1997)
76. J. H. Oaknin, L. Martin-Moreno, C. Tejedor: Phys. Rev. B **54**, 16850 (1996)
77. P. Hawrylak, C. Gould, A. Sachrajda, Y. Feng, Z. Wasilewski: Phys. Rev. B **59**, 2801 (1999)
78. X.-G. Wen: Phys. Rev. B **41**, 12838 (1990)

    C. de C. Chamon, X.-G. Wen: Phys. Rev. B **49**, 8227 (1994)
79. J. J. Palacios, A. H. MacDonald: Phys. Rev. Lett. **76**, 118 (1996)
80. M. Ciorga, A. Wensauer, M. Pioro-Ladriere, M. Korkusiński, J. Kyriakidis, A. S. Sachrajda, P. Hawrylak: Phys. Rev. Lett. **88**, 256804 (2002)
81. H. Drexler, D. Leonard, W. Hansen, J. P. Kotthaus, P. M. Petroff: Phys. Rev. Lett. **73**, 2252 (1994)
82. K. H. Schmidt, G. Medeiros-Ribeiro, P. M. Petroff: Phys. Rev. B **58**, 3597 (1998)
83. R. J. Warburton, C. Schäflein, D. Haft, F. Bickel, A. Lorke, K. Karrai, J. M. Garcia, W. Schoenfeld, P. M. Petroff: Nature **405**, 926 (2000)
84. J. J. Finley, A. D. Ashmore, A. Lemaitre, D. J. Mowbray, M. S. Skolnick, I. E. Itskevich, P. A. Maksym, M. J. Hopkinson, T. F. Krauss: Phys. Rev. B **63**, 073307 (2001)

    J. J. Finley, P. W. Fry, A. D. Ashmore, A. Lemaitre, A. I. Tartakovskii, R. Oulton, D. J. Mowbray, M. S. Skolnick, M. Hopkinson, P. D. Buckle, P. A. Maksym: Phys. Rev. B **63**, 161305 (2001)
85. M. Fricke, A. Lorke, J. P. Kotthaus, G. Medeiros-Ribeiro, P. M. Petroff: Europhys. Lett. **36**, 197 (1996)
86. L. Gravier, M. Potemski, P. Hawrylak, B. Etienne: Phys. Rev. Lett. **80**, 3344 (1998)
87. J. A. Brum, P. Hawrylak: Comments Cond. Matt. Phys. **18**, 135 (1997)
88. R. J. Warburton, B. T. Miller, C. S. Dürr, C. Bödefeld, K. Karrai, J. P. Kotthaus, G. Medeiros-Ribeiro, P. M. Petroff, S. Huant: Phys. Rev. B **58**, 16221 (1998)
89. P. Hawrylak, D. Pfannkuche: Phys. Rev. Lett. **70**, 485 (1993)
90. S. Sauvage, P. Boucaud, J. M. Gerard, V. Thierry-Mieg: Phys. Rev. B **58**, 10562 (1998)
91. L. Chu, A. Zrenner, G. Böhm, G. Abstreiter: Appl. Phys. Lett. **75**, 3599 (1999)
92. H. C. Liu, M. Gao, J. McCaffrey, Z. R. Wasilewski, S. Fafard: Appl. Phys. Lett. **78**, 79 (2001)
93. P. Hawrylak, M. Korkusiński, S. Fafard, R. Dudek, H. C. Liu: Physica E **13**, 246 (2002)

94. P. Hawrylak: NATO Science Series **3**, 81 (Kluwer, 2000) p. 319
95. P. Hawrylak, M. Korkusinński: Nonlinear Optics **29**, 329 (2002)
96. P. Hawrylak: Solid State Commun. **127**, 793 (2003)

# Exciton Complexes in Self-Assembled In(Ga)As/GaAs Quantum Dots

Manfred Bayer

Experimentelle Physik II, Universität Dortmund
44221 Dortmund, Germany
manfred.bayer@physik.uni-dortmund.de

**Abstract.** In this contribution we will present the results of our recent spectroscopic studies of single In(Ga)As/GaAs self-assembled quantum dots. We will develop the relation between the excitonic states in quantum dots and the confined
single particle shell structure, for which we have studied the absorption spectrum of
excitons by photoluminenscence excitation spectroscopy. We find that the absorption varies strongly with the symmetry of the dot structures: It becomes increasingly
complicated the more the symmetry of the quantum dot is reduced. For dots of high
symmetry we demonstrate that a non-trivial mixing of quantum configurations of
the interacting electron–hole complex leads to distinct absorption spectra which
are controlled by the number of confined electronic shells. We will also discuss the
impact of exciton–phonon interaction on the optical spectra. Most importantly, for
temperatures $T \to 0$ the width of the emission lines seems to be limited by the
radiative lifetime of the excitons only. However, already at moderate temperatures
the coupling to phonons has a considerable influence. In particular, at room temperature the linewidth is on the order of several meV, so that the term 'artificial atom'
is no longer justified. Then we will turn to fine structure effects of excitons confined
in quantum dots. We will demonstrate that the fine structure is a very sensitive tool
to obtain insight into the symmetry of the quantum dots. An intentional symmetry breaking can be induced by a magnetic field with an orientation different from
the quantum dot symmetry axis. This allows for a detailed study of dark excitons.
Further, we show how charged excitons can be distinguished from neutral excitons
by looking at the fine structure. Finally, we will discuss the simplest functional unit
that can be assembled from quantum dots, the 'two-atomic' quantum dot molecule.
We will demonstrate a tunneling induced splitting of the energy levels which can be
as large as 30 meV for barrier widths below 5 nm. Looking for these systems at the
exciton fine structure gives a unique proof of the coherent coupling of the electronic
states of the two quantum dots.

## 1 Introduction

The miniaturization of semiconductor devices during the last few decades has
resulted in many applications revolutionizing mankind's life [1]. Any successful implementation of quantum structures relies on a solid understanding of
their basic physical properties. A particularly important aspect of miniaturization is the reduction of the dimensionality of structures, which has paved
the way from bulk crystals to two-dimensional systems (quantum wells), and

P. Michler (Ed.): Single Quantum Dots, Topics Appl. Phys. **90**, 93–146 (2003)
© Springer-Verlag Berlin Heidelberg 2003

further to one-dimensional systems (quantum wires) [1]. This development has in some sense come to its 'natural end' with the fabrication of high quality quantum dots [2], systems with an effective dimensionality zero, due to which they have been often termed 'artificial atoms'. These structures might soon be standard building blocks in commercial devices such as light emitting diodes, lasers or transistors. The 'end' of development could, however, mark the start of a new phase, in which quantum structures are assembled to obtain new functional units, in which eventually quantum effects are exploited. For example, such units might form the basis for applications in quantum information processing such as quantum encryption or quantum computing which attract considerable interest currently [3,4].

In this chapter we want to discuss the electronic properties of one of the potential constituents of such functional units, quantum dots. Among all the different fabrication techniques for these systems, self-assembly has been proven to be particularly successful because of the high optical quality of the resulting structures. We will summarize here the results of our recent experiments on In(Ga)As/GaAs self-assembled quantum dots with two main objectives:

- To demonstrate that optical spectroscopy allows us to obtain a subtle understanding of the electronic properties of these structures.
- To demonstrate that the self-assembled growth technique allows for a rather precise tailoring of quantum dot properties.

Here we will concentrate on nominally undoped dot structures. Figure 1 shows a scanning electron micrograph of an array of self-assembled InAs/GaAs quantum dots [5]. Unlike structures for optical studies, the dots have not been covered by a GaAs layer for the microscopic investigations. From the micrograph we see that the quantum dots are approximately rotationally invariant around the heterostructure growth direction (which we take as the $z$-direction). Their shapes can be well described by lenses. The average dot density is about $10^{10}\,\mathrm{cm}^{-2}$. We note the inhomogeneity of the dots in the array with respect to their sizes.

First the three-dimensional confinement of carriers in the quantum dots has to be demonstrated. This can be done, for example, rather simply by state filling spectroscopy. Here arrays containing millions of quantum dots are illuminated by an intense, non-resonant laser beam so that many electron–hole pairs are created. The carriers relax into the quantum dots where they occupy the discrete energy levels according to the Pauli principle: Since each dot level can be populated by a finite number of carriers only, several levels will be occupied. After relaxation the electron–hole pairs will recombine according to the selection rule that electrons in a given conduction band state can recombine only with holes in valence band states having the same quantum numbers (which holds at least in the lowest approximation). This gives rise to characteristic emission patterns: we expect several distinct features in

**Fig. 1.** Scanning electron micrograph of an array of self-assembled InAs/GaAs quantum dots. For the microscopic studies the sample has been left uncovered, for spectroscopic studies the samples are covered by a GaAs cap layer [5]

0.1 μm

the spectrum that originate from recombination of the optically generated electron–hole pairs in the discrete dot shells.

An example for such experiments is shown in Fig. 2, in which photoluminescence spectra of an array of InAs/GaAs quantum dots are shown for varying excitation powers. At low excitation only one or two electron–hole pairs are created, which relax into the dot ground states, so that only emission from the lowest confined levels is observed. Independent of the dot geometry, the ground state can be occupied by two carriers only due to its spin degeneracy. When increasing the excitation, additional carriers are generated which have to occupy higher lying shells due to Pauli blocking and the corresponding emission from these shells appears in the spectrum. At the highest excitation powers four distinct emission features are observed for the quantum dots under study giving clear proof of the three-dimensional confinement of the carriers.

From these array spectra we already obtain important hints about the geometric confinement potential of the particles in the dot. Since we are studying quantum dots in the strong confinement regime, one might neglect Coulomb interaction effects in lowest approximation. We note, however, already here that these interaction effects are essential for understanding a number of peculiar features in the optical spectra of single quantum dots (see below). Still we will use this crude approximation as a starting point. For self-assembled dots the structure height is significantly smaller than the lateral size. The vertical and the in-plane motions of the carriers can thus be separated, and the energies of the single particle states are given by two contributions:

$$E = E_z + E_{\text{in-plane}} \,. \tag{1}$$

Due to the strong vertical confinement the excitation along the $z$-direction is always that of the ground state. Therefore higher lying lateral excitations are observed in the spectra of Fig. 2. For them, one notes that they are approximately equidistantly spaced in energy. This equidistance indicates that the in-plane confinement potential (which has rotational symmetry due

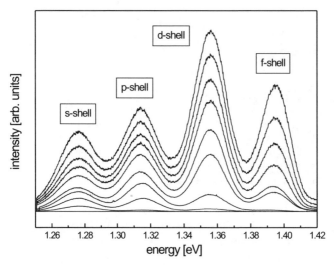

**Fig. 2.** Photoluminescence spectra of an array of self-assembled InAs/GaAs quantum dots recorded at $T = 2\,\mathrm{K}$ using varying optical excitation power

to the lens shape of the present dots) can be described by a parabola $V(r) = \frac{1}{2}\,\mathrm{m}\,\Omega^2 r^2$.

The single particle states of the carriers are described by a radial quantum number $n_r$ and an azimuthal quantum number $n_\varphi$, where the latter quantum number gives the angular momentum around the symmetry axis of the quantum dots. The energy eigenvalues of a two-dimensional harmonic oscillator are given by

$$E_{\text{in-plane}} = \hbar\,\Omega\,(s+1)\,, \qquad (2)$$

where $s = 2n_r + |n_\varphi| + 1 = 0, 1, 2, \ldots$ is the shell index. The shells can also be characterized according to the orbital angular momentum quantum numbers $n_\varphi = 0, \pm 1, \pm 2, \ldots$ as $s, p, d, \ldots$. States with positive and negative angular momenta are degenerate. The degeneracy of a shell is given by $2\,(s+1)$ where the factor of 2 comes from the spin degree of freedom. The angular momentum notation has already been used in Fig. 2 to label the different transitions.

The equidistance of the emission features is observed as a general feature for InAs/GaAs quantum dot samples, as shown in Fig. 3, where high excitation spectra for dots with different dot sizes are shown [5]. The energy separation between the emission lines varies from about 60 to 90 meV. Overall, splittings varying from $\sim 25$ to more than 100 meV have been demonstrated showing the high degree of control that has been achieved in engineering the confinement potential of self-assembled quantum dots.

As mentioned above, the motion of electron–hole pairs which are optically generated in the quantum dots is influenced not only by geometrical quantization but also by Coulomb interaction. Generally the behavior of carriers

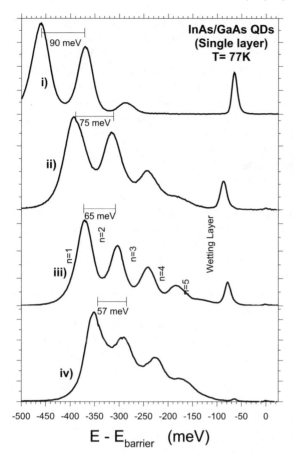

**Fig. 3.** High excitation spectra of different InAs/Gas quantum dot samples with energy splittings between the confined quantum dot shells varying from about 60 to 90 meV. The experiments were performed at a temperature of 77 K [5]

in self-assembled quantum dots is determined by a variety of effects. In dot structures based on GaAs, fortunately, these effects can be categorized by characteristic interaction energies for which a distinctive hierarchy exists, as shown in Table 1.

**Table 1.** Hierarchy of the energies that determine the behavior of carriers confined in GaAs based self-assembled quantum dots

| | Energy | Order of magnitude |
|---|---|---|
| I | Kinetic energy | $\sim (50\text{--}100)$ meV |
| II | Exciton binding energy | $\sim (25\text{--}50)$ meV |
| III | Fine structure effects | $\sim 1$ meV |

The largest energy is the kinetic energy resulting from the geometric confinement of the carriers in the dots and has been discussed already above. The exciton binding energy by which we mean the Coulomb interaction energy of an electron and a hole pair confined in a quantum dot is typically a factor of 2 smaller than the geometric quantization energy. The smallest energies result, for example, from the electron–hole exchange interaction and are typically on the order of a meV. Also the energies for the interaction of carriers with external electromagnetic fields belong to this category, at least at the moderate field strengths available in today's standard experiments. Such effects are the Zeeman interaction of the carrier spins with a magnetic field or the Stark shift of the excitonic levels in an electric field.

When comparing the magnitude of the energies in Table 1 with the inhomogeneous broadening of the spectra in Figs. 2 and 3 (which is typically a few tens of meV) we find that whereas single particle quantization can be readily resolved in the spectra of high quality quantum dot arrays, the resolution of effects which occur on the energy scales II and III is not possible because these energies are comparable or even much smaller than the inhomogeneous broadening. To learn about these effects therefore requires spectroscopy of single quantum dots.

The description of the results that we have obtained during the last few years by optical spectroscopy of single In(Ga)As/GaAs quantum dots and quantum dot molecules is the central topic of this chapter. Before coming to their discussion, we want to stress one important point: despite the constant improvement of the technological schemes for self-organization, there is an unavoidable, fabrication-inherent variation of the quantum dot properties such as composition or geometry. All spectroscopic studies of single quantum dots done up to now suffer from the lack of knowledge about the structural properties of the studied dot structures. This lack of knowledge hampers the comparison of the experimental data with detailed theoretical calculations severely. It will be crucial for future studies to improve this situation.

This contribution is divided into the following sections: in Sect. 2 we will describe the technique which allows us to isolate a single quantum dot for optical studies. Further, a short description of the spectroscopic techniques will be given. In Sect. 3 we will develop the relation of the (charge neutral) excitonic states in quantum dots to the confined single particle shell structure, for which we have studied the absorption spectrum of excitons by photoluminenscence excitation spectroscopy. We find that the absorption varies strongly with the symmetry of the dot structures: it becomes increasingly complicated the more the symmetry of the quantum dot is reduced. For dots of high symmetry we demonstrate that a non-trivial mixing of quantum configurations of the interacting electron–hole complex leads to distinct absorption spectra which are controlled by the number of confined electronic shells. Section 4 describes the impact of the exciton–phonon interaction on the optical spectra. We will show that for temperatures $T \rightarrow 0$ the width of the emission lines seems to be limited by the radiative lifetime of the excitons. However, already at moderate temperatures the coupling to phonons has a considerable influence. In particular, at room temperature the linewidth is on the order of several meV, so that the term 'artificial atom' is no longer justified. Section 5 describes studies of fine structure effects of excitons, also in an external magnetic field. We will demonstrate that the fine structure is a very sensitive tool to obtain insight into the symmetry of the quantum dots. In particular, an intentional symmetry breaking can be induced by a magnetic field with an orientation different from the quantum dot symmetry axis. This allows for a detailed study of dark excitons. Further, we will study how charged excitons can be distinguished from neutral excitons by looking at the fine structure. In Sect. 6, finally, we will discuss the simplest functional unit that can be assembled from quantum dots, the 'two-atomic' quantum dot molecule. We will demonstrate a tunneling induced splitting of the energy levels which can be as large as 30 meV for barrier widths below 5 nm. Looking for these systems at the exciton fine structure gives a unique proof of the coherent coupling of the electronic states of the two quantum dots.

## 2    Experimental Technique

Recently, several experimental techniques have been developed which allow for a study of single quantum dots. These techniques can be divided into two main categories: (a) spectroscopy with a high spatial resolution on as-grown quantum dot samples such as near-field scanning optical microscopy [6] or confocal microscopy [7]; (b) spectroscopy on samples that underwent further technological processing after the growth [8,9]. All the data described here were obtained on such samples, which will be discussed therefore in more detail.

In most cases single quantum dots were isolated by fabricating small mesa structures with lateral sizes down to less than 100 nm [8]. Figure 4 shows a scanning electron micrograph of such a mesa structure. On top the square-shaped gold mask obtained by high resolution electron beam lithography is seen. This pattern is transferred into the semiconductor material by wet chemical etching. The etching depth typically is 100 nm. Due to the wet chemical process the mask is underetched. The quantum dots are located $\sim 50$ nm below the sample surface. In combination with a dot density of $\sim 10^{10}$ cm$^{-2}$ we obtain an average occupation of less than one quantum dot in the smallest mesa structures. Single dot resolution could also be obtained by placing a shadow mask [9] made from aluminium or gold on the quantum dot samples, as shown in the micrograph in Fig. 5. The thickness of the mask is $\sim 100$ nm so that it is not optically transparent. Holes have been prepared in it with diameters as small as 50 nm, through which both optical excitation and detection of the signal is done.

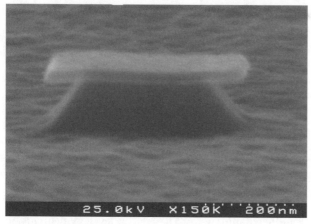

**Fig. 4.** Scanning electron micrograph of a mesa structure with a lateral size of $\sim 100$ nm that has been fabricated by lithography on a planar quantum dot sample. On *top* the gold mask is seen which has been left for the microscopic study. For spectroscopic purposes it is removed

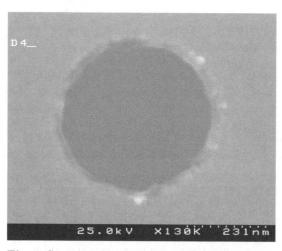

**Fig. 5.** Scanning electron micrograph of a section of a shadow gold mask containing a small hole with a diameter of $\sim 250$ nm to isolate a single quantum dot underneath

As a criterion for mesa occupancy by a single dot we use photoluminescence spectroscopy at low excitation powers. Figure 6 shows corresponding spectra recorded at $T = 2$ K on mesa structures with different lateral sizes [8]. For the unstructured reference quantum dot sample a broad emission band with a width of about 30 meV reflecting the inhomogeneous broadening is observed. When studying a mesa with a size of 300 nm the broad band splits into a set of sharp emission lines giving further proof of the quasi-atomic character of the dots. With decreasing mesa size the number of spectral lines

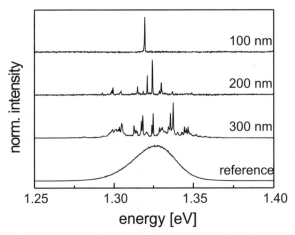

**Fig. 6.** Photoluminescence spectra recorded at $T = 2\,\mathrm{K}$ on a planar, unstructured quantum dot sample (*bottom trace*) and on mesa structures with varying lateral sizes as indicated at each trace [8]. The optical excitation power was limited to $100\,\mu\mathrm{W}$

decreases until we observe a single sharp emission line for the mesa with a lateral size of $\sim 100\,\mathrm{nm}$. We take the observation of a single line as an indication for a single dot in the mesa structure. However, this is certainly not a unique proof. For example, if there were another quantum dot in the structure, it might not be observable at low excitation due to carrier trapping at close-by surfaces. This dot might show up, however, at higher excitation, when the traps at the surface are readily ocupied. Here we will concentrate on low excitation studies, for which this restriction represents no fundamental limitation.

For optical studies, in most cases the structures were held in superfluid helium ($T < 2\,\mathrm{K}$) in an optical cryostat. Also a cryostat with a split magnet coil for fields up to $B = 8\,\mathrm{T}$ was available. The orientation of the magnetic field could be varied relative to the heterostructure growth direction by rotating the sample in the cryostat. For temperature dependent measurements the structures could be placed in the variable temperature insert of an optical cryostat. Laser excitation was done either by Ar- or He-Ne-ion lasers or by a tunable Ti-sapphire laser pumped by a high power Ar-ion laser. The quantum dot emission was dispersed by single or double grating monochromators depending on the required spectral resolution. The polarization of the emission could be analyzed by suitable combinations of linear polarizers and quarter wave retarders. Detection was done either with a Si-charge coupled devices camera cooled by liquid nitrogen or with a Peltier-cooled photomultiplier (with GaAs or S1 photocathodes) interfaced by a photon counting system. To suppress stray light, in some studies the quantum dot was imaged into an intermediate plane, in which a small aperture that could be moved parallel to the plane was placed. Further a linear polarizer with the transmission axis oriented normal to the linear laser polarization was placed in front of the monochromator.

## 3    Excitonic Absorption in Quantum Dots

The absorption of a photon in a direct gap semiconductor creates an electron in the conduction and a hole in the valence band. Due to their mutual Coulomb interaction the carriers bind to form an excitonic complex. The optical absorption spectrum is proportional to the density of states of this composite particle. In three, two, and one dimensional semiconductors, the translational invariance permits us to separate the motion of the two interacting particles into the motions of two independent particles, that of the center of mass and that of the relative coordinate of the electron–hole pair. The resulting equations of motion are in principle exactly soluble. In particular, hydrogen-like solutions are found for the relative motion. The binding energy of the exciton in bulk, for example, is given by:

$$E_X^B = -\frac{Ryd}{n^2} , \tag{3}$$

where $Ryd$ is the Rydberg energy and $n$ is the principal quantum number. For an ideal two-dimensional system, the confinement of the carrier wave functions causes a considerable enhancement of the binding energy which is given by

$$E_X^B = -\frac{Ryd}{\left(n - \frac{1}{2}\right)^2} , \tag{4}$$

while in ideal one dimensional systems the binding energy would be logarithmically divergent. The wave function of the relative motion in these (D-dimensional) structures can be thought of as being constructed by superposing free particle states, that is plane waves. The mixing of the single particle states is forced by the Coulomb interaction:

$$\psi(r) = \frac{1}{(2\pi)^{D/2}} \int A(k) \exp(ik \cdot r) \, d^D k , \tag{5}$$

where the contribution of each wave is weighted by $A(k)$. This superposition is obviously just a Fourier transform. This will serve us as a guide for constructing the exciton wave function in quantum dots from confined free particle states, for which the carrier motion is suppressed.

In contrast to the systems with higher dimensionality, in a system with three dimensional confinement the separation of the two-particle motion is no longer possible. The calculation of the excitonic absorption spectrum requires the solution of the genuine two-body problem of an interacting electron–hole pair. Because the two particles are distinguishable, the Pauli exclusion principle does not operate and the problem is the simplest example of an interacting boson system. The main effect of the two particle nature of the problem is that the number of possible two particle configurations that can be constructed from the single particle orbitals need not be related in a simple way to the number of these orbitals.

There is another important difference which sets quantum dots apart from structures of higher dimensionality. From the representation of the exciton wave function given above we see that in such systems the Hilbert space that is available for constructing an exciton has infinite dimensionality. In contrast, the Hilbert space has only finite dimensionality in quantum dots. Here we neglect the contribution of the continuum states in the barriers surrounding the dot to the exciton wave function. As we will discuss below in more detail, this seems a reasonable assumption for the single quantum dot structures studied here. This contribution might be, however, very important for unpatterned samples.

The goal of our studies is to establish a relation between the exciton states in quantum dots and the available single particle shell structure: how has the exciton wave function to be constructed from the single particle states? We note that this is only one particular aspect which determines the exciton spectrum. For example, also the coupling of the excitons to the phonons might have significant impact, in particular when the splitting between the confined shells is comparable to the optical phonon energy. This additional aspect was not the subject of our studies. In the next section we only study the dephasing of the ground state exciton due to its coupling to phonons. We note, however, that we found no indications for a strong exciton–phonon coupling, which might be related to the considerable variation of the shell splitting from the LO-phonon energy.

Experimentally, two different types of $In_{0.60}Ga_{0.40}As$ self-assembled quantum dots embedded in a GaAs matrix were designed. Figure 7 shows high excitation photoluminescence spectra of mesa structures of the two sample types which contain millions of dots [10]. For the first type of dots (left panel) only two shells are confined: besides emission from the two-fold spin degenerate $s$-shell with orbital angular momentum $n_\varphi = 0$, emission from the first excited $p$-shell is detected. The $p$-shell is four-fold degenerate in structures that exhibit rotational symmetry. In addition to spin degeneracy it has a degeneracy of the orbital angular momentum $n_\varphi$ which can be $+1$ or $-1$. Therefore the total number of confined states adds up to 6 for type 1 dots. For the second type of quantum dots (right panel) one more shell, the six-fold degenerate $d$-shell is additionally confined, for which the orbital angular momentum $n_\varphi$ is $+2, 0$ or $-2$. The total number of confined shells is therefore 12 for type 2 quantum dots. Thus the two quantum dot samples allow us to study the modification of the excitonic states, when the size of the available Hilbert space is doubled.

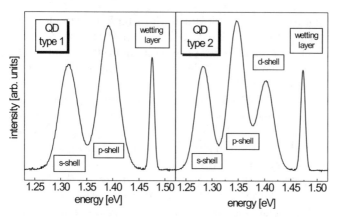

**Fig. 7.** Photoluminescence spectra recorded at high excitation power on arrays of $In_{0.60}Ga_{0.40}As/GaAs$ quantum dots. In the dot type 1 sample (*left panel*) only two electronic shells were confined in the dots; in the second type (*right panel*) three shells were confined [10]

Let us first concentrate on quantum dots of type 1 and type 2 that exhibit a high geometric symmetry. This symmetry is reflected by their optical properties: when studying the fine structure of the ground state exciton, no indications of a fine structure splitting are observed within the experimental accuracy given by the spectral resolution of 25 µeV. The emission seems to be fully circularly polarized at $B = 0$. In general, it seems impossible from the fabrication scheme to find a quantum dot with a geometry that is perfectly invariant under rotations around the growth axis. Still, the asymmetry could be so small that it is hidden underneath the homogeneous broadening due to the finite radiative lifetime of the exciton.

Figure 8 shows photoluminescence (dotted traces) and photoluminescence excitation (solid traces) spectra [10] taken on such single quantum dots of type 1 and type 2. The PLE spectra were recorded by setting the detector wavelength to that of the corresponding s-shell emission and varying the wavelength of the laser that was focussed on the mesa structure. We note that emission at the detection wavelength was also observed when moving the laser away from the mesa and focussing it on a spot between two mesas, where no quantum dots were located. In both types of spectra (laser on and off mesa) broad backgrounds of very similar shapes and intensities appear. We attribute them to a rather high background doping in the particular samples. To get access to the 'pure' quantum dot-related features, we have therefore subtracted the spectra recorded off-mesa from those recorded on-mesa. By doing so, some information about spectral lines involving quantum dot transitions of weak oscillator strength or transitions from confined to wetting layer states might get lost. In any case, no significant absorption continuum was observed for the quantum dots under study (see discussion

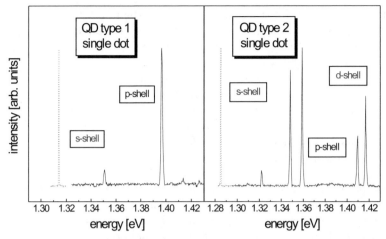

**Fig. 8.** Photoluminescence (*dotted traces*) and photoluminescence excitation (*dotted traces*) spectra of single In$_{0.60}$Ga$_{0.40}$As/GaAs quantum dots of type 1 (*left panel*) and type 2 (*dot B, right panel*) [10]. Both dots exhibit rather high geometric symmetry as evidenced by exciton fine structure investigations

below). For the dots of type 1 we find very simple excitonic absorption spectra which consist of two main intense features only, one in the $s$-shell and one in the $p$-shell. In contrast, for the dots of type 2 five intense features appear, one in the $s$-shell, two in the $p$-shell and two in the $d$-shell. The splitting of the two lines in the $p$-shell which are about of the same oscillator strength is roughly 10 meV.

In both cases, between the $s$- and $p$-shell another weak absorption feature which is located about 36 meV (energy of the GaAs LO-phonon) above the $s$-shell is observed. It is related to phonon-assisted absorption. However, the intensity of this feature is rather weak. Recently there have been several reports of a considerable dipole momentum in quantum dots which originates from a vertical displacement of the electron and the hole wave functions [11]. The weak intensity of the phonon replicum indicates that the excitons in the quantum dots do not show a large dipole momentum. If there were an effective charge in the system, the coupling strength to the LO-phonons would be considerably larger than for a charge neutral system leading to strong phonon replicas in the absorption.

Besides these features there are no indications of further absorption lines with significant oscillator strength. Above the $p$-shell a continuum of states appears which, however, is quite weak as compared to the main absorption features. This continuum becomes important only close to the onset of the wetting layer (not shown). This observation is surprising since it seems to be contradicting other experimental studies [12], which have reported a continuum of states that is extended in energy all the way down to almost the ground state exciton. This continuum was attributed to an efficient coupling of the confined quantum dot states to the states of the wetting layer. These studies were done on unpatterned samples. In this case, the laser beam might be absorbed in the wetting layer, from where carriers might relax into the quantum dot ground state, where they could recombine. In our case the experiments were done on single quantum dots in mesas with very small lateral sizes of $\sim 50\,\mathrm{nm}$. Thus the wetting layer has been chopped off very sharply, which will certainly affect the absorption spectra significantly. The quantum dot can no longer be considered as a zero dimensional system located in a quasi-two dimensional layer but has to be viewed as a quantum dot embedded in a shallow quantum dot. Thus, a continuum of wetting layer states is excluded.

We also note that the studies were performed by using laser excitation with a fixed linear polarization so that excitons of a given spin polarization are created in the quantum dot. If the dots are to exhibit a geometrical asymmetry (which should be, however, small since we do not observe a polarization splitting of the ground state exciton within the experimental accuracy, see above) only states with polarization identical to the laser are excited. Thus, in general some states in the exciton spectrum might have been missed in the spectra. This is, however, not essential for the key point of the studies: demonstrating the connection between the electronic shell structure and the exciton states.

In the following we will show that excitonic effects can explain the different absorption spectra of type 1 and type 2 quantum dots. For this purpose we have to return to the construction of exciton states from single particle levels [10]. Using the representation of the exciton wave function in higher dimensional structures as a guide, we obtain the following exciton representation:

$$|\psi\rangle = \sum_{\substack{s,\,s' \\ n_\varphi}} A_{n_\varphi}^{s,s'} |s, +n_\varphi\rangle |s', -n_\varphi\rangle . \tag{6}$$

Here the first (second) state vector on the right hand side describes the electron (hole). We note that the total orbital angular momentum of the electron–hole pair must be zero. Simply speaking, the angular momentum quantum carried by the absorbed photon is 'used' for the excitation from the $p$-like valence to the $s$-like conduction band. The calculation of the exciton states is performed in two steps:

(1) The consideration starts with working out the appropriate single particle pair states of electron and hole $|s, +n_\varphi\rangle |s', -n_\varphi\rangle$. These states can be divided into two classes: states that are optically active with $s = s'$ – their number in each shell is equal to the degeneracy of the shell $s + 1$; and states that are optically inactive with $s \neq s'$, but which also have zero orbital angular momentum.

While the electron–hole states $|s, +n_\varphi\rangle |s, -n_\varphi\rangle$ correspond to vertical transitions, photons do not create individual pairs of them but rather a linear superposition of them, $\sum |s, +n_\varphi\rangle |s', -n_\varphi\rangle$. Hence only one of the $s+1$ degenerate states couples to the light field while the remaining $s$ states are optically inactive. It is therefore necessary to find a coherent exciton representation which explicitly accounts for this symmetry of coupling to the light field. This is done by a transformation to Jacobi-like coordinates. The one, totally symmetric state among the resulting orthogonal basis states is the bright one. This procedure and the underlying symmetry is analogous to Kohn's theorem for intraband transitions, and can be loosely termed 'Kohn's theorem for excitons'.

(2) The second step in the procedure of finding the exciton states is the superposition of all the electron–hole pair states which physically corresponds to mixing by Coulomb interactions. This is done by diagonalizing the exciton Hamiltonian using the above form for the exciton wave function, i.e. the calculation of the coefficients $A_m^{s,s'}$. Here also the optically inactive pair states will contribute through mixing with the bright states. The most important of the dark states are those with energy resonant to optically active states. For example, by construction the superpositions of electron–hole states from different shells which have the form $(|s-1, +m\rangle |s+1, -m\rangle \pm |s+1, +m\rangle |s-1, -m\rangle)$ are resonant with the optically active, coherent state in shell $s$. These are precisely the nontrivial states which strongly modify the absorption spectrum, because the mixing with bright states will result in a 'normal mode splitting' as observed experimentally.

It is best to illustrate these constructions for quantum dots with two and three confined shells as studied in the present experiments. Let us begin with the dots of type 2, for which we find six optically active configurations, one in the $s$-shell, two in the $p$-shell and three in the $d$-shell reflecting exactly the degeneracy $s+1$ of these shells

$$
\begin{aligned}
|a\rangle &= |0,0\rangle |0,0\rangle \,, \\
|b\rangle &= |1,+1\rangle |1,-1\rangle \,, \\
|c\rangle &= |1,-1\rangle |1,+1\rangle \,, \\
|d\rangle &= |2,+2\rangle |2,-2\rangle \,, \\
|e\rangle &= |2,-2\rangle |2,+2\rangle \,, \\
|f\rangle &= |2,0\rangle |2,0\rangle \,.
\end{aligned}
\tag{7}
$$

Under neglect of Coulomb interaction effects the excitonic absorption would reflect just the single particle denity of states. The spectrum of a type 2 quantum dot would then consist of three spectral features as shown in the upper left panel of Fig. 9. The oscillator strength of each feature would be given by the shell degeneracy.

Besides the optically active states, there are two optically inactive configurations with zero total angular momentum but originating from different

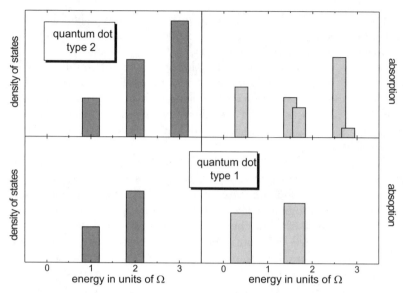

**Fig. 9.** *Left panels:* Density of states for non-interacting electron–hole pairs as a function of energy (in units of the confinement potential frequency) for quantum dots of type 1 and type 2. *Right panels:* Absorption spectrum of an exciton in a quantum dot of type 1 and type 2 [10]

shells

$$|g\rangle = |0,0\rangle\,|2,0\rangle\,,$$
$$|h\rangle = |2,0\rangle\,|0,0\rangle\,. \tag{8}$$

These states involve an electron (a hole) in the zero angular momentum state $|00\rangle$ of the $s$-shell and a hole (an electron) in the zero angular momentum state $|20\rangle$ of the $d$-shell. Hence for the construction of an exciton in a three-shell quantum dot, eight states with zero total angular momentum are available. The next step is the transformation into the coherent exciton basis:

$$|A\rangle = |a\rangle\,,$$
$$|B\rangle = (|b\rangle + |c\rangle)/\sqrt{2}\,,$$
$$|C\rangle = (|b\rangle - |c\rangle)/\sqrt{2}\,,$$
$$|D\rangle = (|d\rangle + |e\rangle + |f\rangle)/\sqrt{3}\,,$$
$$|E\rangle = (|d\rangle - |e\rangle)/\sqrt{2}\,,$$
$$|F\rangle = (|d\rangle - |e\rangle - 2|f\rangle)/\sqrt{6}\,,$$
$$|G\rangle = (|g\rangle + |h\rangle)/\sqrt{2}\,,$$
$$|H\rangle = (|g\rangle - |h\rangle)/\sqrt{2}\,. \tag{9}$$

Only the three symmetric states $|A\rangle$, $|B\rangle$, and $|D\rangle$ are optically active, as can be seen from expanding the interband polarization operator $P^+$ which generates an electron–hole pair in the quantum dot in the new coherent basis

$$P^+ |0\rangle = |a\rangle + |b\rangle + |c\rangle + |d\rangle + |e\rangle + |f\rangle = |A\rangle + \sqrt{2}|B\rangle + \sqrt{3}|D\rangle . \quad (10)$$

State $|A\rangle$ corresponds to an electron and a hole in their $s$-shells. In state $|B\rangle$ the electron and hole are in their $p$-shells, and in $|D\rangle$ they are in the $d$-shells. Further we note that the dark state $|H\rangle$ is degenerate with state $|B\rangle$ (not only in a single particle picture but, surprisingly, even when the Coulomb interaction is taken into account). In addition, state $|F\rangle$ is located energetically close to $|D\rangle$.

Finally, the mixing of these single particle configurations by Coulomb interactions has to be included. Due to this mixing bright configurations share oscillator strength with dark ones which therefore become optically active. From diagonalizing the exciton Hamiltonian in this basis we find that the totally antisymmetric states $|C\rangle$, $|E\rangle$, $|G\rangle$ cannot couple to the bright states and form a dark subspace. On the other hand, the interaction of the states $|F\rangle$ and $|G\rangle$ with the states $|A\rangle$, $|B\rangle$ and $|D\rangle$ will make them optically active. The exciton states are therefore given by $|X_i\rangle = C_A^i |A\rangle + C_B^i |B\rangle + C_C^i |C\rangle + C_D^i |D\rangle + C_E^i |E\rangle$, $i = 1, \ldots, 5$ with energies $E_i$. Thus the excitonic absorption spectrum of a type 2 quantum dot will be given by five excitonic states: one with predominantly $s$-shell character, two with predominantly $p$-shell character, and two with predominantly $d$-shell character, in agreement with the experimental findings. The excitonic absorption can be calculated according to Fermi's golen rule:

$$A(\omega) = \sum_i \left| C_A^i + \sqrt{2}C_B^i + \sqrt{3}C_D^i \right|^2 \delta(\hbar\omega - E_i). \quad (11)$$

The upper right panel of Fig. 9 shows the absorption spectrum for a quantum dot of type 2. The attractive Coulomb interaction leads to a renormalization of the transition energies as compared to the free particle spectrum. Due to the interaction between them the degeneracy of $|B\rangle$ and $|H\rangle$ is lifted and the two lines are split. The oscillator strength is almost equally distributed among them. We want to emphasize again that this splitting is not a consequence of a dot shape asymmetry but is a mere consequence of Coulomb interactions and the induced mixing of quantum configurations. A similar splitting occurs for the $d$-shell excitons. For them, a weak and a strong absorption feature appear in the spectrum.

The single particle configurations $|d\rangle$ to $|h\rangle$ exist in quantum dots with three electronic shells but are obviously absent in dots with two confined shells. Thus for type 1 dots the first three single particle configurations $|a\rangle$, $|b\rangle$ and $|c\rangle$ need to be considered only in the construction of the exciton states. The transformation to Jacobi coordinates gives states $|A\rangle$, $|B\rangle$ and $|C\rangle$. The last state is optically inactive. Therefore only two absorption lines are observed in the spectrum of a two-shell quantum dot. For these dots, only a geometry asymmetry can result in a splitting of the $p$-shell absorption. The lower panels in Fig. 9 show the density of states of a non-interacting electron–hole pair as well as the calculated absorption spectrum of a symmetric quantum dot. Similar to the case of a type 2 dot the red-shift of the absorption features as well as the redistribution of oscillator strength can be seen, when Coulomb-scattering is included.

While these results are qualitatively in good accord with the experimental data, quantitatively there are clear differences. To obtain quantitative agreement was, however, not the scope of the studies and cannot even be expected due to the neglect of effects due to the electron–phonon coupling (polaronic effects) as well as the lack of knowledge of parameters of the quantum dot and its surroundings. It is this latter point to which we want to return briefly at the end of this section. Also other single quantum structures have been studied, which were located in mesa structures with larger lateral sizes (i) and which exhibited geometrical asymmetries (ii). In the following we want to discuss on the basis of spectroscopic data for a type 1 quantum dot some potential consequences of (i) and (ii) on the excitonic absorption spectra.

In Sect. 5 we will see that quantum dots (even of the same sample) can exhibit strong variations of their symmetry. One of the most sensitive tools for studying the symmetry is the exciton fine structure (see below). Figure 10 shows the emission of such a quantum dot of type 1, the emission of which is strongly linearly polarized at $B = 0$ with a polarization splitting of about $120\,\mu\text{eV}$. It was located in a mesa with a lateral size of $\sim 250\,\text{nm}$. Note, that also here a background substraction was done as in Fig. 8. From the comparison with the spectra in Fig. 8 remarkable differences are noted: the absorption spectrum is dominated by not only two, but by three intense features, one in the energy range of the $s$-shell and two lines in the range of the $p$-shell. Besides these features, a large number of additional lines appear. These additional lines can be assigned to the following origins:

(i)  *Effects of a considerable symmetry breaking of the quantum dot*

    1. The two absorption lines with large oscillator strength in the $p$-shell are separated by about $8\,\text{meV}$. A polarization analysis shows that the lines exhibit strong orthogonal linear polarization. Therefore we attribute the splitting of these two lines to a lifting of the $p$-shell degeneracy due to a dot shape asymmetry which breaks the rotational invariance in the dot plane. Consequently angular momentum is no longer a good quantum number. Treating the symmetry breaking as a perturbation, it will lead to a mixing of the two unperturbed $p$-shell angular momentum eigenstates described above (one bright and one dark). The new eigenstates will be linear combinations of them and due to the mixing the dark state will share oscillator strength with the bright one.

    2. There are weak features about $10\text{–}15\,\text{meV}$ above the $s$-shell exciton, which might be related to the nominally forbidden transitions

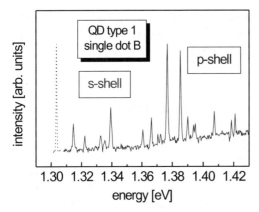

**Fig. 10.** Photoluminescence excitation spectrum of a single $In_{0.60}Ga_{0.40}As/GaAs$ quantum dot of type 1 having low geometrical symmetry [10]

from states in the $p$-shell of the valence band to $s$-shell states in the conduction band. The observability of these features originates from a symmetry breaking which lifts the strictness of the selection rule $\Delta s = 0$ for optical transitions. Vice versa, the features at $1.365\,eV$ could be related to transitions from $s$-shell valence band to $p$-shell conduction band states.

3. Further, one notes that the phonon related features are much more prominent than for the dots in Fig. 8. Besides the GaAs phonon assisted features we also find spectral lines related to the InAs bulk and interface LO-phonons (29 and 32 meV). The strength of all these features indicates that there might be a permanent dipole momentum in the system facilitating the coupling to phonons.

(ii) *Effects of placing the quantum dot into a larger mesa structure*

4. Since the dot of Fig. 10 is located in a much larger mesa structure, it has to be considered as a zero dimensional system embedded in a quasi-two dimensional layer. There is a considerable background, in particular for energies $> 1.36\,eV$ in the absorption spectrum, which increases with increasing energy. We attribute this background to an efficient coupling of the dot structure to the wetting layer. That is, the states of the wetting layer are extended considerably towards the low energy range, where confined quantum dot states are located. Consequently mixing of quantum dot states with states in the wetting layer becomes important. The origin of the features with rather weak oscillator strength around the $p$-shell tentatively lies in mixing of quantum configurations in the $p$-shell with states in the close-by wetting layer.

The data in Figs. 8 and 10 show that single quantum dots with drastically varying absorption spectra can be found within one ensemble of dot structures. To obtain a better understanding, in particular of the impact of the dot geometry as well as of the surroundings of the dot, it will be essential to perform more comprehensive studies, in which for example the mesa size is varied systematically.

# 4    Dephasing of the Ground State Exciton

The excitonic properties are, however, not only determined by the available electronic states in the quantum dot structures, but also by coupling to lattice vibrations. While writing this chapter, this topic has still been the scope of intense studies, both from the experimental and the theoretical side. Conventionally exciton–phonon coupling is treated in a weak coupling picture (at least in III–V semiconductors): the characteristic coupling strength is so small that the exciton and phonon basis are retained as 'eigenstates' and all interaction effects can be treated by means of perturbation theory. Certainly it would be more correct, but also much more cumbersome to diagonalize the full Hamiltonian resulting in polarons as eigenstates. For self-assembled quantum dots it has even been suggested that the coupling strength is so large that such a non-perturbative treatment is essential for a correct treatment of the exciton–phonon interaction [13].

Here we want to present only data for the homogeneous linewidth related to two limiting cases: (a) very low temperatures below $T = 2\,\text{K}$, where phonons are almost completely frozen out and (b) rather high temperatures around room temperature where the behaviour is dominated by optical phonons. To obtain a rigid picture of the homogeneous linewidth in quantum dots is essential for semiconductor quantum dot based quantum information processing. Corresponding proposals are only worth to be pursued if the time to manipulate a quantum bit (based either on charges or spins) is much shorter than the dephasing time (inversely proportional to the homogeneous linewidth). For charge excitations, for example, manipulation times on the order of $100\,\text{fs}$ can be envisioned by using modern techniques of ultrafast spectroscopy. From this time scale we can estimate a lower limit for the dephasing time of about $1\,\text{ns}$ to obtain a ratio of switching to dephasing time of $10^4$. This estimate would exclude right away the use of charge excitations in higher dimensional structures as quantum bits, because much shorter dephasing on the order of a few ps have been reported for them.

Figure 11 shows corresponding spectra of two quantum dots recorded at $T = 2\,\text{K}$. The optical excitation was done quasi-resonantly below the wetting layer into an excited $p$-shell quantum dot state [14]. Extremely sharp Lorentzian emission lines with line widths of $3.4 \pm 0.4\,\mu\text{eV}$ are observed for both dot types. Deconvolution of the spectral resolution of the setup which is $2.0\,\mu\text{eV}$ gives a homogeneous linewidth of $1.8 \pm 0.4\,\mu\text{eV}$. This linewidth corresponds to a dephasing time of about $750\,\text{ps}$. This value is in agreement with recent data from non-linear spectroscopy on quantum dot arrays (four-wave

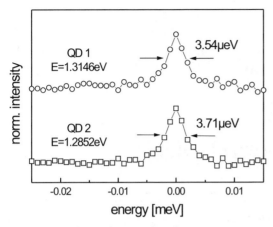

**Fig. 11.** High resolution photoluminescence spectra of two single quantum dots recorded at $T = 2\,\mathrm{K}$ using quasi-resonant laser excitation below the band gap of the surrounding barriers. To faciliate the comparison of the two dots, the energy of the center of the emission has been shifted to zero energy in both cases [14]

mixing and spectral hole burning). It shows that it might be worth considering also excitons in quantum dots as qu-bit candidates, which seemed doubtful, before corresponding results on exciton dephasing were available [15].

Here we want to add an important comment: in the next section we will demonstrate that a quantum dot asymmetry results in a fine structure splitting. In lowest approximation the exciton emission consists not of a single line but of a doublet split by up to $\sim 100\,\mu\mathrm{eV}$. The present observations of single, very sharp emission lines seem to indicate that the emission either originates from charged excitons (see below) or that within the quantum dot ensemble structures can be found which are almost free of any asymmetry. Such an absence of asymmetry would be essential for the creation of polarization entangled photon pair states [16].

We note that even at these very low temperatures the homogeneous linewidth increases with increasing temperature. While the only relevant excitations at these temperatures for exciton scattering are the acoustic phonons, real excitation of carriers from the ground to excited states is not possible due to the considerable energy splittings between the confined states both in the conduction and valence bands which are much larger than the thermal energy in this temperature range. The origin of the temperature dependent linewidth must thus lie in a pure dephasing process. Recent calculations show that it might be related to the finite lifetime of acoustic phonons with non-zero wave number/energy (we note that for $k = 0$ phonons, as lowest energy excitations, the lifetime obviously has to be infinity).

From an extrapolation of the homogeneous linewidth to zero temperature we obtain an upper limit for the homogeneous linewidth of $1.5\,\mu\mathrm{eV}$, resulting in a lower boundary for the dephasing time of more than $900\,\mathrm{ps}$. From time-

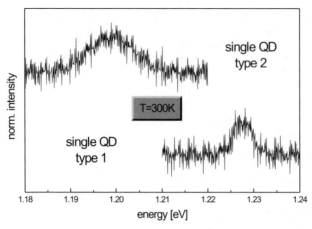

**Fig. 12.** Photoluminescence spectra of two single quantum dots recorded at room temperature [14]

resolved studies performed on arrays of quantum dots an average radiative lifetime of $\sim 1\,\mathrm{ns}$ is obtained. The direct comparison shows that at very low temperatures the exciton dephasing seems to be limited by its radiative decay only, which might be considered as the strongest justification of the term 'artificial atom' for a quantum dot.

While these data clearly are encouraging for quantum information processing, the data obtained at room temperature for the exciton homogeneous linewidth preclude the use of this term for a quantum dot under these circumstances. Figure 12 shows the spectra of the same two single quantum dots as in Fig. 11 at $T = 300\,\mathrm{K}$. Broad emission bands with widths of a few meV are observed then, corresponding to dephasing times of less than a ps. Thus coherent information processing with quantum dots at room temperature seems not to be possible. The location of the dots in the crystal matrix sets a clear limit for the comparison with atoms found in nature, although at low temperatures this fixed location is a big advantage, since the interaction with a light field is much better controllable than for atoms which move freely or are temporarily captured in a trap.

## 5   Fine Structure of Excitons and Excitonic Complexes

In the section on the excitonic absorption spectra in quantum dots we have seen that there are considerable variations with respect to symmetry within a dot ensemble which are reflected by the optical properties of the dots. In this section we want to elaborate sensitive tools for investigating the quantum dot symmetry. We will find that indeed such tools can be developed by addressing the fine structure of the ground state exciton in quantum dots spectroscopically. This fine structure consists of several contributions [17,18,19]:

1. At zero magnetic field it arises from the exchange interaction, which couples the spins of the electron and hole.
2. In an external magnetic field it is extended by the Zeeman interaction of the electron and hole spins with $B$.
3. In general also the interaction of the carrier spins with the spins of the lattice nuclei needs to be included in the discussion. This interaction results in the Overhauser shift of the exciton energy [20]. For its observation, the optically generated excitons have to be spin-polarized to align the spins of the lattice nuclei. This can be obtained by resonant optical excitation of the quantum dots using circularly polarized light. In the experiments described below, however, non-resonant excitation with linearly polarized light was used. Consequently the time averaged spin polarization is zero and the lattice is depolarized. Therefore this part of the fine structure interaction is neglected here.

The exchange interaction consists of a short- and a long-ranged contribution. First we will analyze the short ranged part, for which we will use the methods of invariants [17,18] which can also be used for describing the interaction of the carrier spins with an external magnetic field. In it, for coupling carrier spins and for coupling spins to $B$ the most general inner product forms of the angular momentum operators of electron and hole are chosen that are invariant under the transformations of the symmetry group describing the system. The long-ranged exchange interaction can be easily included in the discussion because it 'exhibits the same spin structure', that is, its contributions can simply be added to the terms of the short-ranged interaction.

The general form of the spin Hamiltonian for the electron–hole exchange interaction of an exciton formed by a hole with spin $J_h$ and by an electron with spin $S_e$ is given by [17,18]

$$H_{\text{exchange}} = - \sum_{i=x,y,z} \left( a_i J_{h,i} \cdot S_{e,i} + b_i J_{h,i}^3 \cdot S_{e,i} \right) . \tag{12}$$

Due to the strain in self-assembled quantum dots the heavy and light hole states are split in energy by at least several tens of meV. This splitting is considerably larger than the fine structure interaction energies, and therefore the light hole states are neglected here. The single particle basis from which the excitons are constructed therefore consists of the heavy hole with $J_h = 3/2$, $J_{h,z} = \pm 3/2$ and the electron $S_e = 1/2$, $S_{e,z} = \pm 1/2$.

From these states four excitons are formed, which are degenerate when the spin Hamiltonian is neglected. These states are characterized by their angular momentum projections $M = S_{e,z} + J_{h,z}$. States with $|M| = 2$ cannot couple to the light field and are therefore optically inactive (dark excitons), while states with $|M| = 1$ are optically active (bright excitons). With these angular momentum eigenstates the matrix representation of $H_{\text{exchange}}$ can be constructed. Using the exciton states $(|+1\rangle, |-1\rangle, |+2\rangle, |-2\rangle)$ as a basis, the following matrix representation is obtained

$$H_{\text{exchange}} = \frac{1}{2} \begin{pmatrix} +\delta_0 & \delta_1 & 0 & 0 \\ \delta_1 & +\delta_0 & 0 & 0 \\ 0 & 0 & -\delta_0 & \delta_2 \\ 0 & 0 & \delta_2 & -\delta_0 \end{pmatrix} \tag{13}$$

Here the following abbreviations have been introduced: $\delta_0 = -3/4 \left( a_z + \frac{9}{4} b_z \right)$, $\delta_1 = 3/8 \left( b_x - b_y \right)$, and $\delta_2 = 3/8 \left( b_x + b_y \right)$. Since the coefficients of the terms linear in the hole momentum are considerably larger than those of the cubic ones, the first term dominantly gives the diagonal matrix elements, while the second term gives the off-diagonal elements. The matrix has block diagonal form. Therefore dark and bright excitons do not mix with each other, and their energies differ by the electron–hole exchange energy $\delta_0$. Due to the off-diagonal matrix elements in the corresponding subblocks, in general the excitons with each $|M| = 1$ and $|M| = 2$ are hybridized: rotational symmetry of the structures studied implies $b_x = b_y$ resulting in $\delta_1 = 0$. In this case the states $|+1\rangle$ and $|-1\rangle$ are eigenstates of $H_{\text{exchange}}$. If, however, the rotational symmetry is broken ($b_x \neq b_y$), angular momentum is no longer a good quantum number, and the $|M| = \pm 1$ excitons are mixed with one another: the new eigenstates $|L_{1/2}\rangle$ are symmetric and antisymmetric linear combinations of the angular momentum states and are split from one another by $\delta_1$. In contrast, the excitons with $|M| = 2$ always hybridize independently of dot symmetry. The splitting patterns are shown schematically in the central part of Fig. 13 for quantum dots with $D_{2d}$ symmetry and with symmetry $\langle D_{2d}$ [22,23]. In general, both asymmetry splittings $\delta_1$ and $\delta_2$ should be rather small compared to $\delta_0$ because they are given solely by the coupling matrix elements that are proportional to $J_h^3$.

The effect of the long ranged part is a splitting of the bright excitons in structures with symmetry $\langle D_{2d}$ into transverse and longitudinal components, while it does not influence the dark states. It has the same spin structure as the short range exchange Hamiltonian. Thus, in the zero-field exchange Hamiltonian it can be simply included by adding the corresponding energies to the off-diagonal matrix elements in the subblock of the $|M| = 1$ excitons. Therefore it does not change the fine structure pattern principally, it only enhances the splitting of the bright exciton doublet.

The general form of the interaction of the electron and hole spins with an external magnetic field $B = (B_x, B_y, B_z)$ of arbitrary strength and orientation is given by [17,18].

$$H_{\text{zeeman}} = \mu_B \sum_{i=x,y,z} (g_{e,i} S_{e,i} - g_{h,i} J_{h,i}) B_i, \tag{14}$$

where $\mu_B$ is the Bohr magneton. Here we want to discuss only two principally different field configurations which are sketched in Fig. 14. The field can be oriented either parallel to the growth direction (Faraday configuration) or

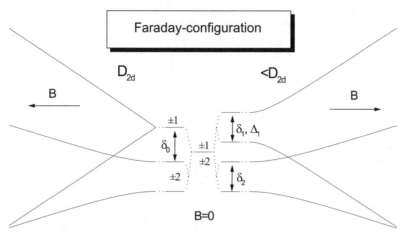

**Fig. 13.** Exciton fine structure in quantum dots with $D_{2d}$ symmetry (*left half*) and in dots with symmetry smaller than $D_{2d}$ (*right half*) at zero field and in a magnetic field. The magnetic field is applied in the Faraday configuration [21,22]

**Fig. 14.** High resolution transmission electron micrograph of a self-assembled InAs/GaAs quantum dot [5]. The *arrows* indicate the two principally different orientations of the magnetic field relative to the heterostructure growth direction ($z$-direction): parallel in Faraday configuration (FC) or perpendicular in Voigt configuration (VC)

perpendicular to it (Voigt configuration). The first configuration does not perturb the rotational invariance around the heterostructure growth direction, the second geometry obviously destroys this symmetry.

### 5.1 Faraday Configuration

First the Faraday configuration will be considered ($B \, || \, z$), for which the Hamiltonian is given by

$$H_{\text{zeeman}} = \frac{\mu_B B}{2} \begin{pmatrix} +(g_{e,z} + g_{h,z}) \, 0 & 0 & 0 \\ 0 & -(g_{e,z} + g_{h,z}) \, 0 & 0 \\ 0 & 0 & -(g_{e,z} - g_{h,z}) \, 0 \\ 0 & 0 & 0 & +(g_{e,z} - g_{h,z}) \end{pmatrix}. \quad (15)$$

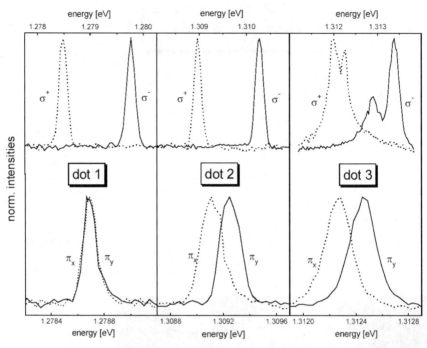

**Fig. 15.** Polarized photoluminescence spectra of three different $In_{0.60}Ga_{0.40}As/$ GaAs quantum dots recorded at zero magnetic field (*lower half*) and at $B = 8\,T$ (*upper half*). At zero magnetic field the circular polarization of the emission was analyzed, at high fields the linear polarization was studied [21,22]

The matrix has diagonal form because the rotational symmetry of the system around the $z$-axis is maintained for this field configuration. The total Hamiltonian of the system is obtained by adding $H_{\text{zeeman}}$ to $H_{\text{exchange}}$. In it, the strength of the diagonal matrix elements can be tuned relative to the strength of the off-diagonal ones through the magnetic field dependence of the Zeeman interaction.

For a quantum dot with $D_{2d}$-symmetry the spin-splitting of the $\pm 1$ states increases linearly with increasing magnetic field. We note that in the model used non-linearities of the spin-splitting (as they are typically observed for quantum wells, for example) cannot occur since band mixing is excluded by restricting to the lowest lying heavy hole bands. For the dark excitons, the spin-splitting shows a non-linear dependence on $B$ because of the hybridization of the $|M| = 2$ excitons at zero field. For an asymmetric quantum dot of lower symmetry the $|M| = 1$ exciton states are linear combinations of the angular momentum eigenstates where the coefficients depend on magnetic field. Their energy splitting deviates from a linear dependence: for low fields it varies quadratically with $B$; only for high fields, for which the Zeeman interaction energies are considerably larger than the exchange energies, is

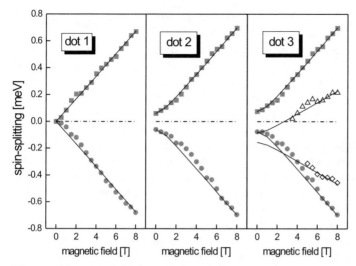

**Fig. 16.** Magnetic field dependence of the exciton transition energies observed for three different In$_{0.60}$Ga$_{0.40}$As/GaAs quantum dots. *Symbols* give the experimental data, *lines* the results of fits to the data using forms obtained from diagonalizing the exciton fine structure Hamiltonian. For simplicity, the diamagnetic shift of the center of the emission has been subtracted in each case [21,22]

a linear field dependence found for the spin-splitting. The left and right parts of Fig. 13 show the evolution of the exciton fine structure splitting in a magnetic field [21,22]. We note that due to their different symmetry the lower energy level of the bright exciton and the upper level of the dark excitons cross each other with increasing magnetic field.

Let us turn now to the discussion of the experimental data obtained in the Faraday configuration. The three panels in Fig. 15 show polarized photoluminesecence spectra of three different In$_{0.60}$Ga$_{0.40}$As/GaAs single quantum dots which were recorded at $B = 0$ (lower traces) and $B = 8$ T (upper traces) [21,22]. The zero magnetic field spectra were analyzed with respect to their linear polarization, while at high fields the circular polarization of the emission was studied. These three quantum dots resemble three different classes of structures observed for the In$_{0.60}$Ga$_{0.40}$As/GaAs quantum dots. The spectra of dots belonging to one class all show the same principal features, although the fine structure parameters, exchange energies and $g$-factors vary from dot to dot.

The emission from dot 1 (left panel) shows no significant linear polarization at zero magnetic field. In contrast, the emission of the two other quantum dots is split into linearly polarized spectral lines at $B = 0$. The splitting between these lines is about 120 μeV for dot 2 (mid panel) and 150 μeV for dot 3 (right panel). In non-zero magnetic field, the emissions split due to the Zeeman interaction of the exciton spin with $B$. As a common feature at

$B = 8\,\mathrm{T}$, the spectra of all three dots show complete circular polarization. The low energy part of the spectrum is $\sigma^+$-polarized, and the high energy one is $\sigma^-$-polarized. However, the number of spin-split lines varies from dot to dot. For the first two dots a splitting into a doublet is observed, while the third dot exhibits a splitting into a quadruplet.

The theoretical analysis of the exciton fine structure for quantum dots of different symmetries given above permits us to understand the observed features in the photoluminescence of the three dot classes [21,22]. From the experimental data we find that there are two characteristic quantities from which information about the quantum dot symmetry can be derived. The first quantity is the magnetic field dependence of the energy splitting between the spectral lines. The second one is the polarization of the emission. Both quantities will be discussed in the following.

The simplest situation is found for dots exhibiting $D_{2d}$ symmetry, for which the exciton angular momentum $M$ is a good quantum number. In this case a single emission line is observed at $B = 0$ due to recombination of the degenerate $M = \pm 1$ excitons, as observed for dot 1. Applying a magnetic field results in a spin-splitting. The symbols in Fig. 16 show the observed exciton transition energies as functions of the magnetic field for the three dots of Fig. 15. To facilitate the discussion of the fine structure effects, we have subtracted the energy of the center of the emission lines for each field strength. This center of emission shifts diamagnetically to higher energies with increasing $B$. The emission of dot 1 splits linearly into a doublet with increasing $B$. $\Delta E = 1.4\,\mathrm{meV}$ at $8\,\mathrm{T}$ corresponding to an exciton $g$-factor of $-3$. The solid lines give the results of fits to the experimental data using the energies calculated by solving the exciton fine structure Hamiltonian.

The situation becomes more complicated for a dot with a shape deformation for which the rotational symmetry is broken. The non-vanishing off-diagonal elements in the subspace of the bright excitons in $H_{\mathrm{exchange}}$ mix the states with $M = +1$ and $-1$. The new coupled eigenstates repel each other, and the emission from these quantum dots is split by the exchange energy $\delta_1$, as observed for dot 2 (middle panel). From the transition energies we obtain a $\delta_1$ of about $120\,\mu\mathrm{eV}$. In contrast to dot 1, at low $B$ the energy splitting between the exciton transitions increases quadratically with $B$ and then transforms into a linear dependence. The transition into a linear dependence occurs at about $2\,\mathrm{T}$ because at these field strengths the diagonal Zeeman interaction terms in the Hamiltonian are already considerably larger than the off-diagonal $\delta_1$. This means that the rotational symmetry is only moderately broken and it can be easily restored by a magnetic field.

Finally, the behavior of dot 3 with a quadruplet splitting in a magnetic field is discussed. The observation of four exciton emission lines implies that the dark excitons become visible and that any possible dot symmetry is lifted. This symmetry breaking can have different origins. First, the quantum dot symmetry can be badly broken so that it exhibits no symmetry at all. In this case all four band edge exciton states should be observable already at zero magnetic field. From the spectroscopic data we obtain no clear proof for such behavior because at $B = 0$ only two emission lines are observed, which are, however, rather broad. In addition, the symmetry breaking could be also magnetic-field induced, for which we envisage the following picture. If the dot structure (and therefore its internal [001] crystal axis) is slightly tilted with respect to the heterostructure growth direction, the spectroscopy is effectively no longer performed in the Faraday configuration because there is a field component in the quantum dot plane. This component which is described by $H_{\text{zeeman}}$ for the Voigt configuration (see below) causes a mixing of $|M| = 1$ and 2 excitons making the dark excitons visible.

Similarly to the case of dot 2, for the bright excitons in dot 3 $\Delta E$ depends quadratically on $B$ for low fields and shows a linear dependence at high $B$. The splitting $\Delta E$ between the 'dark' excitons shows a linear $B$-dependence in the field range in which these states can be resolved. All the results can be described well within the framework for the exciton fine structure developed above, as the comparison of the experimental data (symbols) with the results of the fits (lines) shows.

The different behaviors of the quantum dots show up in the magnetic field dependence of the circular polarization of their emission as well. The degree of circular polarization CP is defined by

$$\text{CP} = \frac{I^+ - I^-}{I^+ + I^-}, \tag{16}$$

where $I^{+/-}$ are the intensities of the $\sigma^{+/-}$ polarized components in the spectra. Figure 17 shows the experimental data for CP as a function of magnetic field for the two $\text{In}_{0.60}\text{Ga}_{0.40}\text{As/GaAs}$ quantum dots 1 and 3 of Fig. 15. For dot 3 we have evaluated only the intensity data of the two emission features of strong intensity. While dot 1 shows circularly polarized emission already for small magnetic fields, the polarization transforms gradually from a linear to a circular one for dot 3. The behavior of dot 2 (not shown) is very similar to that of dot 3. For both these structures complete circular polarization is observed for $B > 3\,\text{T}$ only. This transition reflects the restoration of the rotational symmetry by the magnetic field.

The CP can be calculated from the exciton eigenstates $|L_{1/2}\rangle$ for quantum dots with symmetry lower than $D_{2d}$ [21]:

$$\text{CP}(1/2) = \mp 1 \pm \frac{1}{r^2 + r\sqrt{1+r^2}}. \tag{17}$$

Here we have introduced the ratio $r$ of the spin splitting to the asymmetry energy: $r = (g_{e,z} + g_{h,z})\mu_B B/\delta_1$. At low magnetic fields ($r \to 0$) the circular polarization vanishes, CP $\to 0$, for both structures because the fine structure eigenstates are linear combinations of the circularly polarized excitons. At high fields the Zeeman interaction is much larger than the asymmetry

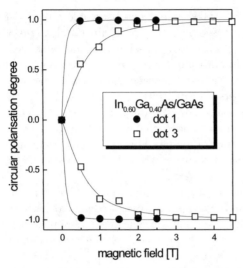

**Fig. 17.** Circular polarization of the emission plotted against magnetic field for two different $In_{0.60}Ga_{0.40}As/GaAs$ quantum dots with different asymmetry induced exchange energy splittings $\delta_1$ at zero magnetic field [22]

exchange interaction ($r \to \infty$). Then the off-diagonal elements can be neglected, and the linearly polarized states $|L_{1/2}\rangle$ transform into circularly polarized states.

$$|L_1\rangle \to |+1\rangle \ ,$$
$$|L_2\rangle \to |-1\rangle \ . \tag{18}$$

Consequently CP $\to \pm 1$. The CP calculated with the quantum dot fine structure parameters from experiment are shown by the lines in Fig. 17, from which good agreement with the experimental data is seen. For the present quantum dots the Zeeman splitting becomes comparable to the asymmetry energy $\delta_1$ in dots 2 and 3 for small magnetic fields < 1 T. For quantum dot 3 having the largest $\delta_1$, CP is already 50% at 0.5 T, 80% at 1 T and approaches unity for $B > 2$ T.

## 5.2 Voigt Configuration

An orientation of the magnetic field perpendicular to the heterostructure growth direction, i.e. in the $x$–$y$ plane, changes the matrix representation of the Zeeman interaction significantly. For simplicity we consider only the case of a magnetic field aligned along the $x$-direction. The matrix representations

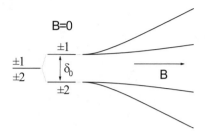

**Fig. 18.** Exciton fine structure of a highly symmetric quantum dot in a magnetic field aligned in the Voigt configuration [23]

of the Hamiltonian can be written as

$$H_{zeeman} = \frac{\mu_B B}{2} \begin{pmatrix} 0 & 0 & g_{e,x} & g_{h,x} \\ 0 & 0 & g_{h,x} & g_{e,x} \\ g_{e,x} & g_{h,x} & 0 & 0 \\ g_{h,x} & g_{e,x} & 0 & 0 \end{pmatrix} . \tag{19}$$

The in-plane magnetic field destroys the rotational symmetry and causes a mixing of bright and dark excitons resulting in observability of the 'dark' states in the spectra. In a classical picture the carriers react to $B_x$ by a precession of the carrier spins around the in-plane field. Due to the precession of the electron spin the $+1$ $(-1)$ exciton couples to the $+2$ $(-2)$ exciton, while the precession of the hole couples the $+1$ and the $-2$ excitons as well as the $-1$ and the $+2$ excitons.

Figure 18 shows a sketch of the exciton fine structure in the Voigt configuration [23]. For simplicity we have assumed that the asymmetry exchange energy splittings $\delta_1$ and $\delta_2$ are negligibly small ($D_{2d}$-symmetry). In contrast to the crossing observed for the Faraday configuration, the spin-splitting now shows a kind of anticrossing behavior. The excitons which are bright at $B = 0$ both shift to higher energies with increasing magnetic field, while the energies of the $|M| = 2$ excitons are lowered by $B$.

Let us turn now to the experimental data. The breaking of quantum dot symmetry by deliberately tilting the magnetic field out of the [001] direction allows us to obtain an unprecedented understanding of the dark exciton states. For these studies we have selected an $In_{0.60}Ga_{0.40}As/GaAs$ quantum dot 1 that shows $D_{2d}$ symmetry. Figure 19 shows photoluminescence spectra of this dot which were recorded in the Voigt configuration for varying magnetic field strengths [23]. With increasing $B$ the center of the emission features moves to higher energies due to the diamagentic shift of the exciton. At $B = 0$ a single emission line corresponding to emission from $|M| = 1$ excitons is observed. On its low energy side an additional spectral line appears at $B = 2\,T$ which originates from the $|M| = 2$ excitons at zero field. For higher fields each of the two emission lines splits into a doublet. The energy splitting between them increases with magnetic field.

Figure 20 shows the observed exciton transition energies versus magnetic field in the Voigt configuration (pointing along the [100] direction) [25]. We

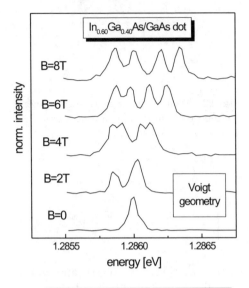

**Fig. 19.** Photoluminescence spectra of a single $In_{0.60}Ga_{0.40}As/GaAs$ quantum dot with high symmetry for different magnetic fields. The field is oriented in the Voigt geometry [23]

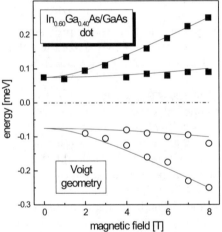

**Fig. 20.** Exciton transition energies observed for a highly symmetric $In_{0.60}Ga_{0.40}As/GaAs$ quantum dot plotted against magnetic field in the Voigt geometry [23]

have also rotated the dot in steps of 45° around the heterostructure growth direction, so that $B$ was aligned along the [110] and [010] directions. Within the experimental accuracy we obtained the same results for these configurations, which gives a strong confirmation of the high in-plane symmetry of the dot. Therefore, we show in Fig. 20 only the data for the [100] orientation. To study only fine structure effects, again the energy of the center of the emission features has been subtracted for each field strength.

The splitting of the energy levels follows the sketch in Fig. 18, as the comparison with the calculated energy splittings (lines) in Fig. 20 shows. The bright and dark excitons show no splittings at $B = 0$ within the experimental error due to the high dot symmetry. Over the whole magnetic field range, the splitting between the two inner emission lines is mainly given by the exchange interaction $\delta_0$, and the Zeeman interaction has little influence on their energies. In contrast, it changes significantly the energies of the outer emission features. Here it should be noted that in quantum wells a quadruplet splitting cannot be observed in the Voigt configuration because experiments show that the in-plane hole $g$-factor is about zero, which leads to a two-fold degeneracy of the exciton states in the Voigt configuration. Furthermore the electron $g$-factors are found to be isotropic in quantum wells, in contrast to the present data. Thus the directionality dependence of $g$-factors in self-assembled quantum dots is considerably different from that in quantum wells.

The presented results have a particularly strong effect on the operation of single photon emitters [24,25] that has been discussed already at the end of Sect. 2. The polarization of the emitted photon will strongly depend on the symmetry of the quantum dot. A solution for this problem might be provided by engineering the polarization of the optical modes of a resonator into which the embedded quantum dot can emit. This polarization might be controlled by the geometrical shape of the resonator. Further, when thinking about using these devices for creating pairs of polarization entangled photons through the radiative decay of a biexciton, it becomes clear from previous considerations that this concept works only for perfectly symmetric quantum dots: only then are the two photons which arise either from the left or the right circularly polarized channel in the first decay in the decay cascade energetically degenerate and therefore indistinguishable. Any asymmetry of the dot lifts this degeneracy, the photons become distinguishable and the entanglement breaks down.

## 5.3   Fine Structure of the Biexciton

After having discussed the exciton fine structure in quite some detail, we want to discuss the fine structure of the emission from the biexciton complex. Figure 21 shows photoluminescence spectra of an $In_{0.60}Ga_{0.40}As$/GaAs quantum dot recorded at different magnetic fields (the quantum dot 1 from above) [8]. The optical excitation power was raised to a level that also emission from the biexciton is observed besides the exciton emission located on the high energy side. The biexciton is shifted by about 3 meV to lower energies. This energy shift corresponds to the biexciton binding energy. With increasing magnetic field both emission features split into doublets. The splitting patterns are very similar, in particular the low energy feature in each doublet is $\sigma^+$-polarized while the high energy one is $\sigma^-$-polarized.

The observed behavior is summarized in Fig. 22 in which we have plotted the exciton and the biexciton transition energies against magnetic field [8]. As discussed before, their field dependences are given by two contributions, the diamagnetic shift due to the orbital magnetic confinement and the spin-splitting due to the Zeeman interaction of the carrier spins with $B$. Except for its low energy shift, the biexciton shows exactly the same diamagnetic

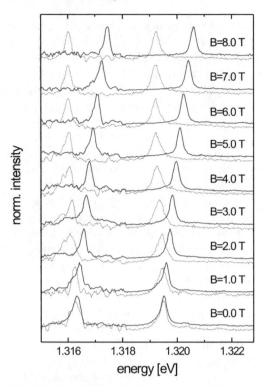

norm. intensity

energy [eV]

**Fig. 21.** Circularly polarized photoluminescence spectra of a single $In_{0.60}Ga_{0.40}As/$ GaAs quantum dot for different magnetic fields. A laser excitation power was chosen that permitted us the observation of both the exciton and the biexciton complex [8]

shift and also the same magnetic field splitting as the exciton. The splittings, for example, are shown in the insert of Fig. 22, from which they are seen to coincide within the experimental accuracy. Let us discuss both aspects of this behavior.

If we neglect Coulomb correlations in the strongly confined quantum dots studied here, the diamagnetic shift of excitonic complexes is given by the sum of the diamagnetic shifts of the electrons and holes that form the complex. Consequently the diamagnetic shift of the biexciton and the exciton emissions have to be identical because they are determined by the shifts of the recombining electron–hole pair.

Now let us turn to the spin-splitting. Due to the Pauli exclusion principle, the biexciton (in its ground state, that is, both electrons and holes occupy their ground $s$-shells) is a spin-singlet state, which does not show a splitting in a magnetic field, as sketched in Fig. 23. However, in the spectra the splitting of the emission is not only determined by the initial state of the recombination process but also by its final state. For the biexciton, this final state is an exciton. Therefore the splitting of the biexciton emission just reflects the

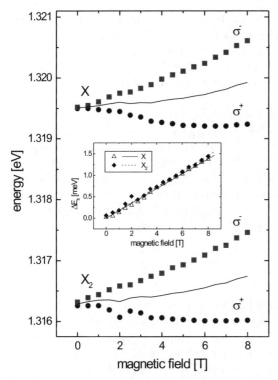

**Fig. 22.** Transition energies of exciton $X$ and biexciton $X_2$ plotted against magnetic field (Faraday configuration). The *inset* shows the field dependence of the spin-splitting for both transitions [8]

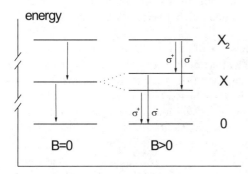

**Fig. 23.** Scheme of the exciton $X$ and biexciton $X_2$ levels at zero and finite magnetic fields. The *arrows* indicate allowed optical transitions

splitting of the exciton in the magnetic field and does not contain additional information about the fine structure.

We note that besides these two basic aspects the results have another important implication. For bulk crystals it is well established that a magnetic field enhances the exciton binding energy considerably. At zero field the electron–hole separation is given by the Bohr radius, whereas at high magnetic field it is given by the magnetic length which is inversely proportional to the square root of $B$. Therefore the exciton binding energy increases approximately proportional to $\sqrt{B}$ in the high field regime. From this dependence one would expect also a strong field dependence of the biexciton binding energy.

Very recently we have shown that this is not the case for quantum wells [26]. Whereas the exciton binding energy is strongly enhanced by a magnetic field, the biexciton binding energy does not depend on $B$ at all (within the experimental error). The same result is found here. Defining the biexciton binding energy as the energy difference between the centers of the exciton and biexciton emission also here it shows no variation with $B$. This independence might prove an important test for the often sophisticated forms of trial wave function that are used for calculating the biexciton binding energy by variational techniques.

## 5.4   Charged Excitons

The possibility to control individual excitations, either charge or spin, in quantum dot structures is an indispensable prerequisite for the new field of quantum information processing in a semiconductor environment. Utilizing ultrafast optical techniques, an exciton might be considered as a proper candidate for such an excitation [27]. However, as a quantum bit it suffers from its relatively short radiative lifetime on the order of a ns, after which the information will be lost. A residual charge in a dot with its spin as quantum bit offers the possibility to overcome this lifetime problem. The spin might be probed and controlled in a coherent fashion by the creation of an electron–hole pair with laser pulses. Even entanglement between two spin states might be created in this way.

In these schemes the intermediate state of processing is given by a charged exciton. Very recently, a number of experimental and theoretical studies of charged excitons confined in single quantum dots have been performed. Often the identification of charged excitons is based on the comparison of their binding energy (the energy separation from the neutral exciton) with the results of detailed calculations. Although sophisticated techniques like numerical diagonalization are used for these calculations, they rely on a number of parameters which are not known with high accuracy, such as the masses of the carriers or shapes of the dots. Therefore we aim in this section at working out a number of experimental criteria, which allow for a unique distinction between neutral and charged excitons. These criteria are (a) the possibility to create a biexciton complex by two-photon absorption and (b) the appearance of an exchange energy splitting between dark and bright excitons.

n-type modulation doped self-assembled $In_{0.60}Ga_{0.40}As/GaAs$ quantum dots have been studied by photoluminescence and photoluminescence excitation spectroscopy. The sample structure is identical to that of the nominally

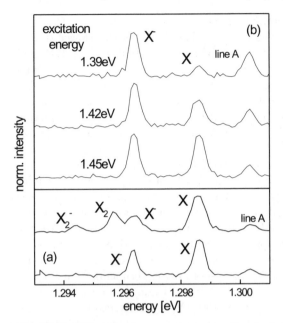

**Fig. 24.** (a) Photoluminescence spectra of a single $In_{0.60}Ga_{0.40}As$/GaAs quantum dot for varying excitation power. The excitation energy was above the GaAs barrier. (b) Photoluminescence spectra of the same energy recorded at low excitation power for varying laser energies indicated at each trace

undoped samples studied up to this point except of a Si $\delta$-doped sheet that is located 20 nm beneath the quantum dot layer. The dopant density was $\sim 10^{10}$ cm$^{-2}$ as compared to a mean dot density of $\sim 2 \times 10^{10}$ cm$^{-2}$. Thus the average number of electrons per dot is $\sim 0.5$ only, assuming a complete transfer of electrons across the barrier.

Figure 24a shows exemplary photoluminescence spectra of a single quantum dot, where the excitation was done non-resonantly into the GaAs barriers ($E_{exc} = 1.56$ eV). At low excitation (bottom trace) two intense emission lines are observed, which are separated by slightly more than 2 meV. Our tentative assignment of these lines is emission from the charge neutral exciton $X$ (for the high energy line) and emission from the negatively charged exciton $X^-$ (for the low energy line). The line A will be discussed later. To obtain more insight, we have increased the optical excitation power (top trace in Fig. 24a). When doing so, two further emission lines appear on the low energy side. The first one is separated 3 meV from the exciton, which is very much the same value that we observed previously for the binding energy of the biexciton $X_2$ for an undoped but otherwise identical sample. The second line is located 2.5 meV below the charged exciton and by analogy we attribute it to the negatively charged biexciton.

Figure 24b shows the evolution of the emission spectra when the laser excitation energy is reduced from above to below the barrier (lower traces) and further to below the wetting layer which is located at $E = 1.416\,\text{eV}$. The laser energies were $E = 1.39\,\text{eV}$ (bottom trace), $E = 1.42\,\text{eV}$ (mid trace) and $E = 1.45\,\text{eV}$ (top trace). When doing so, the ratio of emission intensities of neutral and charged excitons is inverted in favor of the charged complex. This shows that non-resonant excitation results in photodepletion of the free carrier population in the dots, while this population is quite stable for resonant excitation.

Further insight into the correctness of this classification can be taken from two-photon absorption experiments. If the laser energy is tuned midway between the biexciton $X_2$ and the exciton $X$ emission lines $E_{\text{exc}} = 0.5\,[E(X_2) + E(X)]$, the biexciton complex may be created by simultaneous absorption of two photons. The biexciton as a spin-singlet state requires the absorption of two photons of opposite circular polarizations (in symmetric quantum dots, which do not show a linear polarization splitting due to dot asymmetry). Figure 25a shows the corresponding experiments with linearly polarized laser light, which is seen in the spectra due to diffuse scattering. With increasing excitation power ($P_0 \sim 100\,\mu\text{W}$), symmetrically to the laser line, two emission lines appear with equal intensities. Their intensities depend approximately quadratically on excitation power. If, on the other hand, the polarization is changed to a circular one, no biexciton can be generated, whatever level of excitation power we apply.

Turning to the case of the charged complex, a two-photon absorption process is impossible when the laser energy is tuned midway the charged biexciton and the charged exciton lines. Since the $s$-shell may be occupied by two carriers only due to Pauli blocking, the residual charge blocks the absorption of a two-photon pair. This is demonstrated in Fig. 25b, where for $E_{\text{exc}} = 0.5\,[E(X_2^-) + E(X^-)]$ even at very high excitation ($P_{\text{exc}} \sim 100\,P_0$) with linear polarization no indication of an absorption is found. Only scattered laser light is detected. These data are a clear indication that the assignment of Fig. 24a is correct.

Further confirmation is obtained from studies of the fine structure of the exciton emission, which arises from the electron–hole exchange interaction. As discussed above, in leading order the exchange Hamiltonian is given by $H_{\text{exchange}} = -a\boldsymbol{J}_h\boldsymbol{S}_e$ resulting in a splitting of optically active and optically inactive states by the exchange energy $\delta_0 = 1.5a$. A charged exciton in its ground state is formed by two electrons with opposite spins and a hole, all of them in the $s$-shell. In case of the charged exciton, due to the Pauli principle the two electrons form a spin singlet state, so that the total electron spin which is given by the sum $\boldsymbol{S}_e = \boldsymbol{S}_{e,1} + \boldsymbol{S}_{e,2}$ of the spins of the two carriers, is zero. Therefore the exchange disappears and the charged exciton emission shows no energy splitting between the recombinations of the electron–hole pair states with $|M| = 1$ and 2.

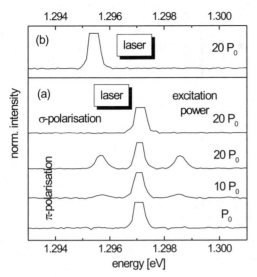

**Fig. 25.** Two-photon absorption experiments on a single $In_{0.60}Ga_{0.40}As/GaAs$ quantum dot for generating a biexciton complex. In (**a**) the exciting laser was located midway between the biexciton and the exciton. Its polarization was either linear ($\pi$-polarisation) or circular ($\sigma$-polarization). At each trace the excitation power level is indicated. $P_0 = 100\,\mu W$. In (**b**) the laser was located between the charged biexciton and the charged exciton

Based on the results described in the previous section, for activating the dark exciton states, we have applied a magnetic field in the Voigt configuration normal to the heterostructure growth direction. In this way, the quantum dot symmetry is broken and the angular momentum $M$ is no longer a good quantum number. Dark and bright excitons become mixed. The observed transition energies are plotted by the symbols in Fig. 26 as a function of magnetic field. Here we have subtracted the diamagnetic shift $\sim B^2$ of the center of emission. At low magnetic fields, an additional feature appears on the low energy side of the exciton which can be attributed to the dark exciton at zero field. For higher fields each of the two features splits into a doublet due to the Zeeman interaction of the carrier spins with the magnetic field. Most importantly, extrapolating the different magnetic field dispersions back to $B = 0$ (as shown by the lines in Fig. 26), a clear splitting between bright and dark excitons is observed which amounts to $\delta_0 = 130\,\mu eV$.

For the charged exciton emission, in effect also a quadruplet splitting is observed at the highest magnetic fields. However, in contrast to $X$, the overall splitting between the different emissions, in particular between the two inner ones, is much smaller as compared to the neutral exciton. Therefore the inner emission lines cannot be separated spectrally. When doing the extrapolation to zero field, all the energies are seen to converge to a single value (see Fig. 26). No splitting between dark and bright states is resolved. As explained above,

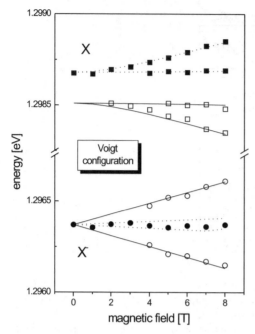

**Fig. 26.** Magnetic field dispersion of the neutral and the charged exciton in a single $In_{0.60}Ga_{0.40}As/GaAs$ quantum dot in the Voigt configuration. *Symbols* give the experimental data, *lines* give the results of fits to the data based on the forms developed in the text

this can only be attributed to interaction of an optically excited hole with an electron spin singlet state as the ground state electron doublet.

Now that we have developed criteria to distinguish between charged and neutral excitons, we are interested in the absorption spectra of these complexes. For this reason, we have performed photoluminescence excitation studies on the quantum dot of Fig. 24. The exciting laser was circularly polarized in these studies. In one case the detection energy was set to the energy of $X$, in another case detection was done at the energy of $X^-$. The results are shown in Fig. 27 for a quantum dot located in a rather wide mesa with a nominal lateral size of 300 nm. The lowest trace shows the absorption of the neutral complex. In agreement with previous studies we find two intense absorption lines, one in the $s$-shell and one in the $p$-shell. In addition, a broad background appears which extends in energy down to the $p$-shell line and is attributed to extended states from the wetting layer.

In the second section we have demonstrated for neutral excitons that the absorption is determined by the number of confined quantum dot shells. For structures with two confined shells, as studied here, only a single feature is found in the $p$-shell. By contrast, for the charged exciton we find two absorption lines in the $p$-shell, which we attribute to an energy splitting between

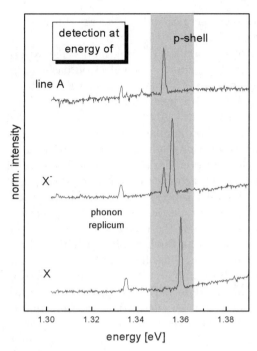

**Fig. 27.** Photoluminescence excitation spectra of a single $In_{0.60}Ga_{0.40}As$/GaAs quantum dot. Detection was done on the neutral exciton (*lowest trace*), on the charged exciton (*mid trace*) and on line A (*top trace*)

different spin configurations. The two electrons can form either a spin triplet or a spin singlet state. The absorption spectrum of the charged exciton differs considerably from that of the neutral one. It does not exhibit a single line in the $p$-shell only but a doublet of lines separated by about 3.7 meV. The oscillator strength of the lower energy line is about a factor of 3 smaller than that of the high energy line. When varying the circular polarization ($\sigma^+$ or $\sigma^-$) of the exciting laser pulse no change in the relative oscillator strength of the two lines is observed. For constructing the absorption spectrum, we first neglect spin and consider all possible electron–hole pair states with zero orbital angular momentum. These states become mixed by Coulomb correlations. The available two-particle states are the same for the two complexes $X$ and $X^-$, so that the different absorption spectra cannot be explained by different orbital configurations.

Therefore different spin configurations must cause the observed doublet splitting of the absorption. The electron in the exciton complex that is optically injected into the $p$-shell can have spin orientation either parallel or antiparallel to the spin of the residual $s$-shell electron. Thus in effect excitation of a spin singlet or a spin triplet state of the electrons can occur. The two complexes have different exchange energies and this energy difference is reflected by the splitting of 3.7 meV observed in the spectra.

Let us also discuss the ratio of the oscillator strengths of the two absorbtion lines. We assume first that the residual $s$-shell electron has spin-up orientation ($S_{e,z} = +1/2$). Excitation by a $\delta^+$-polarized laser will create an electron–hole pair with $S_{e,z} = -1/2$ and a hole with $J_{h,z} = +3/2$. Both carriers will relax fast into the $s$-shell and a ground state trion is formed the emission of which is detected in photoluminescence excitation. Under these experimental conditions, no absorption leading to formation of an intermediate triplet state is possible. However during the recording time of the spectrum (which is typically 30 s for each excitation energy) the spin of the residual electron will undergo spin-flip processes to be in the spin-down configuration ($S_{e,z} = +1/2$). The spin properties of the optically excited electron–hole pair will be the same, but now the relaxation of the electron into the $s$-shell will be blocked as long as no further spin flip (either of the electron in the $s$- or in the $p$-shell) occurs. The hole, on the other hand, will relax fast into the $s$-shell and will recombine with the residual electron. However, this recombination will not be detected in the photoluminescence excitation studies since it will be located at an energy different from that of the ground state trion recombination. Only when the spin flip occurs within the radiative lifetime of the exciton will it be detected in the photoluminescence excitation studies.

The two different spin orientations of the residual electron in the $s$-shell will occur with the same frequency corresponding to a temporal averaging of the two lines. If the flip time were small compared to the lifetime, the two $p$-shell lines would have about the same intensity. The difference in oscillator strength thus reflects quenching of the electron spin relaxation. This is in reasonable accord with the recent report of a quenching of the spin relaxation for ground shell excitons [28]. Since all other involved time scales are short, the ratio of oscillator strengths gives directly the ratio of the spin relaxation time to the radiative decay time. In combination with a radiative lifetime of about a ns (as measured on quantum dot arrays) we obtain in this way a spin-flip time of about 3 ns for the $p$-shell electron.

In the photoluminescence spectra of Fig. 24 we found another faint emission line labelled A about 1.5 meV above the energy of the neutral exciton emission. This emission might be attributed to the recombination of an electron–hole pair in the presence of an electron in the $p$-shell whose relaxation is Pauli-blocked. Therefore we have also performed photoluminescence excitation studies, where the detection energy is set to the energy of the A line. Here we expect detection of the absorption signal only when an electron spin triplet configuration is excited. No signal will be seen for excitation of the spin singlet which permits fast relaxation to the ground state trion. Indeed we observe in the photoluminescence excitation spectrum a single absorption line only. The second line vanishes, fully confirming the above arguments.

# 6    Quantum Dot Molecules

In the preceding section we have demonstrated a solid understanding of excitonic complexes in quantum dots obtained by single dot spectroscopy. In the Introduction we mentioned that this achievement marks the transition from a phase in which the main goal was the fabrication of high quality quantum structures to a new phase in which concepts have to be worked out how these quantum systems can be assembled to obtain new functional units. Here, in the final section, we want to 'close the circle' by discussing the simplest example of such a unit, a quantum dot molecule consisting of two coupled quantum dots.

Currently there is significant interest in quantum information processing [3] using quantum dots [4]. The key building block of a quantum processor is a quantum gate. A quantum gate is used to entangle states of two quantum bits. Quantum gates realized today are based on atomic physics concepts such as ions in traps or nuclear spins in molecules. Recently it has been proposed to use a pair of vertically aligned semiconductor quantum dots as the optically driven solid state quantum gate [4]. In this proposal an electric field applied along the growth direction can localize individual carriers on dot zero or dot one. The different dot indices play the same role as a 'spin', so each particle carries an 'isospin' (qu-bit). When the electric field is turned off the quantum mechanical tunneling rotates the isospin and leads to the superposition of two quantum dot states. The quantum gate is built when two different particles, an electron and a hole, are created optically. In the presence of the electric field the particles are localized on opposite dots. After switching the electric field off, the tunneling and interaction among the two particles will lead to the formation of entangled isospin states. The states can be disentangled at a later time by preventing tunneling through the application of an electric field.

We present here results which support the feasibility of this proposal [29]. This demonstration covers two aspects. First we will demonstrate the possibility of constructing a pair of vertically aligned self-assembled quantum dots with controlled geometries. Second we will demonstrate that electrons/holes tunnel coherently from dot $|0\rangle$ to dot $|1\rangle$ and that the tunneling leads to entangled states of electron–hole pairs. For this reason we isolate a single quantum dot molecule, inject electron hole pairs into it by laser excitation, and study the evolution of the recombination spectrum as a function of the vertical separation between the dots. The remaining task, the demonstration that contacts can be put on a single pair of dots to apply a vertical electric field to control the entanglement, will not be tackled here.

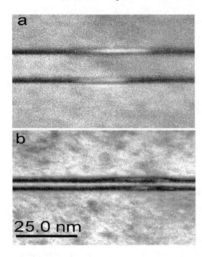

**Fig. 28a,b.** Transmission electron micrographs of InAs/GaAs quantum dot molecules with barrier widths of 16 nm and 4 nm [29]

Promising candidates for coupled quantum dots exhibiting a large splitting are structures fabricated by self-assembly. Indeed, vertical self-alignment has been observed for several years in Stranski–Krastanow epitaxy. The strain field of a quantum dot in a first dot layer facilitates the growth of a second quantum dot above it. Recently, the development of the indium-flush technique [30] allowed the growth of stacks of almost identical coupled quantum dots that display a well-resolved electronic shell structure in state-filling spectroscopy. For the present experiments, the indium-flush procedure was used to grow in close proximity two layers of vertically correlated InAs quantum dots separated by GaAs barriers of varying widths. The quantum dot pairing probability, that is, the probability to find two dots on top of each other and not only a single dot in one of the adjacent layers is almost unity for barrier widths up to $\sim 10$ nm, as high resolution transmission electron microscopy demonstrates. The electron micrographs in Fig. 28 show such quantum dot molecules with barriers of widths $d = 16$ nm (A) and of 4 nm (B) measured from wetting layer to wetting layer [29]. Each dot can be approximated by a disk with a radius $R \sim 8$–12 nm and a height $h \sim 1$–2 nm.

Before we discuss the experiments, let us first develop a simple theoretical model which will help in the interpretation of the spectroscopic data [29]. The model is intended to discuss the essential modifications of the electronic states by coupling of two quantum dots. For simplicity we consider two identical quantum dots (which of course will not be the case for realistic structures leading to a reduction of the tunneling coupling) and we also neglect complications arising from the valence band structure, for example. From coupling of two systems with quasi-atomic densities of states, we expect the formation of 'bonding' and 'anti-bonding' orbitals. The electronic states in each disk can be characterized by a layer index $i = 0$ or $i = 1$ that we term isospin. For weak tunneling and strong lateral quantization, the basic physics is well de-

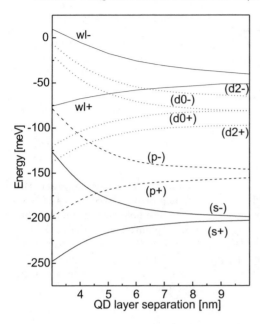

**Fig. 29.** Electron energies in a quantum dot molecule. The $d$-shell is split into the $m = 0$ and $m = \pm 2$ angular momentum channels. Also the wetting layer energies are shown. As parameters for the calculations we used a dot radius of 8 nm, a barrier height of 800 meV and an electron mass of 0.043

scribed by a simple model of $s$-orbitals in each dot. There are two $s$-states $|0\rangle$ and $|1\rangle$ with energy $E_s$ corresponding to an electron on dot 0 and an electron on dot 1. An electron tunnels from dot 0 to dot 1 with a tunneling probability $\langle 0|H|1\rangle = t$ without changing angular momentum. The Hamiltonian $H$ in this limited basis,

$$\begin{bmatrix} E_s & -t \\ -t & E_s \end{bmatrix} \tag{20}$$

is that of a single spin in a magnetic field pointing along the $x$ axis. This similarity is the origin of the notation 'isospin' for the layer index. The off-diagonal tunneling can be eliminated by a rotation to a new basis of symmetric $(+)$ and anti-symmetric $(-)$ orbitals:

$$\begin{bmatrix} |+\rangle \\ |-\rangle \end{bmatrix} = \frac{1}{\sqrt{2}} \begin{bmatrix} |0\rangle + |1\rangle \\ |0\rangle - |1\rangle \end{bmatrix} . \tag{21}$$

The new states are linear combinations of the isospin states 0 and 1. The Hamiltonian in this basis

$$\begin{bmatrix} E_s - t & 0 \\ 0 & E_s + t \end{bmatrix} \tag{22}$$

is diagonal, with $2t$ being the splitting of the two levels which can be interpreted as the 'Zeeman energy'.

In Fig. 29 we show results of a more realistic calculation of the electron energy levels of coupled self-assembled disks as a function of the distance

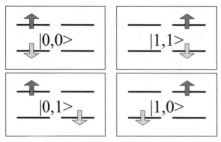

**Energy**

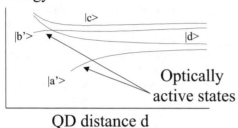

QD distance d

**Fig. 30.** The *upper half* shows the four possible distributions of electron and hole among the two dots in a quantum dot molecule. The *lower half* shows the calculated dependence of the exciton energies on the separation between the quantum dots [29]

between the quantum dot layers $d$. There are three groups of states, the $s$, $p$, and $d$ shells that are confined in the dots. For all the states, the tunneling allows the levels to split into a symmetric $(+)$ and an anti-symmetric $(-)$ state as the quantum dots are brought closer together. In particular, the lowest quantum dot state shows a splitting that increases from only $\sim 5\,\text{meV}$ for $d = 10\,\text{nm}$ to $\sim 40\,\text{meV}$ for $d = 4\,\text{nm}$. For the $p$- and $d$-shells larger splittings between the antisymmetric and symmetric states than for the $s$-shell are found for a fixed barrier width, because these states have a larger penetration into the GaAs barrier resulting in a larger tunnel matrix element.

In a spectroscopic experiment we measure the recombination of an electron–hole complex, so also Coulomb correlations need to be included. We now discuss a simple model of an exciton in a quantum dot molecule in the weak tunneling limit [29]. As shown in the upper part of Fig. 30 there are four isospin states of a single electron hole pair, $|0,0\rangle$, $|1,1\rangle$, $|0,1\rangle$ and $|1,0\rangle$. The first two states correspond to an electron and a hole on dot '0' or dot '1' and can be created optically. The second pair corresponds to indirect electron–hole pairs, e.g., electron on dot '0' and hole on dot '1' and can only be created optically when there is a overlap of the states on the two dots. The interband polarization operator $P^+$ describing the creation of an electron–hole pair can be written as

$$P^+|V\rangle = M_{00}\left[|0,0\rangle + |1,1\rangle\right] + M_{01}\left[|0,1\rangle + |1,0\rangle\right], \qquad (23)$$

where the $M_{ij} = \langle i|j\rangle$ are the overlap integrals of the electron and hole wavefunctions on dots $i$ and $j$. To take the symmetry of coupling to the light field into account, we can write the electron–hole pair states in a new basis of entangled states $|a\rangle, |b\rangle, |c\rangle$ and $|d\rangle$ consistent with the interband polarization operator:

$$\begin{bmatrix} |a\rangle \\ |b\rangle \\ |c\rangle \\ |d\rangle \end{bmatrix} = \frac{1}{\sqrt{2}} \begin{bmatrix} |0,0\rangle + |1,1\rangle \\ |0,1\rangle + |1,0\rangle \\ |0,1\rangle - |1,0\rangle \\ |0,0\rangle - |1,1\rangle \end{bmatrix} . \tag{24}$$

Assuming for simplicity similar tunneling matrix elements for electron and hole, after some algebra the electron–hole pair Hamiltonian in the basis (a–d) can be written as:

$$\begin{pmatrix} E_X - t_1 & -t_2 & 0 & 0 \\ -t_2 & E_X + (V_{00} - V_{01}^D) - t_1 & 0 & 0 \\ 0 & 0 & E_X + (V_{00} - V_{01}^D) + t_1 & 0 \\ 0 & 0 & 0 & E_X + t_1 \end{pmatrix} . \tag{25}$$

Here $E_X = 2E_s - V_{00}$ is the energy of an exciton on a single dot. $-V_{00}$ is the electron–hole attraction for both particles being either on dot '0' or dot '1'. $t_1 = 2tM_{10} + V_{01}^X$ and $t_2 = 2(t + V_0)$. $V_{01}^D$ is the direct interaction for an electron on dot 0 and a hole on dot 1, $V_{01}^X$ is the matrix element responsible for the electron–hole pair scattering from dot 0 to dot 1. $V_0$ is the scattering matrix element involving only one particle, e.g. $\langle 00|V|01\rangle$, and exists only in a finite system such as a quantum dot.

The Coulomb matrix elements $V_{01X}$ and $V_0$ renormalize the tunneling matrix element $t$. Only the optically active states $|a\rangle$ and $|b\rangle$ are still coupled via this renormalized tunneling matrix element. Actually this coupling is important to obtain an entanglement which is independent of the representation basis. The real coupled entangled eigenstates $|a'\rangle$ and $|b'\rangle$ of the exciton Hamiltonian are the excitonic analogs of the symmetric and antisymmetric single particle states, with the splitting controlled by tunneling and by Coulomb interactions. The two entangled states $|c\rangle$ and $|d\rangle$ are exact exciton eigenstates which do not couple to the optically active states and are therefore optically inactive.

The schematic evolution of the four energy levels of the exciton as a function of the distance between the dots is shown in the lower part of Fig. 30 [29]. When the dots are far apart, there are two groups of energy levels separated by the exciton binding energy $V_{00}$. When the dots are brought closer, the optically active states lower their energies and the spacing between them increases in a manner similar to the symmetric–antisymmetric splitting of single particle energy levels. The dark configurations separate from the bright ones and eventually are both located above the optically active ones. The key result is that the eigenstates $a'$ and $b'$ of the electron–hole Hamiltonian are a linear combination of entangled states $a$ and $b$. This guarantees that a state $|10\rangle$, for example, prepared by the application of a strong electric field will evolve into an entangled state once tunneling is allowed. In our simplified treatment,

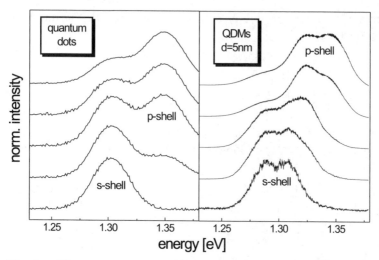

**Fig. 31.** Photoluminescence spectra recorded for varying excitation powers at $T = 2\,\mathrm{K}$ on a single sheet of quantum dots (*left panel*) and on an ensemble of quantum dot molecules with a 5 nm barrier (*right panel*)

the degree of entanglement of the exciton states is close to unity. We note, however, that the degree of entanglement might be considerably reduced by structural asymmetries.

This simple model serves as a guide in understanding the experiments described in the following. First we have studied the quantum dot molecules by photoluminescence spectroscopy recorded on large area mesas. Figure 31 shows the inhomogeneously broadened spectra of a single sheet of InAs/ GaAs quantum dots (left panel) and of an array of InAs/GaAs quantum dot molecules with a GaAs barrier width of 5 nm (right panel). At low excitation, only emission from the $s$-shell is observed for the quantum dots. At higher excitation also emission from the $p$-shell appears 50 meV above the $s$-shell. Turning to the spectra of the molecules, we observe at the lowest excitation power a splitting of the lowest emission line into two features separated by about 25 meV. Note that the higher lying feature cannot be attributed to $p$-shell emission, because in lowest approximation the lateral confinement is not changed by the dot coupling.

$p$-shell emission is therefore expected $\sim 50\,\mathrm{meV}$ above that of the $s$-shell and, indeed, it appears there when increasing the optical excitation. Also for it, a splitting is observed which is larger than that in the $s$-shell ($\sim 30\,\mathrm{meV}$). Thus, these simple array spectra seemingly already confirm our simple expectation of the formation of 'bonding' and 'anti-bonding' orbitals. To obtain further insight, photoluminescence spectroscopy was performed on single quantum dot molecules. For these studies only structures were selected which show at low excitation an emission line doublet in the energy range of the

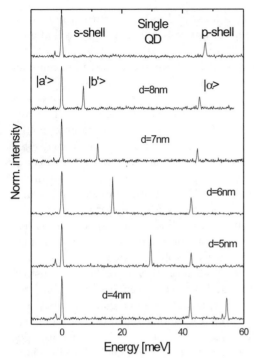

**Fig. 32.** Photoluminescence spectra recorded at ~ 60 K for a single quantum dot (*top trace*) and for quantum dot molecules with barrier widths varying from 8 nm down to 4 nm [29]. To facilitate the comparison of the spectra, the energies of the lowest emission features from the *s*-shell have been subtracted in each case

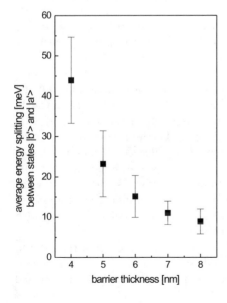

**Fig. 33.** Energy splitting between the entangled exciton states $|b'\rangle$ and $|a'\rangle$ obtained by averaging the data from spectroscopy on a large number of single dot molecules (about 10 for each barrier width) as a function of the barrier width. The 'error bars' indicate the variation within one quantum dot molecule ensemble

s-shell, which we take as an indication for a molecule structure of high symmetry, that is, two almost identical quantum dots which are very well aligned on top of each other.

Figure 32 shows the luminescence of single quantum dot molecules for decreasing barrier width (from top to bottom), from which the evolution of the molecule states with varying separation of the two coupled dot layers is seen [29]. The spectra were recorded at low excitation powers to avoid emission from multi-exciton complexes. The temperature was about $T = 60$ K so that not only ground state exciton emission, but also emission from excited exciton states is observed due to thermal excitation. To avoid variations due to inhomgeneities we have set the energy of the lowest emission feature in the s-shell to zero. The top trace in Fig. 32 shows the emission of a single quantum dot which serves as reference for the emission from the quantum dot molecules. For it, both emission from the s-shell and the p-shell is observed.

For a large dot separation $d = 16$ nm in the reference sample, two emission lines are observed in the s-shell with a splitting varying from < 1 to $\sim 10$ meV. For this distance and temperature, interaction with acoustical phonons prevents coherent tunneling and we expect that the recombination takes place from either state $|0, 0\rangle$ or $|1, 1\rangle$ as in individual quantum dots. The small splitting most probably arises from dot inhomogeneities. For a dot separation $d = 8$ nm in a molecule, the s-shell emission splits into the entangled exciton states $|a'\rangle$ and $|b'\rangle$ (eigenstates of the electron–hole Hamiltonian). The energy splitting between them is about 10 meV. In addition, a further emission line denoted by $|\alpha\rangle$ appears on the high energy side. The energies of all these features depend systematically on $d$: reduction of the quantum dot separation strongly increases the splitting between the two states $|b'\rangle$ and $|a'\rangle$.

This behavior clearly demonstrates the energy level splitting due to the dot coupling. Hence we ascribe the two emission lines as originating from the entangled states of the s-shell exciton. The energy splitting between the two entangled s-shell excitons is shown in Fig. 33 as a function of the separation between the dots. The splitting was obtained by averaging the splitting observed on a large number of single dot molecules for each barrier width. Most importantly, the splitting between the optically active entangled exciton states of the s-shell is more than 25 meV, the thermal energy at room temperature, when the barrier width becomes smaller than 5 nm. The 'error bars' indicate the variation of the splitting for a given $d$. Especially for narrow barriers we find considerable variations. Nevertheless there is the clear trend of an increase of the splitting between the entangled exciton states with decreasing barrier width. In particular, for the smallest barrier width the splitting is always larger than 30 meV.

We want to stress one more time. The theoretical analysis of the exciton states in quantum dot molecules presented here is based on assuming a coupling of two identical quantum dots which are perfectly aligned in the vertical direction. For real structures fabricated by Stransky–Krastanow growth this is obviously a very unlikely situation. Therefore we emphasize that the equality of the dot structures is NOT necessary for observing a coupling induced splitting of the exciton states. Instead such a splitting in the coupled quantum system occurs when the reduction of energy that is obtained from the formation of 'bonding' and 'anti-bonding' orbitals is larger than the energy difference that results from an asymmetry of the two dots.

We also note that the data presented here do not resmble a clear proof of the coherency of coupling of the quantum dot states in the molecule. Such a proof would be possible for example by coherent spectroscopy techniques such as the observation of photon echoes. Very recently, we were able to demonstrate coherent coupling in CW spectroscopy by studying the fine structure of the exciton in quantum dot molecules. For these investigations, structures with a slight asymmetry such as a small lateral displacement or slightly different energies of the dots forming the molecule were chosen. In a magnetic field, pronounced anticrossings are observed when two fine structure levels come into resonance. These anticrossings are possible only due to the coherent coupling of the dots. For details, see [31].

# 7   Conclusions

In summary, we have tried to give an overview of some of the effects that result from Coulomb interactions in excitonic complexes confined in self-assembled In(Ga)As/GaAs quantum dots. As experimental technique we have used optical spectroscopy of single quantum dots. The data obtained in that way have been compared to the results of detailed theoretical studies. One of the main objectives was to develop some generally valid rules which allow us to understand the optical properties in these dot structures, because at the present stage still a veriety of sometimes very different observations are reported. We want to emphasize that our results definitely do not put an end to this very active field. Still we hope that they resemble an important contribution. We also want to stress that the list of references by no means claims completeness. It rather gives the reference to our work. References to the important work done by other groups working in the field can be found in the publications given there.

In detail, we have first studied the excitonic absorption spectrum of self-assembled quantum dots. The absorption is found to depend sensitively on the confined electronic level structure of the quantum dots. For the dependence on the number of confined dot shells a theoretical model has been developed which provides a clear picture of the excitonic states in quantum dots. The results demonstrate that excitons in quantum dots can be understood neither in the frame of a single particle picture nor in the frame of that of an exciton in structures of higher dimensionality. Instead the non-trivial mixing of electron–hole pair quantum configuration needs to be taken into account to explain the experimentally observed features.

Studies of the exciton–phonon coupling in quantum dots reveal two very different regimes of behavior. At very low temperatures when the phonons are frozen out to a good approximation, very sharp excitonic emission lines are observed corresponding to dephasing times of almost a ns. In fact, the line width seems to be limited only by the radiative decay – a result which is encouraging for the application of quantum dots for quantum information processing. At room temperature, on the other hand, the exciton emission is a broad emission band with a halfwidth of several meV resulting in a dephasing on a sub-ps time scale clearly excluding quantum information applications.

Then spectroscopic studies of the exction fine structure in self-assembled quantum dots have been described which allow for obtaining detailed insight into the quantum dot symmetry. Among the spectroscopic tools are linear polarization measurements (without magnetic field). For quantum dots with rotational symmetry the bright exciton emission is circularly polarized, while asymmetric quantum dots exhibit a linear polarization splitting of the ground state exciton luminescence. In highly symmetric dots, the dark excitons can be converted deliberately into optically active states by applying the magnetic field in the Voigt configuration, which destroys the rotational symmetry and the ground state exciton emission consists of a quadruplet.

Further, it was demonstrated that charged excitons can be uniquely identified and distinguished from charge neutral excitons by fine structure studies, because the electron–hole exchange vanishes for them. The identification of charged excitons has been solidified by two-photon absorption experiments. Comparing the absorption spectrum of charged excitons with that of neutral excitons, pronounced differences are observed which arise from the spin structure of the complexes in conjunction with Pauli's exclusion principle.

Finally, in Sect. 6, we have tackled the question whether self-assembled quantum dots indeed have the potential to serve as building blocks for new functional units in which quantum effects are exploited. As an example, we have demonstrated the feasibility of a quantum gate by vertically coupled self-assembled quantum disks. Quantum mechanical coupling of zero dimensional quantum states results in energy level splittings larger than the thermal energy at room temperature. An exciton in the quantum dot molecule was shown to be an entangled state of an electron and a hole. This demonstration opens up possibilities of engineering electronic states at a molecular level for various quantum information processing applications.

## Acknowledgements

The financial support by the Defense Advanced Research Project Agency and the Deutsche Forschungsgemeinschaft (Forschergruppe 'Quantum Optics in Semiconductor Nanostructures') is gratefully acknowledged. The work was also performed under the Canadian European Research Initiative on Nanostructures supported by the European Commission (IST-FET program), by the Institute for Microstructural Sciences, National Research Council of Canada and by the Canadian Natural Sciences and Engineering Research Council. This work would have been impossible without the outstanding theoretical support by the groups of Pawel Hawrylak from the National Research Council in Ottawa and of Tom Reinecke at the Naval Research Laboratory in Washington. The experimental work was based on the excellent quantum dot samples that have been grown both at the Institute of Microstructural Sciences by Simon Fafard and Zbig Wasilewski and at the University of Würzburg by Frank Klopf, Johann-Peter Reithmaier and Frank Schäfer.

I gratefully acknowledge the experimental support by the colleagues with whom I had the pleasure to work at the University of Würzburg: Andreas Kuther, Gerhard Ortner, and Omar Stern (in alphabetical order) in the laboratory of Alfred Forchel. Also the support by the colleagues from the Institute of Solid State Physics (Russian Academy of Sciences) in Chernogolovka, Alexey Dremin, Alexander Gorbunov, Vladimir Kulakovskii, Andrey Larionov, and Vladislav Timofeev is acknowledged.

# References

1. see, for example, C. Weisbuch, B. Vinter: *Quantum Semiconductor Structures: Fundamentals and Applications* (Academic Press, New York 1991)
2. L. Jacak, P. Hawrylak, A. Wojs: *Quantum Dots* (Springer, Berlin, Heidelberg 1998)
3. See, for example: D. Bouwmeester, A. Ekert, A. Zeilinger (Eds.): *The Physics of Quantum Information* (Springer, Berlin, Heidelberg 2000)
4. P. Hawrylak, S. Fafard, Z. R. Wasilewski: Cond. Matter News **7**, 16 (1999)
5. S. Fafard, Z. R. Wasilewski, C. Ni. Allen, D. Picard, M. Spanner, J. P. McCaffrey, P. G. Piva: Phys. Rev. B **59**, 15368 (1999)
6. Y. Toda, S. Shinomori, K. Suzuki, Y. Arakawa: Phys. Rev. B **58**, 10147 (1998)
7. E. Dekel, D. Gershoni, E. Ehrenfreund, D. Spektor, J. M. Garcia, P. M. Petroff: Phys. Rev. Lett. **80**, 4991 (1998)
8. A. Kuther, M. Bayer, A. Forchel, A. Gorbunov, V. B. Timofeev, F. Schäfer, J. P. Reithmaier: Phys. Rev. B **58**, 7508 (1998)
9. D. Gammon, E. S. Snow, B. V. Shanabrook, D. S. Katzer, D. Park: Science **273**, 87 (1996)
10. P. Hawrylak, G. A. Narvaez, M. Bayer, A. Forchel: Phys. Rev. Lett. **85**, 389 (2000)
11. S. Raymond, J. P. Reynolds, J. L. Merz, S. Fafard, Y. Feng, S. Charbonneau: Phys. Rev. B **58**, R13415 (1998)
    P. W. Fry, I. E. Itskevich, D. J. Mowbray, M. S. Skolnick, J. J. Finley, J. A. Barker, E. P. O'Reilly, L. R. Wilson, I. A. Larkin, P. A. Maksym, M. Hopkinson, M. Al-Khafaji, J. P. R. David, A. G. Cullis, G. Hill, J. C. Clark: Phys. Rev. Lett. **84**, 733 (2000)
12. see, for example, Y. Toda, O. Moriwaki, M. Nishioka, Y. Arakawa: Phys. Rev. Lett. **82**, 4114 (1999)
    C. Kammerer, C. Voisin, G. Cassabois, C. Delalande, P. Roussignol, F. Klopf, J. P. Reithmaier, A. Forchel, J. M. Gérard: Phys. Rev. Lett. **87**, 207401 (2001)
    A. Vasanelli, R. Ferreira, G. Bastard: Phys. Rev. Lett. **89**, 216804 (2002)

13. O. Verzelen, R. Ferreira, G. Bastard: Phys. Rev. Lett. **88**, 146803 (2002)
14. M. Bayer, A. Forchel: Phys. Rev. B **65**, 041308 (2002)
15. P. Borri, W. Langbein, S. Schneider, U. Woggon, R. L. Sellin, D. Ouyang, D. Bimberg: Phys. Rev. Lett. **87**, 157401 (2001)
    D. Birkedal, K. Leosson, J. M. Hvam: Phys. Rev. Lett. **87**, 227401 (2001)
16. O. Benson, C. Santori, M. Pelton, Y. Yamamoto: Phys. Rev. Lett. **84**, 2513 (2000)
17. H. W. van Kasteren, E. C. Cosman, W. A. J. A. van der Poel, C. T. Foxon: Phys. Rev. B **41**, 5283 (1990)
18. E. Blackwood, M. J. Snelling, R. T. Harley, S. R. Andrews, C. T. B. Foxon: Phys. Rev. B **50**, 14246 (1994)
19. E. L. Ivchenko, G. E. Pikus: *Superlattices and other Heterostructures*, Springer Ser. Solid-State Sci. **110** (Springer, Berlin, Heidelberg 1997)
20. D. Gammon, S. W. Brown, E. S. Snow, T. A. Kennedy, D. S. Katzer, D. Park: Science **277**, 85 (1997)
21. M. Bayer, A. Kuther, A. Forchel, A. Gorbunov, V. B. Timofeev, F. Schäfer, J. P. Reithmaier, S. N. Walck, T. L. Reinecke: Phys. Rev. Lett. **82**, 1748 (1999)
22. M. Bayer, G. Ortner, O. Stern, A. Kuther, A. A. Gorbunov, P. Hawrylak, S. Fafard, K. Hinzer, T. L. Reinecke, S. N. Walck, J. P. Reithmaier, F. Klopf, F. Schäfer: Phys. Rev. B **65**, 355861 (2002)
23. M. Bayer, O. Stern, A. Kuther, A. Forchel: Phys. Rev. B **61**, 7273 (2000)
24. P. Michler, A. Kiraz, C. Becher, W. V. Schoenfeld, P. M. Petroff, Lidong Zhang, E. Hu, A. Imamolu: Science **290**, 2282 (2001)
25. C. Santori, M. Pelton, G. Solomon, Y. Dale, Y. Yamamoto: Phys. Rev. Lett. **86**, 1502 (2001)
26. T. Baars, M. Bayer, A. A. Gorbunov, A. Forchel: Phys. Rev. B **63**, 153312 (2001)
27. see, for example, G. Chen, N. H. Bonadeo, D. G. Steel, D. Gammon, D. S. Katzer, D. Park, L. V. Sham: Science **289**, 1906 (2000)
28. M. Paillard, X. Marie, P. Renucci, T. Amand, A. Jbeli, J. M. Gérard: Phys. Rev. Lett. **86**, 1634 (2001)
29. M. Bayer, P. Hawrylak, K. Hinzer, S. Fafard, M. Korkusinski, Z. R. Wasilewski, O. Stern, A. Forchel: Science **291**, 451 (2001)
30. Z. R. Wasilewski, S. Fafard, J. P. McCaffrey: J. Cryst. Growth **201**, 1131 (1999)
31. G. Ortner, M. Bayer, A. Larionov, V. B. Timofeev, A. Forchel, Y. B. Lyanda-Geller, T. L. Reinecke, P. Hawrylak, S. Fafard, Z. Wasilewski: Phys. Rev. Lett. **90**, 186801 (2003)

# Optical Spectroscopy on Epitaxially Grown II–VI Single Quantum Dots

Gerd Bacher

Werkstoffe der Elektrotechnik, Fakultät für Ingenieurwissenschaften,
Universität Duisburg-Essen,
47057 Duisburg, Germany
g.bacher@uni-duisburg.de

**Abstract.** Spatially resolved optical spectroscopy applied to single non-magnetic and magnetic II–VI semiconductor quantum dots allows one to study optical, electronic and magnetic properties of single quantum objects. A controlled population of the quantum dot eigenstates by single particles can be obtained by laser excitation, in part combined with electrical current injection. This gives access to intrinsic properties of excitons and biexcitons, like phonon or exchange interaction or coupling to the radiation field. It is demonstrated that the eigenstates in single quantum dots can be manipulated in a well-controlled way simply by applying external electric and magnetic fields. This gives insight into the charge distribution and the g-factor of the particles within the dot. Introducing $Mn^{2+}$ ions into the crystal matrix results in an efficient spin–spin interaction between carriers and magnetic ions. We discuss giant magneto-optical effects like huge effective g-factors and the formation of quasi-zero-dimensional magnetic polarons and demonstrate how the optical response of a magnetic single quantum dot can be used to monitor nano-scale fluctuations of the magnetization.

## 1 Introduction

The huge interest in semiconductor quantum dots (QDs) is driven on the one hand by a variety of promising applications making use of the characteristic $\delta$-like density of states and on the other hand by the possibility to study fundamental physical aspects in these man-made 'artificial atoms'. In contrast to atoms, however, the properties of a semiconductor QD can be tailored in a wide range, e.g. by varying its extension or its composition. For that reason, semiconductor QDs represent a new class of materials with size dependent electronic, optical or even magnetic properties. Besides chemical approaches, the most prominent technique for fabricating semiconductor QDs is self-organized epitaxial growth. While in the past most of these activities have been devoted to III–V compounds (see the Chapter by *Petroff* in this volume), *Xin* and coworkers demonstrated in 1996 for the first time the self-organized growth of CdSe/Zn(Mn)Se QDs [1]. Since this time, the epitaxial growth of II–VI QDs has been demonstrated for a variety of different systems, including CdSe/Zn(S)Se [2,3,4,5,6], CdSe/BeTe [7], Cd(Mn)Se/Zn(Mn)Se [8,9,10], CdTe/ZnTe [11] or CdSe/MgS [12]. Usually,

P. Michler (Ed.): Single Quantum Dots, Topics Appl. Phys. **90**, 147–183 (2003)
© Springer-Verlag Berlin Heidelberg 2003

these QDs are believed to consist of disc-shaped islands with diameters of 10 nm or below and heights of only a few nm. The typical QD densities are on the order of $10^{10}$–$10^{11}$ cm$^{-2}$ [13,14,15], although much larger QDs with a lower density have also been observed in transmission electron micrographs [16].

The physical understanding of these quantum objects has significantly increased since researchers became able to study individual QDs, in particular by means of spatially resolved photoluminescence (PL) spectroscopy [2,15,17] [18,19]. Effects well known from atomic physics, like the Stark effect [20,21,22] [23,24], the Zeeman effect [25,26,27], the Overhauser effect [28] and others have been extensively investigated in single quantum dots (SQDs) during the last few years. An external control of the QD properties can be achieved quite straightforward. This includes on the one hand the manipulation of the QD eigenstates in a well-defined manner by applying external electric or magnetic fields [20,21,22,23,24,25,26,27]. On the other hand, the occupation of QD states with individual charge carriers can be adjusted simply by varying the power of the exciting laser beam. Multiexciton formation has been observed [25,26,29,30,31] and combining optical excitation with electrical carrier injection gives access to charged excitons [32,33,34,35].

Although SQD research on III–V and on II–VI compounds goes hand in hand, two particular properties give prominence to II–VI semiconductor QDs with respect to the more established InAs/GaAs system. First, the high polarity of these materials results in large Coulomb and exchange effects [25,30,36,37] and efficient exciton–phonon coupling [38,39]. Second, the possibility to incorporate magnetic ions isoelectronically into the crystal matrix gives access to magnetic semiconductor QDs with a quantum efficiency high enough for performing SQD spectroscopy [40,41,42]. In particular, adding Mn$^{2+}$ ions has been shown to result in some characteristic magneto-optical effects, like exciton magnetic polaron formation [40,42] or giant g-factors [8] in quasi-zero-dimensional electronic systems.

Not only from the basic physics point of view but also for SQD-based device prospects a detailed understanding and even an external control of optical, electronic or magnetic properties is desirable. The large application potential of single or coupled QDs is evidenced by a variety of novel concepts envisioning devices used for single photon sources and single photon turnstiles [43,44] (see the Chapter by *Michler* in this volume), quantum dot memories [45] (see the Chapter by *Petroff* in this volume) or even quantum computation [46,47]. *Loss* and *DiVincenzo* [46], e.g., suggest to use the spin states of particles in a single QD dot as qubits. In that sense, the role of the spin of electrons, holes or excitons as well as the spin–spin interaction within a QD is of particular interest. Moreover, as the loss of spin information is apparently reduced in QDs [48,49,50,51,52], the usage of (single) QDs should also be promising in the rapidly growing field of spintronics. One example of recent success within this area is the application of (diluted) magnetic semiconductors (DMS) for spin injection into a non-magnetic semiconductor [53,54,55]. Here, one might expect that combining the discrete density of states of zero-dimensional systems with the adjustable magnetic properties known from magnetic semiconductors will open a new path to devices based on the interaction between single carrier spins and the spins of magnetic ions. Examples could be magnetic QD-based spin filters recently suggested by *Efros* et al. [56] or the usage of magnetic SQDs in quantum computation [46].

In this contribution, we review recent results obtained from optical spectroscopy on epitaxially grown non-magnetic and semimagnetic II–VI SQDs, mainly concentrating on aspects typical for these materials. The first part is devoted to the intrinsic properties of self-assembled SQDs, occupied by a well-defined number of particles, i.e. electrons and holes. We discuss the role of electron–hole exchange interaction and exciton–phonon coupling, the coupling of single exciton and biexciton states to the radiation field as well as the formation of charged excitons either by purely optical excitation or by combining laser excitation with electrical current injection. In Sect. 2, we introduce approaches, which allow us to manipulate the SQD eigenstates in a well-controlled way via external electrical or magnetic fields. This will give access to the dipole moments and the g-factors, respectively, of the particles occupying the QD eigenstates. The last part of this contribution introduces magnetic semiconductor QDs. This includes the demonstration of strong magneto-optical effects like giant effective g-factors and the formation of quasi-zero-dimensional magnetic polarons in magnetic SQDs. Moreover it is shown that the optical response of SQDs can be used to monitor magnetization on a nanometer scale, revealing the impact of statistical magnetic fluctuations on the properties of nano-scale magnets.

# 2    II–VI Single Quantum Dots – Intrinsic Properties

The discrete density of states allows us to populate QDs with particles (electrons, holes) one by one. These particles interact with each other, e.g., by Coulomb or exchange interaction, and with their environment, e.g. via phonon scattering or coupling to the radiation field. Due to the high polarity of the II–VI materials studied all these effects are expected to be much more pronounced compared to their In(Ga)As-based counterparts.

## 2.1    Single Excitons in a Quantum Dot

Optical transitions of single electron–hole pairs occupying the QD ground state usually exhibit a spectrally rather narrow emission characteristics. However, this can only be seen if inhomogeneous broadening effects are suppressed, i.e. if one addresses SQDs. This raises on one hand the question on the limitations of the linewidth of optical transitions and on the other

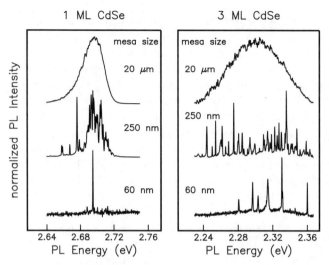

**Fig. 1.** Spatially resolved PL spectra for CdSe/ZnSe quantum dots. SQD selection is obtained by optically addressing etched mesas with diameters down to 60 nm. After [15]

hand allows us to study effects like exciton–phonon coupling or electron–hole exchange interaction, which cannot be resolved in measurements on an inhomogeneously broadened QD ensemble.

### 2.1.1  Single Exciton Transitions and Exciton–Phonon Coupling

The suppression of inhomogeneous broadening effects by reducing the number of QDs contributing to the PL signal can be seen in Fig. 1, where PL measurements on CdSe/ZnSe QDs are depicted for different spatial resolution. Samples containing thin layers of CdSe (nominal thickness of 1 monolayer (ML) and 3 ML, respectively) embedded in ZnSe barriers are compared. To select individual QDs, nano-sized mesas with diameters down to 50 nm have been prepared by electron beam lithography and wet chemically etching [15]. In these experiments, the excitation density was kept low enough to ensure, that at maximum one electron–hole pair occupies the QD ground state. The inhomogeneously broadened PL signal for large mesa sizes, i.e. for simultaneous excitation of $10^5$ to $10^6$ QDs, splits into individual emission peaks, if one reduces the number of QDs addressed by increasing the spatial resolution. E.g. for the 1 ML sample, mainly one peak, which is attributed to the recombination of a single electron–hole pair occupying the ground state of a SQD, dominates the PL spectrum for the highest resolution.

**Emission Linewidth of Single Exciton Transitions.** It is quite interesting to have a closer look at the linewidth of the optical transitions in II–VI

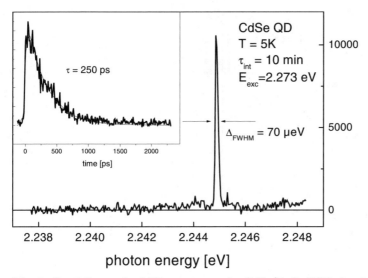

**Fig. 2.** Spatially resolved PL spectrum of a CdSe/ZnSe SQD showing a resolution limited linewidth of 70 μeV. The *inset* shows the transient PL signal of this SQD after pulsed laser excitation. From [51]

SQDs. While in the above presented measurements the linewidth was limited by the spectral resolution of $\approx 400\,\mu\text{eV}$, an emission linewidth of $100\,\mu\text{eV}$ has been found for the CdTe/ZnTe system [57]. *Flissikowski* et al. even report on a *resolution limited* emission linewidth of $70\,\mu\text{eV}$ for CdSe/ZnSe SQDs (see Fig. 2). Apparently, the limitations of this experimental technique do not allow one to extract the intrinsic linewidth of the optical transitions in II–VI SQDs, which for the ground state ultimately has to be limited by the radiative recombination process. Indeed, using time-resolved PL measurements on a CdSe/ZnSe SQD, *Flissikowski* et al. found that except for the damping due to the radiative recombination no further decoherence for ground state excitons occurs [58]. This is in quite good agreement with recent results on self-assembled InAs/GaAs QDs, where it has been demonstrated that the coherence of the ground state exciton is controlled by the radiative recombination process (see the Chapters by *Bayer* and by *Borri* and *Langbein* in this volume).

It has to be noted that similar to what is known for III–V SQDs [59,60,61], some II–VI QD structures exhibit a much larger emission linewidth of SQD transitions. Detailed experimental studies by us [35,62,63] and other groups [36,64] demonstrated that this is due to fluctuating electric fields in the nano-environment. Either by defect states within the epitaxial layer sequence [63] or by surface states [64], excess carriers might be captured and via the quantum confined Stark effect this causes a transient shift of the eigenenergies. Because of the statistical nature of this trapping/re-emission process,

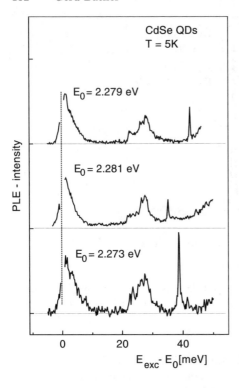

**Fig. 3.** PL excitation spectra of three different CdSe/ZnSe SQDs. The energy zero corresponds to the respective ground state PL position. From [58]

this results in a transient spectral diffusion of the PL signal. Therefore, in time-integrated measurements, where one averages over lots of recombination events, one obtains a broadening of the optical transitions. This is quite similar to what is known from chemically prepared SQDs [65] or single molecule spectroscopy [66]. Vice versa, this spectral diffusion can even be used to identify optical transitions of excitons, biexcitons or charged excitons belonging to the same SQD [36,63,64].

**Excited States and Exciton–Phonon Coupling.** A straightforward experimental access to excited QD states and exciton–phonon coupling is PL excitation (PLE) spectroscopy. While the detection energy is kept constant during the measurements, the energy of the exciting laser beam is tuned over a wide range. Adjusting the detection energy to the QD ground state, a pronounced peak in the PLE spectrum thus indicates an absorption followed by an efficient relaxation of the photoexcited carriers into the QD ground state. Figure 3 reveals PLE spectra measured by *Flissikowski* et al. for different CdSe/ZnSe SQDs [58]. Several features can be extracted from the experimental data.

First, the spectrally narrow peak, which depends on the specific QD, is located between $\approx 35$ meV and $\approx 45$ meV above the ground state and can be attributed to the first excited state of the QD. This is in good agreement with

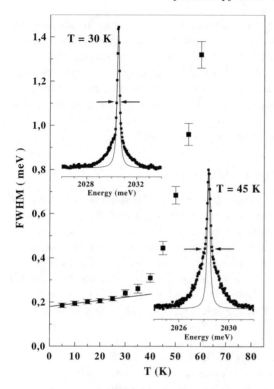

**Fig. 4.** Temperature dependence of the exciton PL linewidth FWHM. The *insets* show PL lines measured at $T = 30\,\text{K}$ and at $T = 45\,\text{K}$ (*symbols*). A strong deviation from a Lorentzian profile (*solid line*) is found. From [38]

recent results of *Rambach* et al., who obtained for the same material system an energy separation between the ground state and the first excited state between about 20 meV and 60 meV varying from dot to dot [67]. Calculations assuming typical QD diameters in the 10 nm range and QD heights of about 1–2 nm as found by TEM measurements [13] underline this interpretation of the data. *Rambach* et al. even demonstrated for some SQDs a splitting of the excited states [67], which apparently orginates from a reduction of the QD symmetry, similar to what was found for InAs/GaAs SQDs [68].

The occurrence of a spectrally broad PLE resonance at around 25–30 meV in Fig. 3 can be attributed to LO phonon replica and indicates an efficient, LO-phonon assisted exciton relaxation to the QD ground state. The large linewidth of the peak demonstrates that in a SQD, the electron–hole pair couples to a distribution of LO phonons, in agreement to what was found by *Gindele* et al. [39]. The increase of the PLE signal close to the detection energy is shown to arise from the exciton coupling to acoustic phonons. *Besombes* and coworkers [38] demonstrated that the acoustic phonon interaction cannot be regarded simply as an inelastic scattering process which induces a loss of phase coherence. As shown in Fig. 4, a strong deviation from a Lorentzian lineshape is instead observed in PL measurements with increasing temperature and sidebands appear progressively around the central zero-phonon line (see inset of Fig. 4). The authors pointed out that in order to account for these experimental findings, stationary eigenstates formed by the mixing of the discrete excitonic states with the continuum of acoustic phonons have to be considered. Interestingly, the thermal broadening of the zero-phonon line is about $1.5\,\mu eV\,K^{-1}$, which is more than a factor of 2 smaller than for comparable quantum well structures. This is apparently a consequence of the three-dimensional exciton confinement, i.e. the absence of final states with suitable energy for the phonon scattering process [69].

### 2.1.2    Electron–Hole Exchange Interaction

Having a closer look on highly spectrally resolved PL spectra of SQDs, one often obtains a doublet structure of the optical transitions. This occurs even in the case of low excitation densities, where definitely only a single exciton occupies the QD ground state, i.e. multiexciton formation can be ruled out. While first reported by *Gammon* et al. for natural GaAs/AlGaAs SQDs in 1996 [19], similar results have been obtained shortly after for CdSe/ZnSe SQDs [2,15,25]. In this system, however, the characteristic energy splitting of the doublet structure exceeds the values found for III–V SQDs by roughly one order of magnitude.

In the left part of Fig. 5, spatially resolved PL spectra of two different CdSe/ZnSe SQDs are plotted for both linearly and circularly polarized detection. As illustrated, QD1 reveals a clear doublet structure with an energy splitting of about 0.8 meV, while for QD2, the energy splitting is quite small and can hardly be resolved in linear spectroscopy. We found that the energy splitting varies from dot to dot in magnitude and sign [70]. However, the two components of the doublet are always linearly polarized and the polarization in most cases is oriented along the [110] and [1$\bar{1}$0] crystal orientation, respectively.

In order to understand these experimental findings, one first has to be aware that due to strain and quantum confinement, the light holes are strongly shifted in energy with respect to heavy holes and consequently the PL signal arises from heavy hole excitons. The ground state of a heavy hole exciton in a SQD is a spin quadruplet, which can be characterized by the $z$-component (= component in growth direction) of the total exciton spin, $M$. If the $z$-component of the electron spin, $s_z = \pm 1/2$, and the $z$-component of the total angular momentum (for simplicity also called 'spin' in the following) of the heavy hole, $j_z = \pm 3/2$, are antiparallel, we get $M = s_z + j_z = \pm 1$ (the so-called 'bright' exciton states), while for parallel spins of the particles $M = s_z + j_z = \pm 2$ (the 'dark' exciton states). According to the selection rules, only the $|\pm 1\rangle$ states couple efficiently to the light field if no mixing between bright and dark states occurs. The spin of the particles results in an exchange interaction between electrons and holes, which lifts the ground state spin degeneracy dependent on the symmetry of the QD (see right part of Fig. 5).

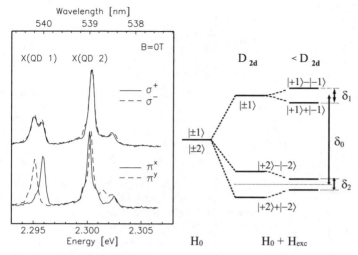

**Fig. 5.** *Left*: Circularly and linearly polarized PL spectra of two different SQDs (QD1 and QD2), respectively. *Right*: Energy level scheme for the exciton ground state in SQDs. The fourfold spin degenerate eigenstate splits into optically active and optically passive states due to the electron–hole exchange interaction $H_{exc}$

It is convenient to write the Hamiltonian $H_{exc}$ for the electron–hole exchange interaction in a matrix representation with respect to the $(|+1\rangle, |-1\rangle, |+2\rangle, |-2\rangle)$ states [70]. Considering first only the short range part of the exchange interaction, one obtains for a QD symmetry of $C_{2v}$ to $D_{2d}$

$$H_{exc} = -\frac{1}{2}\begin{pmatrix} -\delta_0 & \delta_1 & 0 & 0 \\ \delta_1 & -\delta_0 & 0 & 0 \\ 0 & 0 & \delta_0 & \delta_2 \\ 0 & 0 & \delta_2 & \delta_0 \end{pmatrix}. \tag{1}$$

Here, $\delta_0$ respresents the splitting between the bright and the dark excitons, while $\delta_1$ ($\delta_2$) accounts for the splitting of the bright (dark) exciton doublet. From the Hamiltonian it can be seen that no coupling between bright and dark states occurs. It has been shown that $\delta_0$ in self-assembled CdTe/ZnTe QDs almost reaches 1 meV [36], while for CdSe/ZnSe QDs even values clearly above 1 meV have been found [25,71]. Thus, the bright-dark splitting in II–VI QDs is significantly enhanced (by up to an order of magnitude) as compared to InAs/GaAs QDs [72], a consequence of the strong exchange interaction in these materials.

As discussed above, the II–VI QDs are most likely disc-shaped and because of the zincblende structure of the materials investigated, the symmetry for QDs with cylindrical shape is $D_{2d}$. In this case, $\delta_1$ is zero [25] and no splitting of the bright exciton doublet is expected. In contrast, the spin degeneracy of the dark state doublet is already lifted (see Fig. 5). However, a reduction of the symmetry either due to an anisotropic shape and/or due

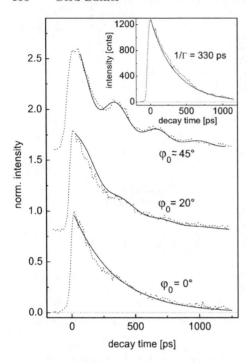

**Fig. 6.** PL transients of a single CdSe/ZnSe QD for quasi-resonant excitation at the 1-LO phonon resonance. Three different polarization configurations are compared, where $\phi_0$ represents the angle between the linear polarization of the laser beam and the [110] crystal orientation. In the *inset*, the PL transient after excitation in the wetting layer is shown for comparison. From [58]

to an interface anisotropy [73] results in $\delta_1 \neq 0$ and the eigenstates are then given by [74]

$$|X^{1,2}\rangle = (|+1\rangle \pm |-1\rangle)/\sqrt{2}. \tag{2}$$

In that case, the optical transitions of ground state excitons are expected to be linearly polarized, which is exactly what is seen in the experiment (see Fig. 5). A quite impressive experiment of *Flissikowski* et al. supports this interpretation [58]. As shown in Fig. 6, photon beats are observed in time-resolved PL spectroscopy on a SQD, if one excites a superposition of the two spin-split eigenstates of the bright exciton. The beating period of 330 ps corresponds to an energy splitting of 13 μeV, not resolvable in the spectral domain.

At this point, one issue has to be discussed. As discussed by *Puls* et al. [71], the energy splitting of the bright exciton states seems to be mainly given by the long range part of the exchange interaction. This effect can be easily included in our matrix representation simply by adding the components of the long range exchange interaction to the subblock of the $|M| = 1$ excitons, while it does not influence the dark states. This contribution mainly increases the bright-dark splitting and, in case of a QD symmetry $< D_{2d}$, the energy splitting of the bright exciton doublet. It is experimentally quite difficult to unambiguously separate the contributions of the long and the short range part of the exchange interaction. However, the experimental data presented by *Puls* and coworkers [71], in particular the fact that the splitting of the bright excitons exceeds the dark exciton splitting significantly is an indication of a contribution of the long range part of the exchange interaction to the fine structure splitting of optically allowed exciton states in SQDs. This, however, does not change the eigenstate symmetry and therefore the fact that the optical transitions are linearly polarized. Note that the exchange splitting of SQD transitions has been found also for CdTe/CdMgTe [57] or CdTe/ZnTe [11] SQDs as well as in the InAs/GaAs system [75], which emphasizes the quite general aspect of these findings.

## 2.2   Biexcitons and Charged Excitons

The ability to vary the power of the exciting laser on one hand and to use (modulation) doped QDs or to inject electrons or holes electrically into a SQD on the other hand allows one to change the population of SQDs with individual carriers in a well-defined way. Here, we concentrate on the formation of biexcitons and charged excitons in II–VI SQDs and discuss their binding energy as well as the coupling to the radiation field.

### 2.2.1   Biexciton Binding Energy in II–VI Single Quantum Dots

As soon as the number of optically generated electron–hole pairs is comparable to or even larger than the number of QDs excited in the sample, the population of the QD eigenstates by more than one electron–hole pair becomes likely. The formation probability of biexcitons (= two electron–hole pairs) is strongly non-linear [76], in accordance to what is known from higher dimensional systems. This can be clearly seen in Fig. 7, where PL spectra of two different CdSe/ZnSe SQDs (QD1 and QD2, respectively) are plotted for various excitation densities. In case of low excitation, only single electron–hole pairs occupy the QD ground state and the exchange doublet (see above) of the single exciton transition dominates the PL spectrum. If one increases the excitation density, new lines appear around 2.27–2.275 eV, red shifted by about 23–25 meV with respect to the single exciton transitions $X$. From the power dependence as well as from the polarization properties (see below), these emission peaks can be attributed to optical transitions from the biexciton state, $X_2$. Interestingly, the same doublet structure occurs for exciton and biexciton transitions of a given SQD.

Because of the Pauli exclusion principle, the QD ground state can be occupied by maximal two electron–hole pairs and the biexciton has to be a spin singlet state as both the electrons and the holes must have antiparallel spins. This has a rather important consequence for optical transitions of biexcitons. As a direct recombination of the biexciton to the ground state is forbidden, the exciton is the final state of recombination for biexciton transitions. Consequently, the symmetry of the exciton eigenstate controls the fine structure splitting of both the biexciton as well as the exciton transition.

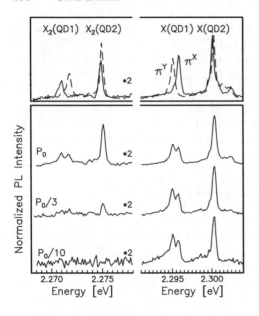

**Fig. 7.** Unpolarized PL spectra from two different SQDs (QD1 and QD2) for different excitation powers ($P_0 = 10\,\mathrm{W/cm^2}$). $X$ denotes the exciton emission while $X_2$ corresponds to the biexciton transition. The *upper part* of the figure shows linearly polarized PL spectra from the two SQDs at $P_0$. After [25]

This becomes even more clear if one looks at the polarization resolved PL spectra (Fig. 7, bottom). E.g. for QD1, both the exciton and the biexciton lines are linearly polarized. Remarkably, the energy sequence of the polarized components is different for the exciton and the biexciton transitions. While the $\pi^y$ component is the low energetic component of the exciton doublet, the low energy component of the biexciton transition is $\pi^x$ polarized and vice versa. This is exactly what we expect from the selection rules of optical transitions in SQDs, if we take into account the exchange splitting of the single exciton state and the fact that neither the biexciton state nor the ground state is split due to their spin singlet structure (see also Fig. 14).

The fine structure of the exciton and the biexciton emission has an important impact on the determination of the biexciton binding energy in QDs. The energetic distance $\Delta_M^*$ between the emission of the single exciton and the two exciton complex apparently depends strongly on the PL polarization. As a consequence, in QDs with a large exchange splitting of excitons, $\Delta_M^*$ and the biexciton binding energy $\Delta_M$ are different entities. The biexciton transition corresponds to the transition from the spin singlet biexciton state to the excited ($|\pm 1\rangle$) rather than to the ground ($|\pm 2\rangle$) exciton state. Thus, in QDs, $\Delta_M$ should be defined by the following equation

$$\Delta_M = \Delta_M^* - 2\delta_0 \mp \delta_1 - \delta_2 \tag{3}$$

for $\pi^x$ and $\pi^y$ polarization, respectively. E.g., for QD1 we obtain $\Delta_M^* = 24.9\,\mathrm{meV}$ (23.3 meV) for $\pi^x$ ($\pi^y$) polarization, which, however, is much larger than $\Delta_M$. Thus, the commonly used energetic distance between the exciton

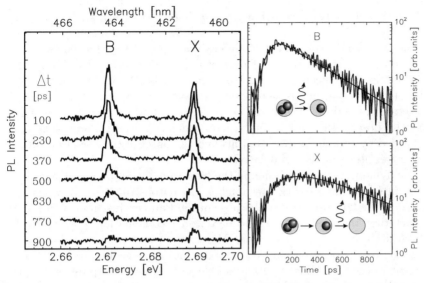

**Fig. 8.** *Left*: Transient PL spectra from a single CdSe/ZnSe QD showing the single exciton $X$ and the biexciton $B$ transition. *Right*: Decay curves for the exciton and the biexciton PL signal. The *solid line* is a fit according to (4). After [30]

and the biexciton emission results in a strong overestimate of the biexciton binding energy by several meV [25]. Nevertheless, the biexciton binding energy for II–VI QDs is strongly enhanced as compared to their III–V counterparts, a direct consequence of the larger Rydberg energy in these materials. Slightly smaller biexciton binding energies have been found for self-assembled CdTe/ZnTe QDs [36] and CdTe/CdMgTe QDs grown on vicinal surfaces [57], while in the CdS/ZnS system biexciton binding energies even far above 30 meV have been obtained [37].

## 2.2.2 Coupling to the Radiation Field – Exciton Versus Biexciton

The fact that the final state of biexciton recombination is the single exciton state can be directly seen in time-resolved experiments. In analogy to what is known from nuclear physics as a consecutive decay, the recombination of the biexciton should feed the population of the exciton state.

In the left part of Fig. 8, transient PL spectra are plotted for a single CdSe/ZnSe QD showing both the exciton $X$ (at $E = 2.690\,\text{eV}$) and the biexciton $B$ (at $E = 2.6715\,\text{eV}$) contribution to the PL signal. Here, an etched mesa with a diameter of only 50 nm has been used for SQD selection. This has allowed us to address only one individual QD, i.e. any interaction between particles in different dots can be neglected. No transient change of the lineshape and the emission energy is observed for both lines, indicating the absence of any marked filling of excited QD states. However, the relative contribution of the two peaks changes in a characteristic way with increasing time.

The time dependence of the PL intensity of the exciton and the biexciton transition are shown in the right part of Fig. 8. The decay of the biexciton signal is monoexponential with a time constant of 310 ps. In contrast, the onset of the exciton line is significantly delayed, resulting in "plateau-like" characteristics of the exciton decay curve. This behavior is in contrast to what is usually observed in bulk or high quality quantum wells, where biexcitons are formed by combining two excitons and the onset of the biexciton rather than the exciton emission is delayed.

The dynamics of the exciton and the biexciton in a SQD can be described taking into account the discrete number of excitons (0, 1, or 2) which are allowed in the QD ground state because of the Pauli exclusion principle, and the fact that the final state of the biexciton recombination is an optically allowed state of the single exciton. Moreover, neither biexciton formation nor biexciton dissociation have to be taken into account due to the 3D confinement in the QD. This results in an elegant analytical expression for the time evolution of the probabilities for the population of the biexciton and the exciton states, $w_2$ and $w_1$, respectively:

$$\frac{d\,w_2}{dt} = -\frac{w_2}{\tau_2} \quad \text{and} \quad \frac{d\,w_1}{dt} = -\frac{w_1}{\tau_{1t}} + \frac{w_2}{\tau_2} \tag{4}$$

with $1/\tau_{1t} = 1/\tau_1 + 1/\tau_{1s}$, where $\tau_1$ and $\tau_2$ are the radiative recombination times of the bright exciton and the biexciton, respectively, and $\tau_{1s}$ is the spin flip time from bright to dark excitons states.

This equation system results in an exponential decay of the biexciton line and a nonmonotonic time dependence for the exciton line, in full accordance with the experimental findings. This clearly indicates a cascaded decay of biexcitons and excitons, in agreement with the conclusions drawn from the optical selection rules. Taking into account the fact that $\tau_{1s}$ exceeds the radiative lifetime of bright excitons significantly [30,77], one obtains $\tau_1 \approx \tau_{1t}$. From a fitting of the experimental data with this simple model (see solid line of Fig. 8), we surprisingly obtain a radiative lifetime of the biexciton which is comparable to the single exciton one [30].

In general, the ratio between the biexciton and the exciton lifetime, $\tau_2/\tau_1$, should depend on both the spin structure of the particles and their spatial wavefunctions which govern the coupling to the photon field, i.e. the dipolar superradiance effect. It has been shown that in the limit of very small QDs (dot radius $r_{\text{dot}}$ much smaller than the Bohr radius $a_B$), Coulomb interaction induced corrections to the electron and hole wavefunctions in the confining potential are negligible. In that case, the biexciton lifetime should be half the exciton one [30]. Apparently this contradicts our experimental findings. However, model calculations show an increase of the spatial separation of the hole wavefunctions with increasing dot radius, while in contrast the electron wavefunctions are almost homogeneously distributed within the dot [78]. In close analogy, e.g. to the spatial separation of the protons in an $H_2$-molecule, this is a direct consequence of the different masses of electrons and holes and has an important impact on the radiative lifetimes of biexcitons. As the annihilation of a $|+\frac{3}{2}\rangle$ $(|-\frac{3}{2}\rangle)$ hole with a $|+\frac{1}{2}\rangle$ $(|-\frac{1}{2}\rangle)$ electron is dipole forbidden, a spatial separation of the holes results in an increase of the biexciton lifetime, because only closely lying electron–hole pairs with antiparallel spins are able to couple to the radiation field, i.e. the biexciton lifetime is expected to approach the exciton one in the case of QDs with sizes comparable to the Bohr radius. In the limit of extremely large 'dots' (i.e. $r_{\mathrm{dot}} \gg a_B$), a biexciton lifetime even larger than the exciton one is expected [30,79]. This simple model idea agrees quite nicely with the experimental findings, suggesting that both the spatial distribution of the carrier wavefunctions as well as their spin structure have to be taken into account when discussing the coupling of excitons and biexcitons to the radiation field [30].

### 2.2.3  Charged Excitons

Laser excitation usually generates electron–hole pairs, i.e. the number of electrons populating the QDs is expected to be identical to the number of holes. In semiconductors, however, both (modulation) doping and electrical injection can be used for an efficient injection of *either* electrons *or* holes. Combining these techniques with laser excitation in QDs, one may expect the formation of charged excitons occupying the QD eigenstates.

**Modulation Doped Quantum Dots.** In a variety of II–VI quantum wells it was demonstrated that laser excitation above the barrier bandgap results in the occurrence of a new PL line located at the low energy side of the single exciton signal. This includes experiments on modulation doped CdTe-quantum wells [80] and nominally undoped as well as doped ZnSe-quantum well heterostructures [81,82]. Apparently excess electrons (holes) from the barrier are captured by the quantum well and in conjunction with the optically excited electron–hole pair form negatively (positively) charged excitons.

Quite recently, *Besombes* and coworkers [36] applied this experimental technique to CdTe/ZnTe SQDs. As can be seen in Fig. 9, a new line $X^-$ appears in the SQD PL spectra, energetically located between the single exciton $X$ and the biexciton $X_2$ line. The authors conclude that this line arises from negatively charged excitons. Although the trion and the biexciton signal coexist in the SQD PL spectrum, one should note that the QD is either populated by a biexciton or by a charged exciton at a time. Only due to the statistical nature of the QD population one observes both lines in time-integrated PL measurements, where one usually averages over lots of recombination events. Depending on the specific SQD studied, typical trion binding energies range from 6 to 8 meV and a strong correlation between the binding energies of the charged exciton and the biexciton is observed [36]. Similar findings have been reported in CdSe/ZnSe SQDs grown by metal

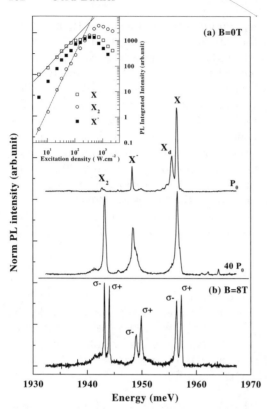

**Fig. 9.** (a) PL spectra of a SQD taken under two different excitation densities ($P_0 = 5\,\mathrm{W/cm^2}$). (b) Non-polarized PL spectrum at $B = 8\,\mathrm{T}$. The *inset* shows the integrated PL intensity of the emission lines of the exciton $X$, the charged exciton $X^-$ and the biexciton $X_2$ plotted versus excitation density. From [36]

organic vapor phase epitaxy using cathodoluminescence experiments. Here the binding energy of the trion even reaches values in the order of $10\,\mathrm{meV}$ [83]. Both groups used the correlated spectral diffusion observed for the trion and the neutral exciton PL signal to demonstrate that both transitions originate from the same SQD, similar to what was observed for exciton–biexciton pairs in CdSe/ZnSe SQDs [63].

**Electrical Injection of Electrons into Single Quantum Dots.** The purely optical approaches to manipulate the charge inside a quantum dot are based on statistical emission/re-emission processes of impurities in the semiconductor matrix embedding the dots. This all-optical technique will not be sufficient to *switch* the charge state of a QD by external means. To obtain a direct control over the charge inside a SQD electrical injection of single carriers is desirable [33,34].

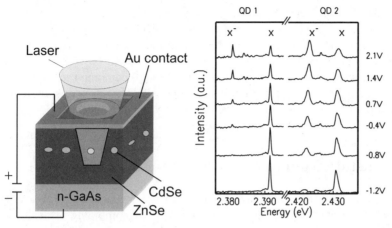

**Fig. 10.** *Left*: Sample geometry used for injecting electrons into SQDs. *Right*: Characteristic PL spectra from two different SQDs for different voltages, showing the emission of neutral $(X)$ and negatively charged excitons $(X^-)$. After [35]

In Fig. 10, the sample geometry used for a controlled injection of single electrons into a II–VI SQD is shown schematically. Self-assembled CdSe QDs embedded in ZnSe barriers have been grown on a highly $n$-doped (doping level $\approx 2 \times 10^{18}\,\mathrm{cm}^{-3}$) GaAs substrate. To address SQDs a 100 nm thick chromium mask was deposited on the sample surface containing lithographically defined nanoapertures of various sizes (diameters down to 80 nm). The metal mask was also used as a top contact for the application of a vertical electric field. By applying a bias between the top contact and the substrate the Fermi level of the heterostructure can be tuned relative to the SQD ground state allowing electrons from the substrate to enter the SQD.

The right part of Fig. 10 shows the PL emission of two individual QDs, labelled as QD1 and QD2, for various bias voltages applied to the top contact. For a high negative bias the PL spectra consist of a single line due to exciton $(X)$ recombination from the ground state. In this case the Fermi level of the heterostructure is below the QD ground state and electrons from the substrate cannot occupy QD states. For increasing positive bias, a second line, denoted as $X^-$, emerges on the low energy side. This spectral feature reflects the emission from the negatively charged exciton state, since electrons from the substrate can reach the QD ground state. The binding energy of the negatively charged exciton is 10.4 meV for QD1 and 8.0 meV for QD2. Interestingly, both the lineshape and the linewidth of the neutral and the charged exciton emission are remarkably similar for $X$–$X^-$ pairs belonging to the same SQD. In fact, these data show that the charge state of a SQD can be switched by applying a vertical electrical field.

It is well worth comparing the results found for II–VI SQDs with the ones obtained in GaAs-based SQDs. The main difference here seems to be of a quantitative nature: the binding energies of trions in II–VI SQDs are much higher than in GaAs/AlGaAs [84] or InAs/GaAs [33,34] SQDs. This is similar to what is known for the biexcitons and is related to the larger Coulomb effects in the II–VI materials. However, it might be very attractive for spintronics or quantum information technology to combine the technique of electrical single electron injection into a SQD with the usage of spin aligners consisting of magnetic semiconductors for injecting single, spin polarized electrons into a SQD or even a SQD molecule.

# 3    Eigenstate Manipulation by External Fields

Up to now, we have discussed how to control the occupation of SQDs with individual carriers either by optical excitation or by combining optical and electrical carrier injection. For the usage of SQDs for device applications in addition an external control of the SQD eigenstates would be desirable. Here, we show how the eigenstates in II–VI SQDs can be manipulated in a well-controlled way by applying either external electric or magnetic fields.

## 3.1    Applying Electrical Fields – The Stark Effect

By applying electrical fields to low dimensional semiconductor structures the eigenstates can be changed via the quantum confined Stark effect. This is well-known for quantum wells or for QD ensembles. Exposing a SQD to a well-defined external electrical field, however, is quite challenging and requires a highly developed nanotechnological approach. In the framework of the quantum confined Stark effect, the energy shift $\Delta E$ of the QD eigenstates in an external electric field $F$ is for small fields usually described by

$$\Delta E = \mu_{el} F + \alpha F^2 , \tag{5}$$

where $\mu_{el}$ and $\alpha$ are the components of the permanent dipole moment and the polarizability, respectively, in the direction of the electric field $F$. Thus, a careful analysis of the field-induced energy shift of the SQD PL signal can be used to extract detailed information about the permanent and the induced dipole moment of the recombining particles in the QD.

### 3.1.1    Lateral Electrical Fields

In CdSe/ZnSe QDs the dot height is usually much smaller than its lateral extension [13,15]. Thus, the electronic eigenstates should be much more sensitive to lateral electric fields as compared to vertical ones. In order to apply well-controllable lateral electric fields, a capacitor-like geometry as shown in the left part of Fig. 11 was developed. Self-assembled CdSe/ZnSe have been grown on an insulating GaAs substrate. For SQD selection, etched mesas

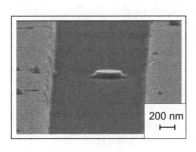
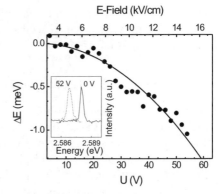

**Fig. 11.** *Left*: Scanning electron micrograph of an etched mesa structure used for SQD selection embedded between two Au contacts. *Right*: Energy shift $\Delta E$ versus lateral electric field [*Symbols*: Experiment, *solid line*: Fit according to (5)]. The *inset* shows normalized PL spectra for two different voltages

acting as nanoprobes have been prepared lithographically while in a second lithography step, pairs of Au electrodes with a gap between $d = 2\,\mu m$ and $d = 8\,\mu m$ have been defined [24]. Assuming that the mesa containing individual CdSe/ZnSe QDs is placed close to the center of the capacitor and the mesa diameter is small compared to the distance between the electrodes, the electric field $F$ applied to the SQDs is given in a good approximation by

$$F \approx \frac{U}{\varepsilon d}. \tag{6}$$

Here $U$ is the applied voltage, $\varepsilon = 9.3$ is the dielectric constant of the CdSe/ZnSe QD and $d$ is the distance between the electrodes. Thus, the electric field acting on the QD can be calculated quite accurately, in contrast to other approaches, where simply metal stripes have been defined on top of the samples [21,23].

As can be seen in the right part of Fig. 11, applying a voltage to the contacts results in both an energy shift and a linewidth broadening of the single exciton luminescence signal (see inset of Fig 11). E.g. for this particular SQD, a voltage of $U = 52\,V$ (corresponding to an electric field of about $13\,kV/cm$ with $d = 4\,\mu m$) causes a redshift of the PL peak of about $1\,meV$. While the linewidth broadening at high electric fields is a consequence of a field-induced tunneling of the carriers out of the SQD [21,85], the energy shift $\Delta E$ is due to the quantum confined Stark effect.

The experimental data can be described by a pure quadratic dependence of the energy shift on the applied electric field (solid line in Fig. 11). No indication of a significant contribution from the linear Stark effect due to a permanent dipole moment is found. For this SQD, we obtain an exciton polarizability of $\alpha = 4.9 \times 10^{-3}\,meV\,cm^2\,kV^{-2}$. This value is in quite good agreement with theoretical calculations, assuming a disc-shaped QD with a diameter of 8 nm and a lateral potential depth for the electrons and holes of 100 meV [24], which seems to be quite reasonable for these kinds of CdSe/ZnSe QDs [13,15].

### 3.1.2   Vertical Electrical Fields

In order to apply electrical fields in the vertical direction, the technological approach depicted in Fig. 10 is used. At first glance, one might expect that in this geometry the sensitivity of the eigenstates to electrical fields is smaller than in the case of lateral fields, as the extension of the QD in the vertical direction is much smaller than in the lateral one. However, two important points have to be emphasized. First, the intrinsic zone of the Schottky diode can be kept small, i.e. the electric field mainly drops in the vicinity of the QD region. Second, there is no air bridge between the contacts and the QD resulting in higher effective electrical fields acting on the SQD for a given voltage.

Applying a vertical voltage of only a few volts to the Schottky diode results in a measurable energy shift of the QD PL signal. This is shown in Fig. 12, where the Stark shift $\Delta E$ is depicted for the neutral and the charged exciton for two different SQDs. Although the Stark shift seems to be linear throughout the investigated voltage range, a clear separation of the linear and the quadratic contribution to the Stark effect is quite difficult due to the small energy shifts obtained. As can be seen from the data the Stark shift of the charged exciton emission for QD2 amounts to about 1.0 meV for the highest applied bias value of $U = 3.0$ V, whereas the neutral exciton emission from the same SQD shows a considerably smaller energy shift of only 0.4 meV. This might be due to the fact that the number of negatively charged particles is twice as large for the trion than for the neutral exciton. The same ratio of the dipole moments is found for the emission lines of QD1, which generally show a weaker dependence on electric field than that of QD2,

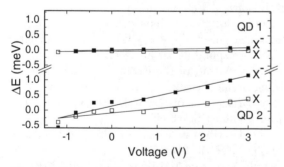

**Fig. 12.** Energy shift versus applied voltage in vertical direction for the neutral and the charged exciton emission from two SQDs (QD1 and QD2, respectively). The *solid lines* are guides to the eye

most likely due to a stronger confinement and/or a smaller QD size in the growth direction.

## 3.2   Applying Magnetic Fields – The Zeeman Effect

**Single Exciton Transitions.** Due to the spin of the particles, the QD excitons cannot only couple to electric fields, but also to magnetic ones. This gives the opportunity to manipulate the eigenstates of these 'artifical atoms' using the Zeeman effect, similar to what is known for 'real' atoms.

In the matrix representation introduced above (1) the Zeeman interaction can be included quite easily in the Hamiltonian, which is for the Faraday geometry given by

$$
H_{\text{exc},B} = -\frac{1}{2} \cdot \begin{pmatrix} -\delta_0 + g_1\mu_B B & \delta_1 & 0 & 0 \\ \delta_1 & -\delta_0 - g_1\mu_B B & 0 & 0 \\ 0 & 0 & \delta_0 - g_2\mu_B B & \delta_2 \\ 0 & 0 & \delta_2 & \delta_0 + g_2\mu_B B \end{pmatrix} . (7)
$$

Here, $g_1 = g_e + 3g_h$ and $g_2 = g_e - 3g_h$ ($g_e$ = electron g-factor and $g_h$ = g-factor of the heavy hole) represent the effective g-factors of the bright and the dark excitons, respectively. Solving the Schrödinger equation, the magnetic field $B$ induced change of the eigenstates can be derived. Limiting ourselves to the optically active bright excitons we obtain for the eigenenergies $E_{1,2}$ and the eigenstates $|X^{1,2}\rangle$

$$
E_{1,2} = 1/2 \left( \delta_0 \mp \sqrt{(g_1\mu_B B)^2 + \delta_1^2} \right) , \tag{8}
$$

$$
|X^{1,2}\rangle = \frac{|+1\rangle \pm \left(\sqrt{1+\alpha^2} \mp \alpha\right)|-1\rangle}{\sqrt{2}\sqrt{1+\alpha^2 \mp \alpha\sqrt{1+\alpha^2}}}, \alpha = g_1\mu_B B/\delta_1 . \tag{9}
$$

From these calculations it can be seen that both the energy splitting as well as the symmetry of the eigenstates are expected to depend on $\alpha$, which is given by the ratio between the Zeeman and the exchange energy. In order to demonstrate this experimentally, a vertical magnetic field (Faraday geometry) was applied to self-assembled CdSe/ZnSe SQDs.

In Fig. 13, magneto-PL spectra for two different QDs are plotted for various magnetic fields. Low excitation density ensures only recombination from neutral single excitons. The solid lines correspond to $\sigma^+$-, the dashed line to $\sigma^-$-polarized light. Two important features can be extracted from the data. First, the energy splitting between the upper and the lower spin state of the exciton increases with $B$, and second, the polarization of the optical transitions changes from purely linear at zero field to almost completely circular for the highest magnetic field of 8 T applied. This is exactly what one expects from the magnetic field induced change of the eigenenergies and the eigenstates (8,9).

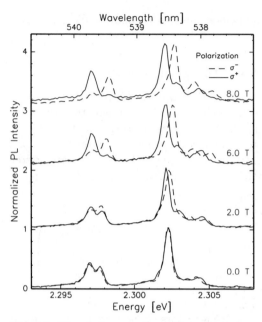

**Fig. 13.** Circularly polarized PL spectra from two individual QDs for different magnetic fields applied in the Faraday geometry

**Biexciton Transitions.** Because of the Pauli exclusion principle, a pair of excitons (= biexciton) occupying the QD ground state is a spin singlet state. In analogy to what was found for the exchange splitted exciton and biexciton transitions, one would expect that the Zeeman effect will modify the biexciton transitions in a quite similar way to the exciton ones.

This can be nicely seen if one looks at the polarization properties of optical transitions in a SQD as a function of magnetic field. In Fig. 14, linearly polarized zero field PL spectra of two SQDs, QD1 and QD2, are compared to circular polarized ones at $B = 8$ T. It is interesting to follow the polarization sequence for biexciton and exciton transitions. While at $B = 0$ the $\pi^x$ polarized component of the exciton transition in QD1 is the higher energetic part of the doublet, the situation is exactly the opposite for the biexciton, where the $\pi^x$ polarized component forms the low energy part of the doublet. As discussed in Sect. 2.2 (see, e.g. Fig. 7), this can be understood by looking at the symmetry of the exciton eigenstates at zero field, which, e.g. for the asymmetric dot is a linear combination of $|+1\rangle$ and $|-1\rangle$ states. For high magnetic fields, the $|+1\rangle$ and the $|-1\rangle$ states become eigenstates of the system, resulting in circular polarized transitions for excitons and biexcitons. Interestingly, the energy sequence of the $\sigma^+$ and $\sigma^-$ transitions at $B = 8$ T is the same for excitons and biexcitons, i.e. for both transitions the $\sigma^-$-polarized component is the high energetic one of the Zeeman doublet. Comparing the zero field data and the experiments at high magnetic fields,

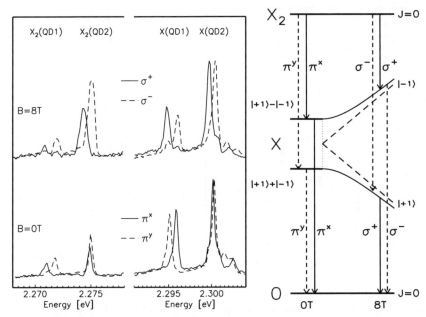

**Fig. 14.** *Left*: PL spectra from two SQDs showing the single exciton ($X$) and the biexciton ($X_2$) transition. Linearly polarized data are plotted for $B = 0$ and circularly polarized spectra are shown for $B = 8\,\mathrm{T}$. *Right*: Energy level scheme for the bright exciton and the biexciton states in a QD. The optical transitions and the corresponding polarizations are included as *solid* and *dashed lines*, respectively

it can be clearly seen that if the $\pi^x$ ($\pi^y$) components of the exciton line transform with increasing magnetic field B into the $\sigma^-$ ($\sigma^+$) lines, the biexciton emission component $\pi^x$ transforms into $\sigma^+$ and, vice versa, $\pi^y$ into $\sigma^-$ lines, in full agreement with the theoretical expectation (see right part of Fig. 14).

In Fig. 15, the energy splitting $\overline{\delta_1}(B) = E_2 - E_1$ is plotted for both the exciton and the biexciton transition for QD1 and QD2. Due to the fact that no Zeeman splitting occurs for the biexciton state, the magnetic field induced energy splitting should be identical for the exciton and the biexciton transition. This is exactly what is seen in the figure. As in QD1 the zero field exchange splitting $\delta_1 \approx 0$, $\overline{\delta_1}(B)$ increases almost linearly with $B$ with a g-factor of $g_1 = 1.6$. For the asymmetric QD2, in contrast, $\overline{\delta_1}(B)$ strongly differs from the linear dependence. This deviation is related to the large zero-field splitting $\delta_1$ of the radiative exciton state due to electron–hole exchange interaction. Taking the exchange splitting $\delta_1$ from the zero-field data, the energy shift of both QDs can be described quite nicely according to (8) with the effective g-factor $g_1$ of the bright exciton as the only fitting parameter. It is worth noting that no sigificant variation of the g-factor was found for the SQDs under investigation.

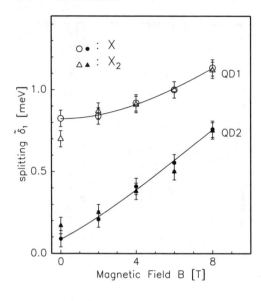

**Fig. 15.** Energy splitting $\overline{\delta_1}(B)$ of the spin doublet versus magnetic field for two different QDs. Exciton $(X)$ and biexciton $(X_2)$ transitions are compared. From [86]

From the variation of the eigenstates with magnetic fields, one can easily see that the interplay between exchange and Zeeman interaction results in a change of the eigenstate symmetry with increasing field. The linear polarization obtained for zero-field transitions transforms into circular polarized transitions for high magnetic fields (see Fig. 14). From the eigenstates given by (9), one can define a degree of circular polarization $P$ given by

$$P = \frac{I^{\sigma+} - I^{\sigma-}}{I^{\sigma+} + I^{\sigma-}} = 1 - \frac{1}{\alpha^2 + \alpha\sqrt{1 + \alpha^2} + 1}. \tag{10}$$

In order to compare this magnetic field-induced change of the eigenstate symmetry with experimental data, one has to be aware that analyzing exciton transitions would not give the correct result, as in that case the measured polarization degree not only depends on the eigenstate symmetry, but also on the spin flip process between the exciton eigenstates. For that reason, we plotted in Fig. 16 the polarization degree of the biexciton transition versus $\alpha$. The transition from the spin singlet biexciton state to the radiative doublet of the single exciton is not affected by any spin flip processes. Consequently, the polarization degree does purely reflect the symmetry of the exciton eigenstates. As can be seen in Fig. 16, indeed the experimental data can be well described by the theory. Thus, the interplay between exchange and Zeeman interaction gives a coherent description of the fine structure of single exciton and biexciton transitions in CdSe/ZnSe SQDs. Recently, similar results have been obtained for CdTe/CdMgTe SQDs [57].

Finally, we would like to comment on the dark states of the exciton. As mentioned above, in CdSe/ZnSe QDs the splitting between dark and bright states exceeds 1 meV and for a QD symmetry of $C_{2v}$ or above, a magnetic field

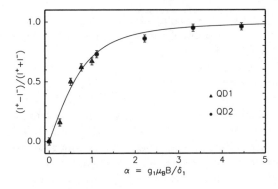

**Fig. 16.** Degree of polarization $P$ taken for the $X_2$ transition. The *symbols* characterize experimental data, the *solid line* is a fit according to (10). From [86]

in Faraday geometry cannot mix bright and dark states. For localized states in CdZnSe/ZnSe quantum wells, it was shown by *Puls* et al., that a magnetic field in Voigt geometry can cause mixing of bright and dark states [77], in accordance to what is expected from theory [75]. In addition, even for Faraday geometry a contribution of dark states to the single exciton PL spectrum was observed in InGaAs/GaAs [72] and in CdTe/CdMgTe [57] SQDs. This is attributed to a further reduction of the QD symmetry, where a pronounced mixing of bright and dark states will become possible even for a magnetic field applied in the growth direction.

## 4   Single Magnetic Semiconductor Quantum Dots

In II–VI semiconductors, magnetic ions like, e.g. $Mn^{2+}$ can be incorporated isoelectronically into the crystal matrix. Lots of studies on such magnetic semiconductors have been reported so far for bulk and quantum wells. Only recently, people succeeded in the epitaxial growth of quantum dots containing magnetic ions [8,9,10]. II–Mn–VI QDs with a PL efficiency large enough to perform optical studies on SQDs became accessible. This is in contrast to III–Mn–V compounds, where the $Mn^{2+}$ ions act as p-dopants and poor quantum efficiencies are obtained so far. First measurements have been performed on 'natural' CdMnTe/CdMgTe SQDs [42] or lithographically defined CdTe/CdMnTe SQDs [87]. Nowadays, the research efforts mainly concentrate on self-assembled Cd(Mn)Se/Zn(Mn)Se QDs and recently, optical spectroscopy on CdSe/ZnMnSe SQDs was reported for the first time [40,41].

### 4.1   Giant Magneto-Optical Effects in a Zero-Dimensional System

In magnetic semiconductor SQDs, exchange interaction not only occurs between electrons and holes, but also between charge carrier spins and the spins of magnetic ions of the crystal matrix. This has some important consequences, like the occurrence of giant magnetooptic effects. This includes the giant Faraday and the giant Zeeman effect as well as the formation of exciton magnetic polarons (EMPs) [88]. An EMP is a small area in the crystal where the spins of the charge carriers and the spins of the $Mn^{2+}$ ions are strongly correlated due to sp-d exchange interaction.

The formation of quasi-zero-dimensional magnetic polarons as bound states becomes evident if temperature dependent measurements are performed on a diluted magnetic SQD. At low temperatures, the $Mn^{2+}$ spins are aligned in the quantum mechanical exchange field of the charge carriers. For increasing temperature the magnetic ordering inside the EMP is reduced and thus the emission line should shift to higher energy. It is important to note here that in a QD, the electronic localization does not change with temperature, provided the temperature is low enough to suppress a significant occupation of excited QD levels or extended states in the wetting layer or the barrier. Thus, in the equilibrium case the variation of the PL energy $E(T)$ with temperature directly reflects the thermally induced change of the $Mn^{2+}$ spin alignment and can be well described by a Brillouin-like behaviour characteristic of a paramagnetic spin system [42,89]:

$$E(T) = E_{\text{gap}}(T) + C \; \mathcal{B}_{5/2}\left(\frac{5\mu_B g_{\text{Mn}} B_{\text{MP}}}{2k_B(T+T_0)}\right),\tag{11}$$

where $C = -\frac{1}{2}\gamma N_0(\alpha-\beta)xS_0$ and $B_{5/2}(x)$ is the Brillouin function for spin S = 5/2. Here, $E_{\text{gap}}(T)$ describes the temperature dependence of the bandgap, $g_{\text{Mn}}$ the g-factor of the $Mn^{2+}$ ions and $C$ is a temperature independent constant depending on the well-known exchange parameters for the electron $N_0\alpha$ and the hole $N_0\beta$, the Mn concentration $x$, the effective spin of the localized $Mn^{2+}$ ions $S_0$, and the overlap $\gamma$ of the exciton wavefunction with the $Mn^{2+}$ ions. The exchange field $B_{\text{MP}}$ describes the interaction between the carrier spins and the spins of the magnetic ions and the phenomenological parameters $S_0 < 5/2$ and $T_0 > 0$ take into account the antiferromagnetic interaction of neighboring Mn ions.

In diluted magnetic semiconductors the magnetization $M$ can be described by a modified Brillouin function [89]

$$M(B,T) = xN_0 g_{\text{Mn}}\mu_B S_0 \; \mathcal{B}_{5/2}\left(\frac{5\mu_B g_{\text{Mn}} B}{2k_B(T+T_0)}\right).\tag{12}$$

For zero external magnetic field the magnetic field $B$ is given by the exchange field $B_{\text{MP}}$ and combining (11) and (12) one immediately obtains a nice relation between the temperature dependent PL energy shift $\Delta E$ and the magnetization $M$ in a magnetic semiconductor SQD:

$$\Delta E(T) = E(T) - E_{\text{gap}}(T) = -\frac{\gamma(\alpha-\beta)}{2\mu_B g_{\text{Mn}}} M(B_{\text{MP}},T).\tag{13}$$

Note that (13) only holds in the equilibrium case, i.e. if the spins of the magnetic ions reach their equilibrium state prior exciton recombination.

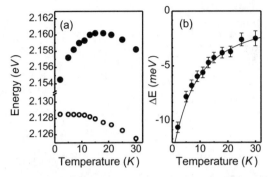

**Fig. 17.** (a) Temperature dependence of the PL energy for non-magnetic (*open symbols*) and magnetic (*solid circles*) SQDs. (b) Experimental (*symbols*) and calculated (*solid line*) magnetic contribution to the PL energy shift of a magnetic SQD versus temperature. The data are plotted in a way that $\Delta E$ reaches zero in the limit of high temperatures. From [42]

In Fig. 17a, the PL energy of a 'natural' CdMnTe SQD is plotted in comparison to its non-magnetic CdMgTe counterpart. The QDs are formed by embedding nominally 3 MLs of $Cd_{0.93}Mn_{0.07}Te$ ($Cd_{0.93}Mg_{0.07}Te$) within $Cd_{0.60}Mg_{0.40}Te$ barriers and SQD selection is obtained by using metal nano-apertures with diameters down to 100 nm [42]. As expected, the PL line of the non-magnetic SQD shows a monotonic shift to smaller energies according to the temperature induced band gap shrinkage. In contrast, the energy of the PL line in the magnetic SQD first increases with $T$. This can be attributed to the thermal suppression of EMPs and thus to a reduction of the EMP binding energy. Only for $T > 20\,K$ does the influence of the band gap shrinkage dominate. By comparing the PL shift of the non-magnetic SQD and the shift found for the magnetic SQD (see symbols in Fig. 17b), we can extract $\Delta E$, i.e. the part of the energy shift which is of purely magnetic origin and thus reflects the change of the *magnetic* localization within the EMP with temperature. As can be seen in the figure, this temperature dependence can be well described by a Brillouin-like behaviour according to (11). The excellent agreement between experiment and calculations strongly supports the validity of the equilibrium model used. Moreover, an EMP binding energy of $E_{MP} \approx 10.5\,meV$ at $T = 2\,K$ as well as an internal exchange field of $B_{MP} = 3.5\,T$ can be directly exctracted from the data.

Due to the magnetic ions involved, the electron–hole pairs do not only couple to an external magnetic field via their band g-factor $g_1$, but in addition via the $Mn^{2+}$ system. In fact, the energy splitting $E_Z$ of the exciton eigenstates of magnetic QDs in an external magnetic field is given by [42]

$$E_Z = g_1 \mu_B B_{ext} + (\alpha - \beta)\gamma N_0 x S_0 \, \mathcal{B}_{5/2}\left(\frac{5\mu_B g_{Mn} B}{2k_B(T + T_0)}\right) , \qquad (14)$$

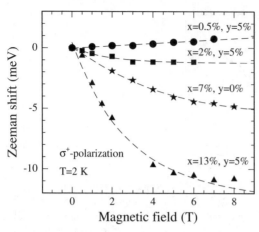

**Fig. 18.** Zeeman shift of the $\sigma^+$-polarized component of the PL signal from natural $Cd_{1-x-y}Mn_xMg_yTe/Cd_{0.6}Mg_{0.4}Te$ SQDs at $T = 2\,K$

The magnetic field $B$ contains both the exchange field $B_{MP}$ and the external field $B_{ext}$ and is given by $B = B_{MP} + B_{ext}$. It becomes clear that the energy splitting $E_Z$ between the $|+1\rangle$ and the $|-1\rangle$ exciton states is strongly enhanced as compared to non-magnetic QDs. Moreover it can be adjusted by changing the concentration of the magnetic ions in the SQD. This is shown in Fig. 18, where for 'natural' $Cd_{1-x-y}Mn_xMg_yTe/Cd_{0.6}Mg_{0.4}Te$ SQDs the energy shift of the $\sigma^+$ polarized PL component is plotted versus external magnetic field. For a $Mn^{2+}$ concentration of $x = 0.005$, the PL signal shifts to higher energies, which corresponds to a negative g-factor of $-3.2$, mainly given by the band g-factor of the exciton. With increasing manganese concentration, a more and more pronounced red shift of the $\sigma^+$ polarized PL peak is obtained and it can easily be seen that this energy shift exceeds the values found in non-magnetic SQD (see Fig. 13) by 1–2 orders of magnitude. This effect is usually called the giant Zeeman effect. Although due to the Brillouin-like energy shift the definition of an effective g-factor may not be appropriate, one can use the linear part of the Brillouin function, i.e. the limit of small magnetic fields, to estimate an effective g-factor $g_{eff}$ in these magnetic SQDs. Depending on $x_{Mn}$, values even larger than $g_{eff} = 100$ can be extracted from the data and just recently, *Kratzert* et al. obtained for self-assembled CdMnSe/ZnSe QDs effective g-factors on the order of 200 [8].

## 4.2    Statistical Fluctuations of the Nano-Scale Magnetization

The unique possibility to tailor electronic and magnetic properties independently has triggered a variety of visionary device concepts, based on controlling the interaction of zero-dimensional carrier spins and the spins of magnetic ions [46,56]. However, as one proceeds to ultrasmall magnetic systems, there arises a *fundamental* limit for operating such devices: statistical magnetic fluctuations are expected to severely restrict the device functionality. As we have shown above, the PL signal of a magnetic SQD can be used to monitor changes of the magnetization. The question arises, whether nano-scale magnetic fluctuations are accessible by SQD spectroscopy.

For that purpose, self-assembled CdSe/ZnMnSe SQDs have been studied by magneto-PL spectroscopy. A nominal CdSe thickness of 2.5 monolayers and a Mn content of $x = 0.25$ have been chosen. In such a wide-bandgap self-assembled II–VI QD, the exciton is three-dimensionally confined. For magnetic fields below $\approx 10$ T, the QD radius is smaller than the typical magnetic length of $l = \sqrt{\hbar/eB}$. In addition the energy spacing between the ground state and excited states is in the range of several tens of meV [58,67], i.e. much larger than the thermal energy for temperatures below $\approx 100$ K. Within these limits, neither increasing the temperature nor applying a magnetic field is expected to change the exciton wavefunction significantly. Thus, the PL energy of a magnetic QD is expected to directly reflect changes in the magnetization via the *sp-d* exchange interaction. To get a sufficient spatial resolution for the selection of individual QDs metal apertures with diameters down to 80 nm have been defined lithographically on top of the samples [40,42].

In Fig. 19, the magnetic field dependent (left) and the temperature dependent (right) PL spectra of CdSe/ZnMnSe SQDs are compared to their non-magnetic CdSe/ZnSe counterparts. The energy shifts versus temperature and versus magnetic field have been discussed above and were related to the suppression of the EMP and the giant Zeeman effect, respectively. The most striking result of the experiments is the large emission linewidth of SQD tran-

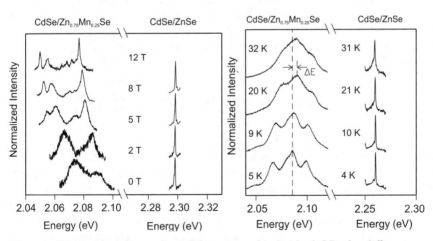

**Fig. 19.** Highly spatially resolved PL spectra of individual QDs for different magnetic fields (*left*) and different temperatures (*right*). Note the distinct difference in the energy shift and the emission linewidth of non-magnetic CdSe/ZnSe SQDs and magnetic CdSe/ZnMnSe SQDs

sitions in the CdSe/ZnMnSe system. Moreover, the linewidth strongly narrows with increasing magnetic field and a distinct broadening with increasing temperature, which cannot be explained by phonon scattering, is found. This behaviour, which is characteristic for all the magnetic SQDs studied, is obviously related to the influence of the magnetic ions located within the spatial extent of the exciton wavefunction. Using time-resolved PL spectroscopy it has been shown that in this sample the EMP formation time is much smaller than the recombination lifetime [90], i.e. the equilibrium case of the EMP is being probed in case of continuous wave SQD spectroscopy.

At first glance one would expect that in the equilibrium case the EMP in a given SQD is characterized by a well-defined energy – i.e., the PL signal should be spectrally quite narrow. However, as the SQD is being probed repeatedly in our experiment, statistical variations of the orientation of the magnetic ion *spins*, i.e., fluctuations of the magnetization within the exciton wavefunction, should result in a broadening of the SQD emission peak in time-integrated experiments [41,91]. This is quite similar to what is known from fluctuating *charges* in the environment of a non-magnetic SQD, which is found to result in a distinct linewidth broadening in CW PL measurements (see Sect. 2) [60,62,64,65].

In order to get a theoretical approach to the nano-scale fluctuations of the magnetization one can assume a Gaussian distribution of the magnetization, which is characterized by a broadening of $\Delta M = \sqrt{\langle M^2 \rangle 8 \ln 2}$. Being aware that the exchange interaction between the carrier spins and the magnetic ions is dominated by the heavy hole – $Mn^{2+}$ spin interaction and the in-plane hole g-factor is expected to be small in our QDs, we monitor for Faraday geometry, predominantly the longitudinal magnetic fluctuations. The broadening of the magnetization distribution $\Delta M$ thus transfers into a broadening of the SQD emission signal FWHM according to $FWHM = \Delta M B_{MP} V_{eff}$. Using the well-known fluctuation-dissipation theorem $\langle M^2 \rangle = (dM/dB) k_B T / V_{eff}$, one obtains [91]

$$FWHM = B_{MP} V_{eff} \sqrt{\langle M^2 \rangle 8 \ln 2} = \sqrt{B_{MP} \cdot 8 \ln 2 \cdot k_B T \left( -\frac{dE}{dB}\bigg|_B \right)}. \quad (15)$$

The effective volume is given by $V_{eff} = \gamma V$, where $V$ is the volume occupied by the exciton wavefunction, and $\gamma$ takes into account the fact that only a part of the exciton wavefunction actually overlaps with the $Mn^{2+}$ ion spins. This leads us to a very fundamental result: the emission linewidth FWHM is proportional to $\sqrt{\langle M^2 \rangle}$, i.e. the optical response of the SQD is directly correlated to the statistical magnetic fluctuations on a scale defined by the spatial extent of the exciton wavefunction. Even more, the emission linewidth should be proportional to $(-dE/dB)^{1/2}$. As the emission peaks narrow with increasing magnetic field, the optical response of individual QDs can be easily resolved, and the Zeeman shift (and therefore $dE/dB$) can be determined very accurately.

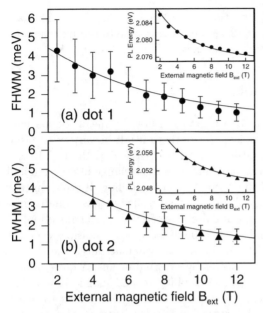

**Fig. 20a,b.** PL linewidth FWHM as a function of $B_{ext}$ for two different SQDs. The *insets* show the magnetic-field induced energy shifts of the PL lines for the same SQDs. *Symbols* correspond to experimental data, *solid lines* are model calculations according to (14) and (15). From [41]

In the insets of Fig. 20, the Zeeman shift for two SQDs with strongly different PL energies is depicted. The solid lines represent fits using (14) showing rather excellent agreement with the experiment. As the internal exchange field $B_{MP}$ can be obtained by analyzing the temperature dependent energy shift of the PL signal at zero field [40], we are in a position to describe quantitatively the linewidth broadening due to magnetic fluctuations. This can be seen in Fig. 20, where the FWHM is plotted for the same two SQDs as a function of $B_{ext}$. It is evident that the value of the PL linewidth as well as the magnetic-field-induced linewidth narrowing, which is ascribed to the suppression of magnetic fluctuations, is quantitatively reproduced by our model. Thus, the PL linewidth from a single magnetic semiconductor QD can be used to analyze statistical magnetic fluctuations on a scale of a few nanometers. It has to be emphasized that this is only true if the EMP formation time is much smaller than the recombination lifetime. Otherwise, the transient energy shift during EMP formation as well as memory effects due to an incomplete magnetic relaxation between two recombination events have to be taken into account when discussing the linewidth of optical transitions in magnetic SQDs [42]. It is interesting to note that investigating a magnetic SQD corresponds to detecting a magnetic moment of only $\mu \approx 100\mu_B$ at $B_{ext} = 0$, as estimated using the relation $\mu \approx E_{MP}/B_{MP}$. The linewidth broadening at low temperatures corresponds to fluctuations of $\pm20\mu_B$ at $B_{ext} = 0$, which are reduced to about $\pm3\mu_B$ at $B_{ext} = 12\,T$ – i.e., we are able to resolve magnetic moment changes in the magnetic ion ensemble of only a few $\mu_B$.

## 5  Summary and Outlook

The ability to address individual II–VI quantum dots by means of spatially resolved optical spectroscopy opened the path to study a variety of fascinating aspects of individual quantum objects. In particular, the II–VI compounds are characterized by a strong Coulomb and exchange interaction and by a pronounced light–matter and exciton–phonon coupling. It is shown that the quantum dots can be populated in a very controlled way with individual electrons and holes by combining optical and electrical carrier injection. In addition, applying magnetic and electric fields becomes accessible even for single II–VI quantum dots, which allows one to systematically tune the energy and the symmetry of the eigenstates by external means. A quite unique feature of II–VI quantum dots is the possibility to add magnetic ions like $Mn^{2+}$ isoelectronically to the crystal matrix. This, e.g. allows us to investigate fundamental aspects of the interaction between carrier spins and the spins of magnetic ions on a length scale in the order of $10\,nm$.

At this point, let us just briefly outline some possible applications, where the II–VI quantum dots may be superior to the well-known InAs/GaAs system due to their specific properties. First, the deep carrier confinement, the large Coulomb interaction as well as the strong exciton–photon coupling seems to be quite promising for the development of single photon sources operating at elevated temperatures, possibly up to room temperature (see the contribution by *Michler* in this volume). Second, combining (electrical) single electron injection into a II–VI single quantum dot with the concept of (diluted) magnetic semiconductor spin aligners may give the chance to inject single spin-polarized electrons into a quantum dot, which should be a rather important step for spintronics or quantum computation. Third, including magnetic ions into the crystal matrix of individual quantum dots adds a new degree of freedom for quasi-zero-dimensional systems. This provides the possibility to combine the discrete density of states of quasi-zero-dimensional systems with adjustable magnetic properties, a prerequisite for preparing devices based on the interaction between single carrier spins and the spins of magnetic ions. In addition, combining two single quantum dots with largely different effective g-factors will allow one to define a single quantum dot molecule with adjustable coupling as recently suggested by us [92].

However, preparing and studying individual 'man-made' quantum objects like single quantum dots, magnetic single dots or single quantum dot molecules will certainly exhibit a variety of new and unexpected results in the future. Moreover, even completely new applications might be developed or surprising physical aspects may be discovered.

## Acknowledgements

I am indebted to J. Seufert, H. Schömig, M. Obert, M.K. Welsch, T. Kümmell, R. Weigand, M. Emmerling and A. Forchel from the Technische Physik at the University of Würzburg and to our colleagues A. A. Maksimov, V. D. Kulakovskii, P. S. Dorozhkin and A. V. Chernenko from the Institute of Solid State Physics at Chernogolovka for their collaboration. Special thanks to J. Seufert and H. Schömig for critically reading the manuscript. The high quality growth of the CdSe/ZnSe quantum dots by Th. Passow, K. Leonardi and D. Hommel, of the CdSe/ZnMnSe quantum dots by S. Lee, M. Dobrowolska and J. K. Furdyna and of the CdMnMgTe/CdMgTe samples by C. R. Becker, G. Landwehr and L. W. Molenkamp is gratefully acknowledged. The research has been supported by the Deutsche Forschungsgemeinschaft under the grants BA1422-1 and SFB 410.

# References

1. S. H. Xin, P. D. Wang, A. Yin, C. Kim, M. Dobrowolska, J. L. Merz, J. K. Furdyna: Appl. Phys. Lett. **69**, 3884 (1996)
2. V. Nikitin, P. A. Crowell, J. A. Gupta, D. D. Awschalom, F. Flack, N. Samarth: Appl. Phys. Lett. **71**, 1213 (1997)
3. M. Rabe, M. Lowisch, F. Kreller, F. Henneberger: phys. stat. sol. (b) **202**, 817 (1997)
4. K. Leonardi, H. Heinke, K. Ohkawa, D. Hommel, H. Selke, F. Gindele, U. Woggon: Appl. Phys. Lett. **71**, 1510 (1997)
5. M. Strassburg, V. Kutzer, U. W. Pohl, A. Hoffmann, I. Broser, N. N. Ledentsov, D. Bimberg, A. Rosenauer, U. Fischer, D. Gerthsen, I. L. Krestnikov, M. V. Maximov, P. S. Kop'ev, Zh. I. Alferov: Appl. Phys. Lett. **72**, 942 (1998)
6. H.-C. Ko, D.-C. Park, Y. Kawakami, S. Fujita, S. Fujita: Appl. Phys. Lett. **70**, 3278 (1997)
7. T. V. Shubina, S. V. Ivanov, A. A. Toropov, S. V. Sorokin, A. V. Lebedev, R. N. Kyutt, D. D. Solnyshkov, G. R. Pozina, J. P. Bergmann, B. Monemar, M. Willander, A. Waag, G. Landwehr: phys. stat. sol. (b) **229**, 489 (2002)
8. P. R. Kratzert, J. Puls, M. Rabe, F. Henneberger: Appl. Phys. Lett. **79**, 2814 (2001)
9. Y. Oka, J. Shen, K. Takabayashi, N. Takahashi, H. Mitsu, I. Souma, R. Pittini: J. Lumin. **83-84**, 83 (1999)
10. C. S. Kim, M. Kim, S. Lee, J. Kossut, J. K. Furdyna, M. Dobrowolska: J. Cryst. Growth **214/215**, 395 (2000)
11. L. Marsal, L. Besombes, F. Tinjod, K. Kheng, A. Wasiela, B. Gilles, J.-L. Rouviere, H. Mariette: J. Appl. Phys. **91**, 4936 (2002)
12. M. Funato, A. Balocchi, C. Bradford, K. A. Prior, B. C. Cavenett: Appl. Phys. Lett. **80**, 443 (2002)
13. N. Peranio, A. Rosenauer, D. Gerthsen, S. V. Sorokin, I. V. Sedova, S. V. Ivanov: Phys. Rev. B **61**, 16015 (2000)

14. C. S. Kim, M. Kim, J. K. Furdyna, M. Dobrowolska, S. Lee, H. Rho, L. M. Smith, Howard E. Jackson, E. M. James, Y. Xin, N. D. Browning: Phys. Rev. Lett. **85**, 1124 (2000)

15. T. Kümmell, R. Weigand, G. Bacher, A. Forchel, K. Leonardi, D. Hommel, H. Selke: Appl. Phys. Lett. **73**, 3105 (1998)

16. M. Strassburg, Th. Deniozou, A. Hoffmann, R. Heitz, U. W. Pohl, D. Bimberg, D. Litvinov, A. Rosenauer, D. Gerthsen, S. Schwedhelm, K. Lischka, D. Schikora: Appl. Phys. Lett. **76**, 685 (2000)

17. A. Zrenner, L. V. Butov, M. Hagn, G. Abstreiter, G. Böhm, G. Weimann: Phys. Rev. Lett. **72**, 3382 (1994)

18. H. F. Hess, E. Betzig, T. D. Harris, L. N. Pfeiffer, K. W. West: Science **264**, 1740 (1994)

19. D. Gammon, E. S. Snow, B. V. Shanabrook, D. S. Katzer, D. Park: Phys. Rev. Lett. **76**, 3005 (1996)

20. S. A. Empedocles, M. G. Bawendi: Science **278**, 2114 (1997)

21. W. Heller, U. Bockelmann, G. Abstreiter: Phys. Rev. B **57**, 6270 (1998)

22. S. Raymond, J. P. Reynolds, J. L. Merz, S. Fafard, Y. Feng, S. Charbonneau: Phys. Rev. B **58**, R13415 (1998)

23. H. Gotoh, H. Kamada, H. Ando, J. Temmyo: Appl. Phys. Lett. **76**, 867 (2000)

24. J. Seufert, M. Obert, M. Scheibner, N. A. Gippius, G. Bacher, A. Forchel, T. Passow, K. Leonardi, D. Hommel: Appl. Phys. Lett. **79**, 1033 (2001)

25. V. D. Kulakovskii, G. Bacher, R. Weigand, T. Kümmell, A. Forchel, E. Borovitskaya, K. Leonardi, D. Hommel: Phys. Rev. Lett. **81**, 1780 (1999)

26. A. Zrenner, F. Findeis, E. Beham, M. Markmann, G. Böhm, G. Abstreiter: in Adv. Solid State Phys. **40**, B. Kramer (Ed.) (Vieweg, Braunschweig/Wiesbaden 2000) pp. 561–576

27. A. Kuther, M. Bayer, A. Forchel, A. Gorbunov, V. B. Timofeev, F. Schäfer, J. P. Reithmaier: Phys. Rev. B **58**, R7508 (1998)

28. S. W. Brown, T. A. Kennedy, D. Gammon, E. S. Snow: Phys. Rev. B **54**, 17339 (1996)

29. L. Landin, M. S. Miller, M.-E. Pistol, C. E. Pryor, L. Samuelson: Science **280**, 262 (1998)

30. G. Bacher, R. Weigand, J. Seufert, V. D. Kulakovskii, N. A. Gippius, A. Forchel, K. Leonardi, D. Hommel: Phys. Rev. Lett. **83**, 4417 (1999)

31. M. Bayer, O. Stern, P. Hawrylak, S. Fafard, A. Forchel: Nature **405**, 923 (2000)

32. R. J. Warburton, C. Schäflein, D. Haft, F. Bickel, A. Lorke, K. Karrai, J. M. Garcia, W. Schoenfeld, P. M. Petroff: Nature **405**, 926 (2000)

33. F. Findeis, M. Baier, A. Zrenner, M. Bichler, G. Abstreiter, U. Hohenester, E. Molinari: Phys. Rev. B **63**, 121309(R) (2001)

34. J. J. Finley, P. W. Fry, A. D. Ashmore, A. Lemaître, A. I. Tartakovskii, R. Oulton, D. J. Mowbray, M. S. Skolnick, M. Hopkinson, P. D. Buckle, P. A. Maksym: Phys. Rev. B **63**, 161305 (2001)

35. G. Bacher, H. Schömig, J. Seufert, M. Rambach, A. Forchel, A. A. Maksimov, V. D. Kulakovskii, T. Passow, D. Hommel, C. R. Becker, L. W. Molenkamp: phys. stat. sol. (b) **229**, 415 (2002)

36. L. Besombes, K. Kheng, L. Marsal, H. Mariette: Phys. Rev. B **65**, 121314(R) (2002)

37. U. Woggon, K. Hild, F. Gindele, W. Langbein, M. Hetterich, M. Grün, C. Klingshirn: Phys. Rev. B **61**, 12632 (2000)

38. L. Besombes, K. Kheng, L. Marsal, H. Mariette: Phys. Rev. B **63**, 155307 (2001)
39. F. Gindele, K. Hild, W. Langbein, U. Woggon: Phys. Rev. B **60**, R2157 (1999)
40. G. Bacher, H. Schömig, M. K. Welsch, S. Zaitsev, V. D. Kulakovskii, A. Forchel, S. Lee, M. Dobrowolska, J. K. Furdyna, B. König, W. Ossau: Appl. Phys. Lett. **79**, 524 (2001)
41. G. Bacher, A. A. Maksimov, H. Schömig, V. D. Kulakovskii, M. K. Welsch, A. Forchel, P. S. Dorozhkin, A. V. Chernenko, S. Lee, M. Dobrowolska, J. K. Furdyna: Phys. Rev. Lett. **89**, 127201 (2002)
42. A. A. Maksimov, G. Bacher, A. McDonald, V. D. Kulakovskii, A. Forchel, Ch. Becker, G. Landwehr, L. W. Molenkamp: Phys. Rev. B **62**, R7767 (2000)
43. J. M. Gerard, B. Sermage, B. Gayral, B. Legrand, E. Costard, V. Thierry-Mieg: Phys. Rev. Lett. **81**, 1110 (1998)
44. P. Michler, A. Kiraz, C. Becher, W. V. Schoenfeld, P. M. Petroff, Lidong Zhang, E. Hu, A. Imamoglu: Science **290**, 2282 (2000)
45. T. Lundstrom, W. Schoenfeld, H. Lee, P. M. Petroff: Science **286**, 2313 (1999)
46. D. Loss, D. P. DiVincenzo: Phys. Rev. A **57**, 120 (1998)
47. A. Imamoglu, D. D. Awschalom, G. Burkard, D. P. DiVincenzo, D. Loss, M. Shervin, A. Small: Phys. Rev. Lett. **83**, 4204 (1999)
48. M. Paillard, X. Marie, P. Renucci, T. Amand, A. Jbeli, J. M. Gerard: Phys. Rev. Lett. **86**, 1634 (2001)
49. J. A. Gupta, D. D. Awschalom, X. Peng, A. P. Alivisatos: Phys. Rev. B **59**, R10421 (1999)
50. H. Gotoh, H. Ando, H. Kamada, A. Chavez-Pirson: Appl. Phys. Lett. **72**, 1341 (1998)
51. A. Hundt, T. Flissikowski, M. Lowisch, M. Rabe, F. Henneberger: phys. stat. sol. (b) **224**, 159 (2001)
52. M. Scheibner, G. Bacher, A. Forchel, Th. Passow, D. Hommel, J. Supercond.: Incorp. Novel Meagnetism **16**, 395 (2003)
53. M. Oestreich, J. Hübner, D. Hägele, P. J. Klar, W. Heimbrodt, W. W. Rühle, D. E. Ashenford, B. Lunn: Appl. Phys. Lett. **74**, 1251 (1999)
54. R. Fiederling, M. Keim, G. Reuscher, W. Ossau, G. Schmidt, A. Waag, L. W. Molenkamp: Nature **402**, 787 (1999)
55. Y. Ohno, D. K. Young, B. Beschoten, F. Matsukura, H. Ohno, D. D. Awschalom: Nature **402**, 790 (1999)
56. Al. L. Efros, E. I. Rashba, M. Rosen: Phys. Rev. Lett. **87**, 206601 (2001)
57. L. Besombes, K. Kheng, D. Martrou: Phys. Rev. Lett. **85**, 425 (2000)
58. T. Flissikowski, A. Hundt, M. Lowisch, M. Rabe, F. Henneberger: Phys. Rev. Lett. **86**, 3172 (2001)
59. C. Obermüller, A. Deisenrieder, G. Abstreiter, K. Karrai, S. Grosse, S. Manus, J. Feldmann, H. Lipsanen, M. Sopanen, J. Ahopelto: Appl. Phys. Lett. **74**, 3200 (1999)
60. H. D. Robinson, B. B. Goldberg: Phys. Rev. B **61**, R5086 (2000)
61. P. G. Blome, M. Wenderoth, M. Hübner, R. G. Ulbrich, J. Porsche, F. Scholz: Phys. Rev. B **61**, 8382 (2000)
62. J. Seufert, R. Weigand, G. Bacher, T. Kümmell, A. Forchel, K. Leonardi, D. Hommel: Appl. Phys. Lett. **76**, 1872 (2000)
63. J. Seufert, M. Obert, R. Weigand, T. Kümmell, G. Bacher, A. Forchel, K. Leonardi, D. Hommel: phys. stat. sol. (b) **224**, 201 (2001)

64. V. Türck, S. Rodt, R. Heitz, R. Engelhardt, U.W. Pohl, D. Bimberg, R. Steingrüber: Phys. Rev. B **61**, 9944 (2000).
65. S.A. Empedocles, D.J. Norris, M.G. Bawendi: Phys. Rev. Lett. **77**, 3873 (1996)
66. W.P. Ambrose, W.E. Moerner: Nature **349**, 225 (1991)
67. M. Rambach, J. Seufert, M. Obert, G. Bacher, A. Forchel, K. Leonardi, T. Passow, D. Hommel: phys. stat. sol. (b) **229**, 503 (2002)
68. P. Hawrylak, G.A. Narvaez, M. Bayer, A. Forchel: Phys. Rev. Lett. **85**, 389 (2002)
69. M. Grundmann, J. Christen, N.N. Ledentsov, J. Böhrer, D. Bimberg, S.S. Ruvimov, P. Werner, U. Richter, U. Gösele, J. Heydenreich, V.M. Ustinov, A.Yu. Egorov, A.E. Zhukov. P.S. Kop'ev, Zh.I. Alferov: Phys. Rev. Lett. **74**, 4043 (1995)
70. R. Weigand, J. Seufert, G. Bacher, V.D. Kulakovskii, T. Kümmell, A. Forchel, K. Leonardi, D. Hommel: J. Cryst. Growth **214/215**, 737 (2000)
71. J. Puls, M. Rabe, H.-J. Wünsche, F. Henneberger: Phys. Rev. B **60**, R16303 (1999)
72. M. Bayer, A. Kuther, A. Forchel, A. Gorbunov, V.B. Timofeev, F. Schäfer, J.P. Reithmaier, T.L. Reinecke, S.N. Walck: Phys. Rev. Lett. **82**, 1748 (1999)
73. R.I. Dzhioev, H.M. Gibbs, E.L. Ivchenko, G. Khitrova, V.L. Korenev, M.N. Tkachuk, B.P. Zakharchenya: Phys. Rev. B **56**, 13405 (1997)
74. V.D. Kulakovskii, R. Weigand, G. Bacher, J. Seufert, T. Kümmell, A. Forchel, K. Leonardi, D. Hommel: phys. stat. sol. (a) **178**, 323 (2000)
75. For a recent review on In(Ga)As/Ga(Al)As QDs see e.g. M. Bayer, G. Ortner, O. Stern, A. Kuther, A.A. Gorbunov, A. Forchel, P. Hawrylak, S. Fafard, K. Hinzer, T.L. Reinecke, S.N. Walck, J.P. Reithmaier, F. Klopf, F. Schäfer: Phys. Rev. B **65**, 195315 (2002)
76. K. Brunner, G. Abstreiter, G. Böhm, G. Tränkle, G. Weimann: Phys. Rev. Lett. **73**, 1138 (1994)
77. J. Puls, F. Henneberger: phys. stat. sol. (a) **164**, 499 (1997)
78. G. Bacher, R. Weigand, J. Seufert, N.A. Gippius, V.D. Kulakovskii, A. Forchel, K. Leonardi, D. Hommel: phys. stat. sol. (b) **221**, 25 (2000)
79. D. Citrin: Phys. Rev. B **50**, 17655 (1994)
80. K. Kheng, R.T. Cox, Y. Merle d'Augnigne, F. Bassani, K. Saminadayar, S. Tatarenko: Phys. Rev. Lett. **71**, 1752 (1993)
81. O. Homburg, K. Sebald, P. Michler, J. Gutowski, H. Wenisch, D. Hommel: Phys. Rev. B **62**, 7413 (2000)
82. G.V. Astakhov, D.R. Yakovlev, V.P. Kochereshko, W. Ossau, W. Faschinger, J. Puls, F. Henneberger, S.A. Crooker, Q. McCulloch, D. Wolverson, N.A. Gippius, A. Waag: Phys. Rev. B **65**, 165335 (2002)
83. V. Türck, S. Rodt, R. Heitz, O. Stier, M. Strassburg, U.W. Pohl, D. Bimberg: Physica E **13**, 269 (2002)
84. A. Hartmann, Y. Ducommun, E. Kapon, U. Hohenester, E. Molinari: Phys. Rev. Lett. **84**, 5648 (2000)
85. J. Seufert, M. Obert, M. Rambach, G. Bacher, A. Forchel, T. Passow, K. Leonardi, D. Hommel: Physica E **13**, 147 (2002)
86. R. Weigand, G. Bacher, V.D. Kulakovskii, J. Seufert, T. Kümmell, A. Forchel, K. Leonardi, D. Hommel: Physica E **7**, 350 (2000)
87. G. Bacher, T. Kümmell, D. Eisert, A. Forchel, B. König, W. Ossau, G. Landwehr: Appl. Phys. Lett. **75**, 956 (1999)

88. J. K. Furdyna: J. Appl. Phys. **64**, R29 (1988)
89. J. A. Gaj, R. Planel, G. Fishman: Solid State Commun. **29**, 435 (1979)
90. J. Seufert, G. Bacher, M. Scheibner, A. Forchel, S. Lee, M. Dobrowolska, J. K. Furdyna: Phys. Rev. Lett. **88**, 027402 (2002)
91. T. Dietl: J. Magn. Magn. Mater. **38**, 34 (1983)
92. M. K. Welsch, G. Bacher, H. Schömig, A. Forchel, S. Zaitsev, C. R. Becker, L. W. Molenkamp: phys. stat. sol. (b) **238**, 313 (2003)

# Quantum-Dot Lasers

Heinz Schweizer, Michael Jetter, and Ferdinand Scholz

Stuttgart University, 70550 Stuttgart, Germany
h.schweizer@physik.uni-stuttgart.de

**Abstract.** The reduction of the active region of semiconductor lasers to quasi-zero dimensions has different effects on static and dynamic laser parameters as already discussed in the last two decades. Most prominent effects due to thermodynamics of low dimensional electron–hole plasmas are the threshold reduction and improved temperature stability of lasers with low dimensional active regions. Based on the changed density of states also some interesting device properties with respect to modulation response and beam quality of a laser are expected but not yet reached. This obvious discrepancy between theoretical expectations and experimental results must be related to additional effects such as carrier transport and relaxation. After a brief review of nano-fabrication aspects of dot lasers by self assembled methods as well as by lithography based implantation and etching methods, also laser device properties based on thermodynamics in low dimensional systems will be discussed. Strong emphasis will be put on carrier dynamics (transport, recombination, and relaxation). On the footing of a rate equation approach we discuss the static and dynamic properties of quantum dot lasers as a function of the dot array filling factor. Also some applications and device properties will be discussed. Based on periodic dot arrays, gain coupled dot distributed feedback (DFB) lasers can be realized which result in an improved side mode suppression ratio of the laser emission. From modulation experiments, extremely low dynamic chirp of dot lasers can be observed making quantum dot lasers promising candidates for high speed communication in the long wavelength range, if modulation response limitations can be solved.

## 1 Introduction

Dot lasers are nanometer optical devices which combine both length scales defining the confinement of photons (100 nm) and length scales defining the confinement of electrons and holes (10 nm) in the solid. Both scales are strongly different but should be controlled to better than 10 nanometers to realize low dimensional electronic systems and simultaneously macroscopic filter structures with dot arrays.

### 1.1 History of Dot Lasers

In Fig. 1 resonator structures with active regions consisting of wires and dots and passive regions serving as waveguide and cladding region are shown schematically. The picture shows that there are two independent design rules possible. As photons and electrons have fundamentally different dispersion

P. Michler (Ed.): Single Quantum Dots, Topics Appl. Phys. **90**, 185–235 (2003)
© Springer-Verlag Berlin Heidelberg 2003

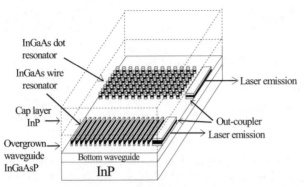

**Fig. 1.** Principle of a low dimensional laser structure of InGaAs wires and dots (active region) embedded in an InGaAsP waveguide and an InP cladding layer. The active wire and dot region can be realized by different fabrication techniques (see text)

relations, the design of laser structures can be decoupled (see seperate confinement heterostructure lasers [1,2]) i) with respect to questions on photon waveguiding and ii) with respect to carrier transport as well as to density of state engineering. Waveguide design meets, nearly independently of the applications (high speed lasers, single frequency lasers, or high power lasers), mainly macroscopic physical aspects (Maxwell's equations, diffusion equations). An exception may be if optical transition energies of a wire or dot array coincide with filter energies of the macroscopic/mesoscopic resonator resonantly. On the other hand, density of state engineering can be realized nearly independently of the photon aspects in separate confinement heterostructure lasers.

The history of quantum dot lasers is strongly correlated with the history of the general development of semiconductor lasers, beginning from bulk lasers, continuing with heterostructure lasers and ending with dot lasers. According to the above mentioned design principles of photon and carrier confinement, the laser structures have been permanently improved from the first bulk lasers [1,2] to the highly sophisticated quantum dot separate confinement heterostructure lasers [3]. Finally with the advent of quantum film lasers the carrier confinement principle not only includes the macroscopic design rules of carrier transport but also the possibility of density of states engineering in a semiconductor material by realizing nanometer structures (films, wires, and dots) in the active region of a laser. Very early, the advantages of density of state engineering were recognized by *Arakawa* and *Sakaki* [4] with respect to a laser's characteristic temperature $T_0$. First experimental indications of improvements could be obtained from laser experiments in high magnetic fields [4]. From the analysis of the density of states in low dimensional structures, a strong improvement of the optical gain as well as of the differential optical gain in going from 3D to 0D semiconductor sytems [5] has been stated. Especially higher modulation frequencies were expected due to higher differential gain in low dimensions. These expectations, however, must be modified by taking into account carrier transport [6,7,8,9] and carrier relaxation effects [10], which can counterbalance the pure density of state effects. With the continous improvement of processing technology, lasers with artificially patterned wires [11,12,13,14] and dots could be realized [15,16]. The advantage of those structures is the possibility to combine optical filter effects (DFB, DBR, complex coupling) with density of states effects. A strong disadvantage in artificially patterned nano-structures is the incompletely suppressed process induced defect level, which counterbalances in most cases density of states effects. A possible solution are self assembling growth techniques to realize quantum dots (SAD). With these techniques different material systems have been investigated (InAs/GaAs [3],[17]–[35], GaInP/InP [36]–[45], InGaAs/GaAs [46]–[60], CdSe/ZnSe [61]–[66]) (see also the Chapter by *Petroff* in this volume. Especially in the material system InAs and GaInAs, the expected improvement of laser operation parameters with respect to threshold and quantum efficiency could be observed. The also expected higher modulation response could not be observed and will be adressed in the sections below. Moreover high output powers could be obtained. However, the also expected improvement in beam quality due to the low optical confinement factor and the mostly passive optical resonator could not be demonstrated up to now and will also be discussed below.

## 1.2   Motivation/Applications/Material Systems

Key parameters in laser applications are the laser threshold $I_{th}$ and its characteristic temperature $T_0$, the quantum efficiency $\eta$ and its characteristic temperature $T_1$. These parameters determine the temperature range in which a laser device can be used. Furthermore, modulation speed and maximum output power and last but not least beam quality are important parameters for communication applications. These requirements favour semiconductor lasers with active regions, whose band dispersion is especially designed (strained structures, quantum films, and quantum dots). According to applications also specific emission wavelengths are required: long wavelengths ($\geq 1300\,\mathrm{nm}$) for optical communication lasers and shorter wavelengths below $1300\,\mathrm{nm}$ for material processing lasers. Especially for CD- and DVD-applications the laser emission should be short enough to ensure high data storage capability and the laser should be stable against light back-reflection to minimize noise. Due to its low threshold and therefore low power consumption as well as higher back-reflection stability, again, dot lasers are in the discussion for these applications.

From published dot laser data, actually we can derive a tendency that dot lasers with larger filling factors of the active dot region show improved threshold current densities [67,68]. This threshold behaviour is remarkably different to the behaviour of 2D lasers, where single well lasers (small confinement factor) for suitably adjusted resonator lengths nearly always show the lowest threshold current densities. Furthermore, besides the thermodynamically expected high characteristic temperature $T_0$ for the laser threshold of a dot laser (see below), nearly all dot lasers show a strong reduction of $T_0$ at higher temperatures combined with a strong reduction of the quantum efficiency (see for instance the InGaAs/GaAs material system [3,69] or the InP/GaInP material system [70,71,72]). These effects might be attributed to the strongly modified carrier transport and carrier relaxation (see also the discussion of carrier relaxation and linewidth by *Borri* and *Langbein* in this volume) in low dimensional lasers and will be discussed in the next section.

Also dot lasers with shorter wavelengths show generally higher threshold values [70,71,72,73,74,75,76] than lasers with longer wavelengths [67,68,69,77] [78,79] indicating that barrier leakage currents play an important role in the (Al)GaInP, (Al)GaInN systems. Of course at the present status also material problems (defect density, doping) may affect the laser performance strongly.

### 1.3    Artificial versus Self Assembly

For the fabrication of low dimensional lasers a paramount number of techniques have been developed. Below we give a list of fabrication techniques to realize wire and dot structures.

1. Lithographical techniques
   - overgrowth of patterned strained layers (buried stressors [80])
   - overgrowth of etched V-grooves (an extension of the buried crescent laser technique to nanometer sizes) [81,82]. V-groove and migration growth (AlGaAs/GaAs material system) along $(111)_A$-planes for V-grooves in the $[01\bar{1}]$-direction [83,84]
   - lithography based patterning (dry etching and overgrowth) techniques [14,16,85,86,87,88]. Dry etching and control of surface recombination and dead layer effects [86,88,89,90]
   - local implantation enhanced intermixing (IEI) techniques [91,92,93,94] [95,96,97,98,99]
   - focused ion beam (FIB) technique (wire and dot structures) [100]
2. Growth techniques
   - growth on vicinal substrates (FLS fractional layer super-lattices [101]) by MOVPE
   - growth on misoriented substrates (tilted super-lattices (TSL) [102]) by MBE
   - SSL (serpentine super lattice) technique [103,104]
3. Self assembled growth techniques (self assembled dots = SAD)
   - growth of dots [17,105,106,107,108,109] (see also Sect. 2)

It turns out that in general, artificially patterned nano-structures based on implantation or etching techniques show a non-vanishing contribution of defect effects. These defects seriously mask quantization effects and therefore clear single dot effects could hardly be observed. Epitaxially grown nano-structures by self organization principles, on the other hand, already showed nice low dimensional effects (see for instance the Chapter by *Petroff* in this volume). Therefore the expected low laser threshold behaviour could be demonstrated [3]. Despite this sucess of low dimensional lasers based on self organized nanometer structures there is an intrinsic disadvantage of SAD structures as size and average distance between dots can be controlled only to a certain degree inside the Gaussian distribution due to the epitaxial growth conditions. However (see the contribution by *Petroff* in this volume and [110]) there are first experimental results of ordered SAD arrays, which overcome these problems.

The Gaussian distribution width shows a twofold effect:

- The distribution width of dots reduces the available gain in a certain energy range and can even result in an inferior device performance of a dot laser compared to a QW laser.
- The distribution of dot–dot distances doesn't allow the realization of an optical filter like a DFB or DBR by SAD growth solely. A post epitaxial realization of a DFB grating is of course always possible, but would not meet the very attractive possibility to realize complex coupled DFB lasers with dot arrays directly.

These restrictions can be overcome in dot and wire lasers with artificially designed nanometer structures, despite the already mentioned limitations by process induced defects. Complex coupled wire and dot CC-DFB lasers in the long wavelength range (1550 nm) could be realized [9,16]. The technoslogical progress triggered by dot and wire DFB laser design has even fertilized further CC-DFB-laser development: CC-DFB lasers with very low threshold and extremely low emission linewidths [111] even for short lasers could be realized.

An approach to combine the benefits of precise optical filter definition with the benefits of the SAD growth technique has been published for the material system GaAs recently [110].

## 2   Epitaxially Fabricated Quantum Dot Lasers

As already mentioned, old crystal growth analysis results have been rediscovered recently for the epitaxial growth of self assembled dots (SAD) as the active region of a semiconductor laser diode. As described in detail elsewhere in this book (see also the contribution by *Petroff* in this volume), the sum of the free enthalpies of layer, interface and substrate

$$\Delta \sigma = \sigma_l + \sigma_i - \sigma_s \tag{1}$$

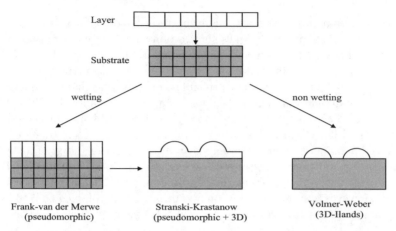

Frank-van der Merwe
(pseudomorphic)

Stranski-Krastanow
(pseudomorphic + 3D)

Volmer-Weber
(3D-Ilands)

**Fig. 2.** Different epitaxial growth modes for strained layer growth. *Left bottom* figure: 2D growth of a strained layer. *Center bottom*: Formation of 3D island arising after wetting layer growth. *Right bottom*: 3D island growth without wetting layer growth

controls the growth mode near the heterointerface, in particular if the epitaxial layer has a lattice constant different to the substrate. It can be classified into three modes depicted in Fig. 2. When the substrate term dominates ($\Delta \sigma < 0$), a two-dimensional epitaxial layer grows pseudomorphically strained to the substrate (Frank–Van der Merwe mode [112]). If, on the other hand, the layer term dominates ($\Delta \sigma > 0$), then the layer does not wet the substrate, but forms three-dimensional islands (Volmer–Weber growth mode [113], right side). For the growth of quantum dots, the most interesting mode turned out to be just in between these two extremes: The growth starts by completely wetting the substrate surface. However, due to the lattice mismatch induced strain, the interface enthalpy $\sigma_i$ increases during growth. After a few monolayers, when reaching a critical thickness, this results in a change of sign of $\Delta \sigma$, i.e. three-dimensional islands start to grow on the wetting layer (Stranski–Krastanov mode [114], middle).

Under optimized growth conditions, the resulting quantum dots show fairly good size and height homogeneity (see below Fig. 9, right) making them useful for basic studies about zero-dimensional systems as well as for applications in device structures like semiconductor laser diodes. Within a short time, most groups working on quantum dot research adapted this simple fabrication method which does not require any lithography or other artificial lateral structurization technique. It has then been applied to many semiconductor material systems including Si/Ge as well as III–V and II–VI compounds.

In the following, after discussing some basic growth considerations, we review the current state of laser diodes employing SAD structures as active region.

## 2.1 Growth Considerations
## for Self-assembled Quantum Dot Laser Structures

As outlined in the contribution by *Petroff* in this volume, the growth of SADs has been achieved mainly by both of today's most sophisticated epitaxial methods: molecular beam epitaxy (MBE) and metalorganic vapor phase epitaxy (MOVPE or MOCVD). In order to achieve self-assembling, a drastic change of the growth conditions established for two-dimensional layer growth is required. First, a large lattice mismatch of several percent between layer and substrate is a prerequisite. Second, a significant lowering of the growth temperature of 100°C or more compared to best two-dimensional layer growth must be employed. The temperature controls directly the dot properties: Lower temperatures lead to larger densities with smaller dimensions owing to the shorter atom diffusion lengths on the growing surface, whereas higher temperatures may cause three-dimensional growth of defect-rich layers because of the large lattice mismatch.

For laser applications, dot densities as large as possible should be obtained in order to increase the filling factor of the gain material (see Sect. 4.3), which is inherently very small for any dot structure compared to a two-dimensional film. Moreover, the dots should be fairly small to show the expected quantization effects, both features asking for low growth temperatures.

A laser structure is built of many layers and the majority of them would require high growth temperatures for best material properties. In order not to destroy the dots during growth of the covering layers, the growth temperature after having deposited the dots should, however, be kept low up to the end of the epitaxial process. This may be in conflict with other needs and design rules, especially for the growth of high quality waveguide layers or optimized doping of the outer barrier layers. Thus, special care must be taken for the temperature profile during the epitaxy of a quantum dot laser structure.

In general, MBE seems to be more favorable for the growth of self-assembled quantum dots. This may be related to the fact that the growth temperatures in MBE are in general lower than in MOVPE. Therefore, the further lowering for dot growth is somewhat less dramatic or leads more easily to the needed higher dot density with still good properties.

## 2.2 Specific Problems
## of Self-assembled Quantum Dot Laser Structures

An ideal quantum dot laser would contain a large number of identical quantum dots in the active region (Sect. 4.2). This basic requirement can hardly be obtained by the Stranski–Krastanov growth mode. The self-assembled dots show a characteristic comparably broad size distribution (Fig. 9) which leads to typical spectral linewidths of several 10 meV. Special techniques such as adequate growth interruptions after depositing the dots [115] or use of miscut substrates [42,59,116] may help to improve the dot size distribution. Still, as mentioned earlier the lateral filling factor of a single quantum dot layer is at least a factor of 10 smaller than that of a quantum well even for highest dot densities which may exceed $5 \times 10^{10}\,\mathrm{cm}^{-2}$. Therefore, the modal optical gain may be quite small [117] and saturate at low emission powers. In consequence, laser action may start from excited states which have a higher density of states compared to the ground state [118], or it will switch from ground state to excited state transitions. Thus, the above mentioned increase of the dot density, in parallel with best size homogeneity, is one of the major challenges for the epitaxial growers. Stacking of quantum dots, as described by *Petroff* in this volume, is just a logical step to further increase the density and decrease dot laser threshold currents [35,119,120] or modal gain, respectively [121]. Additionally, it turned out to improve the size homogeneity of SADs [31,122,123,124] which may lead to extremely low inhomogeneous broadening of the SAD photoluminescence below 20 meV [125].

Obviously, the latter is one of the major characteristics to be considered for optimized quantum dot lasers, as it is directly related with the spectral width of the optical gain curve (see (8) and the Chapter by *Borri* and *Langbein* in this volume). In a quantum well, nearly all carriers excited in the well may finally contribute to the laser action, even if the gain curve is fairly broad, due to the carrier–carrier interaction and redistribution during lasing. On a first glance, this is questionable for a quantum dot structure, which ideally should consist of many electronically noninteracting quantum dots. Thus, only those dots would contribute to lasing which hit exactly the very narrow laser line. This would be a very low number in a structure with a typical photoluminescence line width of some 10 meV resulting from the inhomogeneous broadening of the dot ensemble.

To some extent, this may be counterbalanced by the extremely large specific gain attributed to quantum dots [126]. But fortunately it seems that even a perfect quantum dot system shows some homogeneous line broadening [127,128] (see also the Chapter by *Borri* and *Langbein*). This was explained by a carrier redistribution process between different dots via electronic states in the wetting layer or even the barrier material [129]. By comparing two dot lasers with different barrier heights between vertically stacked dot layers, but otherwise identical, *Groom* et al. [130] found direct evidence for this explanation: the structure with AlGaAs barriers exhibited a broad laser spectrum at lower temperatures attributed to imperfect interdot communication, whereas the device with GaAs barriers showed a much narrower laser spectrum (see Fig. 3). At higher temperatures, the high barrier laser finally outperforms its low barrier counterpart because of reduced thermal carrier escape. Thus, the thermally activated interdot communication pro-

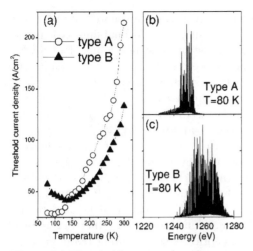

**Fig. 3. (a)** Threshold current density vs temperature for InAs quantum dot lasers with GaAs (type A) or AlGaAs barrier (type B) with five QD layers; **(b)** and **(c)** 80 K lasing spectra for the devices of **(a)**. From [130]

cess makes single energy laser action more probable at higher temperatures, as experimentally observed by many groups.

The broadening mechanism is still a matter of scientific debate. It has been explained by a lattice relaxation mechanism, where the carriers couple to the acoustic phonons during the optical transition [131], or by polarization dephasing, i.e. by interaction between different dots through photons leading to collective lasing of the dot ensemble, whereas coupling via thermally activated carriers would not be required [128].

The inhomogeneous broadening may be taken, on the other hand, as a chance for realizing wavelength tunable lasers, because it is responsible for a broad optical gain spectrum of the dot ensemble. Wavelength tuning can be done either by changing the resonator by external means or by taking adavantage of the fast gain saturation of the quantum dots. Devices starting lasing at a rather long wavelength may gradually shift to shorter wavelengths with increasing injection current, because other quantum dots and/or higher energy levels will take over after saturation of those levels involved in the laser process at low current densities [132]. Indeed, we observed a smooth blue shift of the gain curve of InP-GaInP dots with increasing temperature, which might be explained by this effect [44], whereas an opposite shift would be expected because of the temperature dependence of the band gap.

## 2.3 InAs-GaAs Quantum Dot Lasers

InAs quantum dots on GaAs have become the prototype material system of self-assembled quantum dots and respective lasers. Due to the large lattice mismatch of about 8% between GaAs and InAs, the Stranski–Krastanov growth mode can easily be established. Moreover, the mismatch can be tuned over a wide range by forming the ternary alloy GaInAs enabling basic investigations about the correlations between strain and development of SADs. Finally, this material system has already been the material of choice for most MBE machines, before SAD investigations started, whereas others like phosphides or nitrides impose special growth problems.

At the beginning, scientists have been fascinated by the fact that many basic properties of zero-dimensional systems could be easily studied with quantum dots the fabrication of which was just as easy as that of a quantum well. Of course, it was expected that lasers with such quantum dots would show all those nice properties predicted by theoreticians many years ago. However, these hopes could not be realized yet, especially when studying devices at higher than cryogenic temperatures. This can be mainly attributed to the inherently large dot size distribution mentioned above. Another reason may be the limited quantum dot potential depth. As we are interested in growing very small dots, the quantization energy of electron and hole states becomes fairly large. In addition, the dots are grown under compressive strain shifting the carrier states further to higher energies. Therefore, thermal escape of carriers is still one of the major problems in most quantum dot lasers leading to a strong increase of non-radiative losses and thus of the threshold current when the device temperatures are increased to 300 K or more. Therefore, a very high characteristic temperature close to infinity, as ideally expected for zero dimensional systems and often measured at cryogenic temperatures (c.f. [133,134]), could not yet be observed around room temperature.

However, very soon, another quite important advantage has been discovered. Self-assembled quantum dots are not just a logical step from three-dimensional layers over two- and one-dimensional structures (quantum wells and wires) to zero dimensions, they additionally can be regarded as a new class of materials. In normal two-dimensional film growth of lattice mismatched materials, the layer grows pseudomorphically strained and defect-free on the substrate (see Fig. 2) only up to a critical thickness defined by the built-in strain energy [135]. Then, the layer relaxes by forming a high density of defects and becomes unsuitable for most applications. This limits, as an example, the In content $x$ in $Ga_{1-x}In_xAs$ on GaAs to values below about 15% and thus the emission wavelength of respective laser diodes to values below about 1100 nm.

By the formation of self-assembled quantum dots, these limits can be overcome, opening up the telecom wavelength region of 1300 nm and above to materials grown on GaAs, the technologically best developed and thus most important substrate for compound semiconductors. Today, InAs dot lasers are regarded as the most promising technology for telecom laser diodes on GaAs wafers, besides GaInAsN structures [136] where the lattice mismatch induced strain is reduced by alloying low amounts of nitrogen into GaInAs quantum wells.

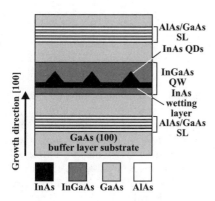

**Fig. 4.** InAs dots embedded in a GaInAs quantum well (schematically). From [138]

Now, SAD lasers did not just compete with quantum well lasers in terms of threshold current or output power, but are potentially enabled to cover emission wavelength regions which otherwise would be impossible or at least difficult to reach. As described later, such a widening of the material range by SADs may also be helpful in other regions of the optical spectrum.

In recent years, tremendous progress has been achieved especially for InAs SAD lasers, published in a vast number of papers. In the following, we can only describe some outstanding results, referring the reader to other review papers [3,127,136] and the original publications.

### 2.3.1  Shift of the Emission Wavelength to the Telecom Windows

As mentioned above, typical SAD systems show fairly large effective band gaps due to quantization and strain effects. Thus, the first InAs quantum dot laser structures emitted at wavelengths shorter than about 1.25 μm. In order to reach the first telecom window at 1.3 μm, several groups [137,138,139] proposed to embed the InAs dots into a GaInAs quantum well (see Fig. 4) which may decrease both the compressive strain in the dot and the barrier height and thus the quantization energy [136]. Moreover, this may help to increase the dot density to values close to $10^{11}$ cm$^{-2}$ [140]. Using these effects, the first CW lasing at 1.3 μm at room temperature (Fig. 5) has been achieved by *Mukai* et al. [141].

A further reduction of emission wavelength up to the second telecom window at 1.55 μm is still a challenge. By growing laterally assosiated InAs dots at very low temperatures (320–400°C) in MBE, *Maximov* et al. could achieve photoluminescence (PL) emission at 1.7 μm [142]. More recently, *Tatebayashi* et al. reported about InAs dots embedded in GaInAs and grown by MOVPE which emitted PL at 1.52 μm [143].

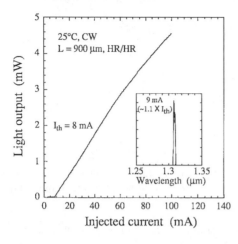

**Fig. 5.** CW light output versus injected current characteristics of an InAs quantum dot laser with a 900 μm cavity length and high-reflectivity-coated facets. Threshold current was 8 mA. Lasing spectra at 9 mA (1.1 × $I_{th}$) is shown in the *inset*, indicating a lasing wavelength of 1.31 μm. From [141] (©1999/2000 IEEE)

## 2.3.2   Low Threshold Current Density and Temperature Sensitivity

As the most obvious figure of merit of a laser diode, the optimization of its threshold current density $j_{th}$ and its temperature dependence expressed by the characteristic temperature $T_0$ are the primary goals of many research attempts. Although already the first SAD lasers showed fairly low $j_{th}$ around $100 \, \mathrm{A/cm^2}$ with excellent temperature stability at cryogenic temperatures (around 70–100 K, see Fig. 6), a drastic increase with temperature was commonly observed [60,133] making the laser action at room temperature less attractive. As mentioned above, high modal gain and the suppression of any losses in the active region or at the mirrors are prerequisites to obtain lasing via the ground state. In order to suppress lasing via excited states, long laser cavities of several millimeters were fabricated [144], often combined with coating the laser facets for higher reflectivities [35,145,146].

Since then, better material quality and basic understanding of SADs have resulted in a drastic improvement of both threshold current and characteristic temperature. Lowest threshold current densities of SAD edge emitter lasers reported to date are around $25 \, \mathrm{A/cm^2}$ [140,146] with best absolute room temperature threshold currents close to 1 mA [147]. These values are better than those of best quantum well lasers, a fact which was never expected by most laser experts because of the random nature of self-assembled quantum dots.

However, the characteristic temperature $T_0$ remained at low values around 70 K, comparable to the competing GaInAsP quantum well lasers on InP substrates. As mentioned above, this is mainly attributed to the thermal escape of carriers out of the ground state involved in the laser process. Very recently, *Shchekin* and *Deppe* pointed out that in particular the valence band structure in the dots may be responsible for the low $T_0$ [148]. Due to the large mass of the holes, the separation between the hole levels is much closer than between

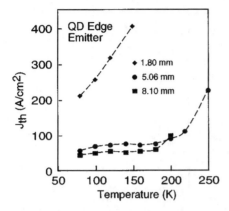

**Fig. 6.** Threshold current density versus temperature for three different cavity lengths for InAs QD laser diodes. Lasing on the $m = 2$ transition for the 1.80 mm long cavity leads to an increased temperature dependence of threshold due to carrier escape from the QDs, while the 5.06 and 8.10 mm long cavities both show regions of temperature-insensitive lasing. From [60]

the electron levels. Whereas the electrons, once arrived in their ground state level, may reside there long enough for subsequent emission processes, the holes may easily thermally escape to higher levels thus reducing the ground state recombination probability. The authors showed both theoretically and experimentally that p-type doping of the dots leads to a higher occupation probability of the dot ground states and thus to improved laser performance in terms of gain and characteristic temperature. $T_0$ could be increased up to 161 K for operating temperatures up to 80°C. Moreover, these effects may result in a significantly higher modulation bandwidth [149].

Another approach to decrease thermal excitation of the carriers and thus increase the characteristic temperature is obviously to increase the band gap of the barrier material. By this technique, *Zhukov* et al. could substantially reduce the threshold current density of their quantum dot lasers at room temperature and achieve fairly high output powers of 1 W [150]. *Schäfer* et al. have embedded their $Ga_{0.4}In_{0.6}As$ SADs in an $Al_{0.33}Ga_{0.67}As$-GaAs graded index separate confinement heterostructure. With this approach, they could achieve laser action via the dot ground state up to 214°C (Fig. 7) with still 7 mW output power per (uncoated) facet [151]. However, a fairly low emission wavelength of about 940 nm at room temperature was obtained.

### 2.3.3  Long Wavelength VCSELs

GaAs based vertical cavity surface emitting laser diodes (VCSELs) emitting around 850 nm have reached a high state of development. In such lasers, the optical feedback is provided by highly reflective epitaxially grown multilayers acting as top and bottom Bragg mirrors around the active zone of a laser.

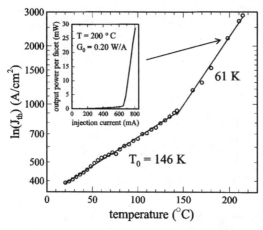

**Fig. 7.** Threshold current density of a 600 μm × 50 μm broad-area quantum dot laser at different temperatures ranging from 20 up to 214°C. The *inset* shows a light output curve at 200°C. From [151]

These devices are very attractive owing to their monomode emission potential, their extremely low threshold currents and the simple device process, where nearly all testing can be done on the wafer, and the difficult and risky production of laser mirrors by facet cleaving becomes obsolete. Similar devices for longer wavelengths around 1.3 or 1.55 μm, however, are just getting to being commercialized, although there is a huge interest to use such devices in glass fibre optics instead of the sophisticated and very expensive edge emitting DFB lasers, as VCSELs should have similar monomode behaviour at much lower price. However, there are basic material problems making the development of such lasers difficult. Structures based on InP wafers can emit the required wavelength fairly well. However, VCSELs made of these materials lattice matched to InP would need a huge number of Bragg mirrors, because the potential mirror materials InP and GaInAsP posses a very small difference of their refractive index.

Therefore, many alternative routes are currently investigated including AlSb-GaSb mirrors, dielectric mirrors, wafer fusing to combine a GaInAs/InP based active region with GaAs/AlAs Bragg mirrors, or special techniques to overcome the problems of the thick Bragg mirrors.

InAs laser structures turned out to be of great benefit here as well as for edge emitters. Again, wavelength regions can be obtained with InAs dots which would not be possible with quantum wells on GaAs. Besides this good reason, already discussed extensively above, an even more important plus is the possibility to use AlGaAs-GaAs layers with a comparably large refractive index difference as Bragg mirrors.

After having achieved laser emission from edge emitters, it was therefore obvious to grow long wavelength quantum dot VCSEL structures. First

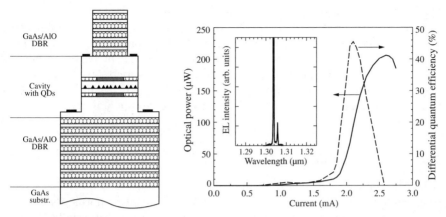

**Fig. 8.** VCSEL structure with self-assembled InAs dots and GaAs/AlO Bragg mirrors (schematically, *left*); light output and efficiency versus driving current for this device (*right*). The laser spectrum is depicted in the *inset* of the right figure. From [155]

devices, emitting at a fairly short wavelength of about 950 nm, have been reported in 1996 [152]. As the Bragg mirror reflectivity still limited the performance, others have used AlO-GaAs mirror stacks by fully oxidizing the epitaxially grown AlAs [153,154]. Extremely low threshold currents below 200 μA have been observed [154] for an emission wavelength around 1000 nm. Only with this mirror technology, a VCSEL emitting at 1.3 μm could be finally realized [155] (Fig. 8).

However, it seems that VCSELs in this wavelength region can be realized more easily with GaInAsN as the active region, the competing material system as pointed out above, where impressive performance has already been achieved [156] with semiconductor Bragg mirrors.

### 2.3.4   Long Wavelength Dot Laser Structures on InP Wafers

As mentioned above, the large lattice mismatch induced strain contributes significantly to the up-shift of the InAs quantum dot band gap. This strain can be easily reduced by growing the dots on InP wafers, where the lattice mismatch to InAs is only about 3.2%. Indeed, such InAs dots show optical emission at much longer wavelength, and lasers emitting at 1.9 μm (although at cryogenic temperatures) have been reported [157]. However, the driving force for developing such lasers is quite small.

### 2.3.5   Are Dot Lasers Mature for Industrial Applications?

Despite these impressive results obtained in many research labs, the semiconductor industry has not yet started to change their long wavelength laser production to quantum dot structures. The main objection is the inherently random growth process of the active region, which is thought to influence negatively the long term reproducibility. Up to now, only a small U.S. company has announced the production of such lasers [132] by MBE on both, GaAs wafers for 1310 nm emission and on InP wafers for 1550 nm emission. The wavelength stability is quoted to be better than 0.25 and 0.3 nm/°C, respectively, whereas quantum well lasers would exhibit about twice these values.

## 2.4     Red Light Emitting Dot Lasers

After the success of InAs self-assembled quantum dots, it is worth studying other similar material systems. Changing arsenic to phosphorus opens another class of compound semiconductors which possess substantially higher band gaps.

### 2.4.1     InP-GaInP Dot Lasers

The perfect analogon would be InP dots in a GaP matrix. However, due to the indirect character of GaP, this material is less attractive for optoelectronic applications. Therefore, research in our labs as well as in others concentrated first on InP dots in a GaInP matrix, a ternary compound which is lattice matched to GaAs for nearly equal parts of Ga and In. Let us come back to the InP/GaP system later.

The larger band gaps of the phosphides open another interesting part of the spectrum for quantum dot laser diodes: GaInP is the material of choice for high brightness red and yellow light emitting LEDs as well as for red light emitting laser diodes with emission wavelengths around 650 nm. As in particular the latter still are inferior compared to GaAs-AlGaAs infrared lasers, quantum dot structures may be a chance to improve them.

Due to the restriction to GaInP on GaAs, the maximum lattice mismatch between the dot material (InP) and the matrix (GaInP) is about 3.7% and thus half the value as in the case of InAs/GaAs. But this is still large enough to give rise to self-assembled dot growth, which has been achieved mainly by MOVPE [124,158,159,160]. The growth of phosphorus containing layers in MBE is much more difficult due to the lack of well controllable solid P sources. *Manz* et al. achieved excellent phosphide and InP dot growth using GaP single crystalline material as evaporation source in their MBE machine [45]. It has been observed that InP dots develop at somewhat higher growth temperatures than InAs dots which is in line with the common trend that arsenides need in general lower growth temperatures than phosphides. We found that dots start to grow after depositing a wetting layer of about 1.6 monolayers at 580°C in low pressure MOVPE [42]. However, two types of dots develop under these growth conditions differing mainly in size. The larger ones (called type B), less interesting in terms of quantization and band

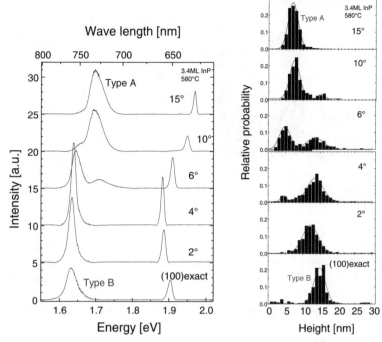

**Fig. 9.** Photoluminescence (PL) signal (*left*) and height probability (*right*), as measured by atomic force microscopy, from InP-GaInP self-assembled quantum dots grown on differently misoriented GaAs wafers. Only on wafers misoriented by more than 10° versus (111)$_B$, could the growth of the larger type B dots emitting at lower energy be completely suppressed. The PL signal at $E > 1.85$ eV stems from the GaInP barrier

gap, soon dominate the sample's characteristics. Lower growth temperatures lead to a larger number of small dots (called type A), which, however, degrade drastically in their spectroscopic properties under these conditions.

Therefore, we have influenced the surface migration properties, which finally are responsible for the dot formation process, by choosing miscut substrates (see Fig. 9). Perfect single mode development of type A dots could be obtained on GaAs wafers miscut by 15° to (111)$_B$ with still very good photoluminescence properties [42] at comparably high growth temperatures. We could establish structures with excellent dot size uniformity as indicated by a small PL linewidth around 40 meV. Also here, the stacking of dots may lead to better size uniformity reducing the inhomogeneous PL linewidth further down to below 30 meV [124]. Similar to InAs dots, the effective band gap of these InP dots lies substantially above that of InP bulk material, due to the large compressive strain and quantization effects. This brings them close to the above mentioned matrix or barrier material Ga$_{0.5}$In$_{0.5}$P, giving rise to strong thermal carrier evaporation and thus to a dramatic decrease

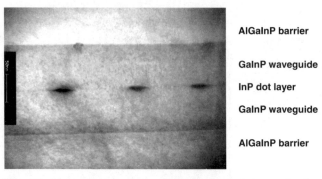

AlGaInP barrier

GaInP waveguide

InP dot layer

GaInP waveguide

AlGaInP barrier

**Fig. 10.** Transmission electron microscopy photograph of a laser diode structure containing InP-GaInP self-assembled quantum dots

of the photoluminescence intensity when increasing the temperature above about 150 K. However, their emission is still at comparably long wavelengths around 700 nm. Either embedding them into a GaInP matrix with higher Ga content or by mixing Ga into the dots themselves, could only slightly change their emission wavelength.

Recent spectroscopic studies have even pointed to another fundamental difference to InAs dots: it seems that the InP valence band lies energetically below the valence band edge of the GaInP matrix material forming a type II band alignment of these dots. Thus, while the electrons may be confined within the InP dots as expected, the holes may just be attracted by Coulomb interaction to the interface between the dots and the matrix. This would reduce the wave function overlap of electrons and holes and thus their recombination probability, as indeed observed by time resolved measurements [161].

Laser structures with InP or GaInP dots could be grown by using the experience of GaInP quantum well laser diodes. In order to provide best dot characteristics, we embedded the dots in a GaInP waveguide layer (Fig. 10). The outer barriers were made of AlGaInP or AlInP. Also here, the growth of the ternary and quaternary p-type doped layers on top of the dots had to be optimized at lower growth temperatures compared to normal GaInP quantum well lasers in order not to destroy the InP dots in the overgrowth process.

These structures showed optically pumped as well as electrically driven laser action [44,124]. Due to the small difference between the quantum dot levels and the band gaps of the surrounding layers, thermal carrier escape was again very severe. The above mentioned indirect band structure in real space may also limit the laser performance. Similar as for InAs dot lasers, we observed excellent temperature stability only at cryogenic temperatures, whereas the characteristic temperature dropped to about 40 K at room temperature, where we, however, still achieved fairly low threshold current den-

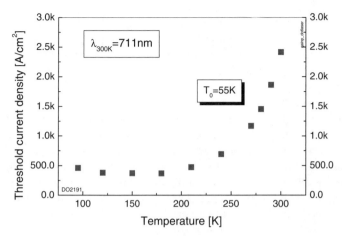

**Fig. 11.** Threshold current density of laser diodes containing InP-GaInP self-assembled quantum dots versus temperature

sities around $1.8\,\mathrm{kA/\,cm^2}$ (Fig. 11) [162]. From the difference between the PL maximum and the laser line, we conclude that the laser emission at room temperature does not occur on the ground state of the dots, but involves excited states.

Similar laser properties have been achieved by *Manz* et al. [45]. Ground state emission could be established, probably due to the fact that by MBE, smaller dots with better size uniformity could be grown.

In order to improve the optoelectronic characteristics of the dots at elevated temperatures, we have investigated quaternary AlGaInP as a matrix material for the InP dots. This material tends to grow with much lower quality than GaInP, in particular if low growth temperatures are required, and thus the self-assembling process needs special optimization. In spite of these problems, we were able to improve the quantum dot PL intensity at room temperature dramatically [124]. Similar results have been obtained by *Ryou* et al. [160] who found largest intensities for barriers with a quaternary composition having the largest direct band gap. Optically pumped lasers grown by them with such structures emitted at somewhat smaller wavelengths of 650–680 nm, and recently, they could achieve electrically driven laser operation down to 637 nm by adding a GaInP auxiliary quantum well which helps to aid carrier collection [163].

### 2.4.2  GaP as Substrate for Dot Lasers?

As mentioned above, InP dots on GaP seem, in terms of lattice mismatch, even more attractive for self-assembled dot growth. GaP was, due to its indirect band structure, never regarded as attractive for laser diode applications. Layers with direct band gap would require some alloying with InP. However, mixing larger amounts of InP to GaP would soon change the lattice constant too far for defect free two-dimensional epitaxial growth. But if we consider quantum dots, we are no longer limited to two-dimensional growth. This potentially opens the chance to use GaP or even AlP as barrier material for laser diodes thus increasing the barrier height for better temperature stability or shorter emission wavelength of red light emitting lasers.

Indeed, InP forms excellently developed self-assembled dots on GaP [43]. However, these dots do not show any photoluminescence, probably due to the fact that their band structure becomes of indirect character due to the large lattice mismatch induced strain [164]. On the other hand, self-assembled GaInP dots on GaP with reduced strain exhibit photoluminescence at wavelengths around 650 nm [39,43] proving that light emission is attainable from material grown on this indirect substrate. Further studies are needed to investigate the potential of these structures for laser action.

### 2.4.3   Red Light Emitting Lasers with AlInAs Dots

An alternative way to shorten the emission wavelength is the use of AlInAs as the quantum dot material, i.e. increasing the band gap of InAs by mixing of AlAs. Although the lattice mismatch to GaAs is reduced, it is still possible to grow nicely developed quantum dots. By embedding them into a well established laser structure containing AlGaAs as large band gap barrier material, laser action at a wavelength of about 720 nm could be obtained [165,166].

### 2.5   Nitride Quantum Dots for Laser Action

Besides the ongoing race towards long wavelength laser diodes on GaAs wafers, an impressively large amount of research focuses on the opposite part of the optoelectronic spectrum. Since the first report of a GaInN laser diode in 1996 [167], many companies have tried to develop such short wavelength lasers which can be used in next generation optical disk memories at an emission wavelength of about 400 nm. In the nitride material system, however, it is difficult to obtain laser action at rather long wavelengths, i.e. in the blue or even green visible part of the spectrum. This is partly true because GaInN cannot be alloyed in the total composition range, but tends to form local composition variations or even shows segregation for In concentrations exceeding 10–15%. These fluctuations, also found in early laser structures, have been considered as quantum dots [168]. However, it became clear that these fluctuations do not really improve the performance of laser diodes, probably because they exhibit a totally random size distribution in strong contrast to typical self-assembled quantum dots.

On the other hand, (Ga)InN possesses a large lattice constant difference to GaN and thus may be an excellent candidate for Stranski–Krastanov grown

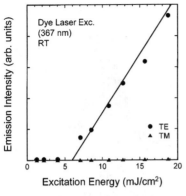 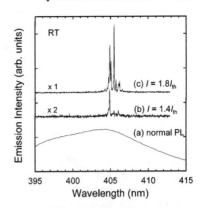

**Fig. 12.** *Left*: Dependence of the polarized emission intensity of a GaInN quantum dot laser on the excitation energy per pulse at room temperature. The excitation source was a dye laser (367 nm). *Right* (**a**): Conventional PL spectrum, excited by a HeCd laser (325 nm), (**b**) and (**c**): Emission spectra above threshold, excited by a dye laser (367 nm). From [175] (©1999/2000 IEEE)

quantum dots. Moreover, such dots may not suffer but may even take advantage of the above mentioned immiscibility and thus open again a spectral range not accessible to quantum well structures.

Indeed, Stranski–Krastanov growth mode of GaN dots on Al(GaN) [169] and of GaInN dots on GaN [170,171] has been successfully obtained. The first optically pumped lasers containing GaInN quantum dots have been reported by *Tachibana* et al. [172]. However, although rather long wavelength photoluminescence emission up to 475 nm has been observed [173], these lasers still emit at the same wavelength (around 405 nm, Fig. 12) as best quantum well lasers, for which, in contrast, the emission could recently be shifted to 460 nm [174] and beyond.

## 2.6  ZnSe Based Quantum Dot Lasers

II–VI compounds offer another wide range of materials which have been studied for many years. In particular, they were regarded as the best materials to realize short wavelength green and blue lasers. Although such lasers could be fabricated with quite good performance in terms of wavelength, threshold current and quantum efficiency, they still suffer from short device lifetimes which could not yet be managed accurately.

Nevertheless, only recently, *Klude* et al. could realize an electrically driven ZnSe based laser by MBE growth with self-assembled CdSe quantum dots as the active region [65]. These lasers emitting at 560 nm exhibited rather large threshold current densities in the range of $7.5 \, \mathrm{kA/cm^2}$ (Fig. 13). However, the authors expect to improve their performance by optimizing the dot size uniformity and by stacking several dot layers. Moreover, the device lifetime

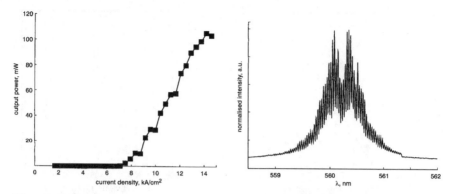

**Fig. 13.** Light output versus current (*left*) and output spectrum (*right*) of a laser containing self-assembled CdSe dots as active region. From [65]

may be increased, because dot structures are expected to be less sensitive to defect formation because of their small lateral filling factor [64] when compared to quantum wells.

# 3   Lithography Based Quantum Dot Laser Fabrication

This section gives a brief overview of lithography based nano-fabrication techniques for wire and dot lasers. The realization of dot and wire lasers requires a flexible definition and fabrication of large scale structures as optical resonators and filters as well as definition of the device periphery for carrier injection. On the other hand on a small length scale, mainly two techniques can be distinguished as mentioned in Sect. 1.3 – ... (i) direct epitaxial growth with reduced flexibility, but in principal improved material quality and (ii) lithography based as implantation and dry etching with enhanced flexibility, but for some cases reduced material quality.

## 3.1   Electron Beam Lithography

The realization of small, arbitrarily shaped crystal structures to obtain quantum size effects can be realized by patterning techniques based on electron beam resist masks in the nanometer range. A quite flexible technique to realize small masks is based on a local variation of the chemistry of the mask material by a well defined local energy supply. The different possibilities to supply energy defines the kind of lithography (photons (UV, and X-rays), electron beams, ion beams):

1. Optical lithography [176,177]
2. X-ray lithography [178,179]
3. Electron beam lithography [180,181]
4. Ion beam lithography [182,183]

Very flexible and commonly used is electron beam lithography (e-beam) to define artificially shaped nano-structures. Very often, except for ion beam lithography, e-beam is also used to define masks for the other above mentioned lithography techniques. The mask materials for e-beam processes consist mostly of polymeric materials. According to the type of resist (negative tone or positive tone) the resists possess long chain lengths (high molecular weights $\approx 10^5$ units) or short chain lengths (small molecular weight). The exposure leads in the case of the positive tone resists to a reduction of the chain length whereas in the case of the negative tone resists a further polymerization of the molecule (increase of chain length) occurs. The development process now uses the solubility of the resist materials which depends on chain length. Therefore, the resolution of the e-beam process not only depends on the minimum electron beam width, but also on the detailed parameters of the development process. Examples for negative tone resists are $\alpha$M-CMS [184], AZPN114 [185], and a common positive tone resist represents PMMA [186]. The contrast $\gamma = 1/[\lg(D_{x,o}/D_{x,i})]$ with $D_{x,i}$ = threshold dose for increase of resist thickness (negative tone resist, polymerization of chains) or for decrease of resist thickness (positive tone resist, chain length reduction), $D_{x,o}$ = dose for full resist thickness (negative tone resist) or zero resist thickness (positive tone resist) is an important key parameter of a resist material and should be large to ensure digital patterns of wires or dots. Furthermore, the resist sensitivity is also an important parameter in defining wire and dot structures. For a resist with sensitivity $D$ the number of electrons defining a pixel with size $l_P^2$ can be calculated according to $N = \frac{1}{q} \cdot D l_P^2$. For PMMA with a sensitivity of $D = 300$–$500\,\mu C/cm^2$ one needs between 2000 and 3000 electrons to define a pixel of $(10\,nm)^2$ corresponding to a statistical fluctuation of 0.018–0.022. On the other hand for a pixel of the same size defined in AZPN114 only 216 electrons are necessary corresponding to a statistical fluctuation of 0.07. Therefore the ideal resist possesses high contrast and high threshold dose $D_i$. Further mechanisms limiting the structure definition are processes of electron back-scattering resulting in a background dose (proximity effect) in the resist. The range of the back-scattered electrons can reach a few $\mu m$ depending on the substrate under the resist [187,188,189]. From dose variation experiments with single wires and dots as well as with wire and dot arrays we found a minimum structure resolution of 20 nm for a PMMA mask and below 50 nm for a AZPN114 mask which is near the minimum focus width (8 nm) of the e-beam lithography system [JEOL JBX-5DII(U)] used. Examples of nanometer optical filters and lasers are given for PMMA in [190] and for AZPN114 in [191]. Some examples of dot arrays are depicted in Fig. 14.

## 3.2 Structurization Technology

After lithography, different technologies can be applied to transfer the lithography pattern to the semiconductor substrate. Most common are dry etching

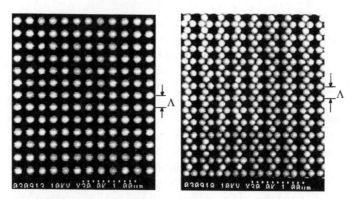

**Fig. 14.** *Left*: SEM picture of a dot array with a simple DFB period ($\Lambda = 240\,\mathrm{nm}$). *Right*: Analogous to left but with filling factor modified DFB period $\Lambda$

techniques or implantation techniques. The properties of dry etching will be briefly discussed. By implantation techniques no devices with real dot features at room temperature could be achieved.

### 3.2.1  Dry Etching

After definition of masks with nanometer sizes the structures can be transfered into the semiconductor by dry etching in conjunction with epitaxial overgrowth (Fig. 15). This technique allows a most flexible wire and dot definition with the additional advantage to use the wire and dot array directly as an optical filter. Many results reported in this contribution are based on this fabrication method. However, this fabrication method is determined strongly by the damage level induced by etching, especially during dry etching [192,193]. Nevertheless, if low damage etching techniques [194,193] in combination with epitaxial overgrowth [195,196,197,198,199,200,193] are applied, structures down to 50 nm with optical functionality can be realized. Examples of single mode lasers based on wires or dots are given in Sect. 5.

## 4  Basic Considerations

The discussion of the advantages of low dimensional systems with respect to device operation parameters are at first glance based on the changed density of states in those systems [4].

### 4.1  Electronic States and Statistics

For a more detailed description, the interaction between the different bands must be taken into account and can change the density of states appreciably (see the contribution by *Hawrylak* and *Korkusiński* in this volume). The

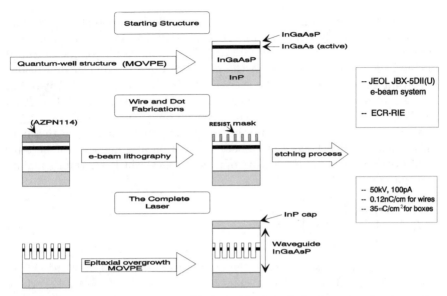

**Fig. 15.** Principal process steps of lithography based dot laser production using dry etching methods for dot array patterning

effects for various dimensions $d$ can be directly seen by studying the density of states $D(E)$. Assuming a spherical dispersion relation $E(k)$, the number of states in $k$-space can be simply derived from the volume $N^d$ of a $d$-dimensional sphere with constant energy $E$ (Eq. 32 in [201]), The density of states $D^d(E)$ follows from the derivative of the volume with respect to energy by

$$D^{(d)}(E) = \frac{\partial N^{(d)}}{\partial E}. \tag{2}$$

Based on the density of states the chemical potential $\mu = (\partial nf/\partial n)_{T,V}$ can be calculated from the free energy $f = F/N$ per particle [201] as a function of carrier density and temperature. The result is the well known density and dimension dependence of the chemical potential

$$\mu \sim n^{2/d}. \tag{3}$$

Using the inverse function of (3), we end up with the well known dependency of the density of transparency on temperature and dimension

$$n_{\text{tr}} \sim T^{d/2}. \tag{4}$$

If we assume in a first approach no further density and temperature dependence of the carrier lifetime in a laser, (4) predicts an extremely weak dependence of the laser threshold current on temperature [4] for low dimensional lasers [202].

The strong structure dependence of the chemical potential according to (3) can be nicely demonstrated in GaAs/AlGaAs quasi-wire structures with different wire widths [195,201,203]. The main result for the chemical potential is a steeper increase of the potential value for smaller dimensions according to the decrease of the average number of states. This behaviour of the chemical potential in low dimensional electron–hole plasmas (LD-EHP) develops continuously to the case of quantum dots and is the reason for the laser threshold density reduction in low dimensional systems. Furthermore this dimension dependence of $k$-space occupation also determines the temperature dependence of the density of transparency, which determines mainly the characteristic temperature $T_0$,

$$T_0^{(d)} = \left( \frac{\partial \ln \left( n_{\mathrm{tr}}^{(d)} \right)}{\partial T} \right)^{-1} \approx \frac{2}{d} \tag{5}$$

if we neglect in a first approach the temperature dependence of the carrier lifetime in the LD-EHP. This results in an increase of $T_0$ for lasers whose active regions have a small dimension $d$.

Based on experimental data of wires with different widths [195], this increase of the characteristic temperature in low dimensional systems can be demonstrated quite systematically by calculating the density of transparency $n_{\mathrm{tr}}$ and its temperature slope (5) [201].

In going to lower dimensions this temperature slope of $n_{\mathrm{tr}}(T)$ becomes continuously larger ending in an infinitely large value of $T_0$ in the 0D-limit [4,201].

## 4.2    Optical Transitions

The optical transitions in a low dimensional gain medium can be described by a wave function

$$\Psi^{(d)i} = u^i \cdot \Phi_z^{(d)i}; d = 3 - 0; i = \mathrm{CB, VB} \tag{6}$$

consisting of a central cell function $u$ for electrons (S-type) or holes (p-type), which describes the polarization of the optical transitions, and an envelope function [204,205,206], which describes the subband selection rule of the optical transitions between different subbands. Together with the density of states (2) the optical gain can be expressed by

$$G(\omega) = \frac{4\pi^2 e^2}{n c m_o^2} \cdot \sum_{c,v} |\, e\boldsymbol{P}_{c,v} \,|^2 \cdot D_{c,v}^d(\omega) \cdot [f^c(\omega) - f^v(\omega)] \,(d = 3, 2, 1, 0). \tag{7}$$

Results on gain calculations can be found in [5].

The matrix elements are given by $\langle S \mid p_x \mid P_x \rangle = \langle S \mid p_y \mid P_y \rangle = \langle S \mid p_z \mid P_z \rangle = P_{cv} = \sqrt{\frac{m_0 E_p}{3}}$, with $E_{\mathrm{p}} = 20.17\,\mathrm{eV}$ for the interband matrix element for InGaAs [207] and $E_p = 22.71$ eV for the interband matrix element for GaAs [205]. Due to the strong $kp$-interaction of the valence bands in 1D and 0D structures, strong mixing occurs between heavy and light hole states. Therefore hole states in wires or dots appear as heavy or light hole like keeping their original pseudo-angular momenta [204,205]. An important contribution to optical gain represents damping mechanisms described by a frequency dependent broadening constant $\Gamma(\omega)$ which broadens according to the relation

$$G(\omega) = \int_{-\infty}^{+\infty} d\omega^* G(\omega, \omega^*) \frac{\Gamma(\omega^*)}{(\omega - \omega^*)^2 + [\Gamma(\omega^*)/2]^2} \tag{8}$$

the originally narrower gain spectrum (7) of the dots. These broadening mechanisms are of homogeneous type (internal fluctuating fields, dot–dot interaction, carrier–carrier scattering) and of inhomogeneous type as dot size fluctuations. For very strong broadening contributions the dot gain can be strongly reduced showing no advantage compared to QW laser gain. This means that the dispersion of dot size widths plays a crucial role in achieving high gain in a dot laser.

## 4.3   Dot Laser Threshold and Laser Dynamics

A simple description of laser operation can be achieved using rate equations for carriers (electrons and holes) and photons [208,209], which describe the interaction of electrons and photons by stimulated and spontaneous emission mechanisms as well as the pump process.

### 4.3.1   Dot Laser Threshold

The stationary solutions of the rate equations lead to a linear dependence of the threshold density $N_{th}$ on losses $\alpha_g$:

$$N_{th} = \frac{\alpha_g}{(\partial G/\partial N)} + N_{tr} \tag{9}$$

and of the output power $S$ on the pump current $I$:

$$S = \frac{1}{G v_{gr} q V} \cdot (I - I_{th}). \tag{10}$$

Assuming a pump term $G_r = I/qV$, we obtain for the threshold current $I_{th}$

$$I_{th} = qV \cdot \frac{N_{th}}{\tau_s}. \tag{11}$$

Assuming a density dependence of the spontaneous lifetime $\tau_s$ as $1/\tau_s = A + B N_{th}$ (with $A$ = linear recombination coefficient, $B$ = bimolecular recombination coefficient) (11) may be replaced by

$$I_{th} = qV N_{th} \cdot (A + B N_{th}). \tag{12}$$

(The meanings of the above used constants are: $\alpha_g$ = optical resonator loss, $G$ = optical gain $\equiv \alpha_g$, $G_r$ = pump rate radiative contribution, $N_{tr}$ = density of transparency, $N_{th}$ = threshold density, $S$ = photon density, $v_{gr}$ = group velocity of light, $V$ = volume of active area.)

The description of the dot or wire laser operation can be obtained by a 2-level (active, passive region) rate equation approach [7,8,9]. Similar to the example of QW lasers the laser threshold current can be expressed in terms of a structural factor $\lambda$ [(13), Fig. 16]:

$$\lambda = N_z \Gamma_x \Gamma_y L \approx N_z L_x L_y \rho_{dot} L \tag{13}$$

with $N_z$ = number of QWs or of dot layer stacks in the growth direction, $\Gamma_x$ = optical filling factor in the $x$-direction, $\Gamma_y$ = optical filling factor in the $y$-direction, $L_x, L_x, L_z$ dot sizes in $x$-, $y$-, $z$-directions, $\rho_{dot}$ = sheet area dot density, $w$ = resonator width, $L$ = resonator length.

The threshold current is then given by

$$I_{th} = q L_z w (C_1 + C_2 \lambda + C_3/\lambda) \tag{14}$$

with constants

$$C_1 = A N_{th1} + 2 B N_{th0} \cdot N_{th1},$$
$$C_2 = A N_{th0} + B N_{th0}^2,$$
$$C_3 = B N_{th1}^2$$

with $N_{th0} = N_{tr} + \alpha_{ac} + \alpha_{wg} \cdot (1-\Gamma)/\Gamma]/(\partial g/\partial N)$ = threshold density contribution (losses and density of transparency) without outcoupling contribution, $N_{th1} = \ln(1/R)/[\Gamma_{zn} \cdot (\partial g/\partial N)]$ (outcoupling), $\Gamma = \Gamma_x \cdot \Gamma_y \cdot N_z \cdot \Gamma_{zn}$ = total optical filling factor, and $\Gamma_{zn}$ = optical filling factor in growth direction ($z$-direction) of one dot layer.

Usually laser thresholds are represented as functions of resonator lengths $L$ and the other structural factors are parametrically used. The plot as a function of the structural parameter $\lambda$ (Fig. 16) gives us a more compact description of the laser threshold $I_{th}$ with respect to geometrical parameters. The threshold current minimum can be obtained in terms of the coefficients $C_1$, $C_2$, and $C_3$. $C_1$ is related to a mixtutre of the outcoupling rate and the internal losses in a resonator. $C_2$ is mainly related to the losses in a resonator and $C_3$ depends mainly on the outcoupling rate in a resonator. From the derivative of $I_{th}$ with respect to $\lambda$, we get the optimum $\lambda_{min}$ value with respect to the minimum threshold current

$$\lambda_{min} = \sqrt{C_3/C_2} \tag{15}$$

and the minimum threshold current by

$$I_{th,min}(\lambda_{min}) = q L_z w \cdot \left( C_1 + 2 \cdot \sqrt{C_2 \cdot C_3} \right). \tag{16}$$

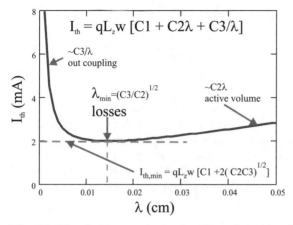

**Fig. 16.** Threshold current as a function of structural parameter $\lambda = N_z \Gamma_x \Gamma_z L$

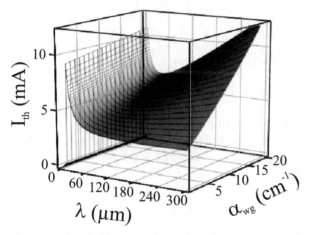

**Fig. 17.** Threshold current as a function of structural parameter $\lambda = N_z \Gamma_x \Gamma_z L$ and optical resonator loss

Besides the monomolecular parameter $A$ and the bimolecular parameter $B$, the optical resonator losses $\alpha_{ac}$, $\alpha_{wg}$ are most prominent material parameters in determining the optimum $\lambda$. For increasing internal losses the minimum value of $\lambda$ decreases and the minimum value of the threshold current increases in a low dimensional laser as shown in Fig. 17. The physical meaning of the minimum threshold current can be seen in more detail if the recombination rates in a laser at the minimum length $L_{min} = \lambda_{min}/(N_z \cdot \Gamma_x \Gamma_y)$ are calculated and give the condition:

$$1 = \frac{B \cdot (N_{th} - N_{th0})^2}{A \cdot N_{th0} + B \cdot N_{th0}^2} \tag{17}$$

with the carrier density at laser threshold given by

$$
\begin{aligned}
N_{th} &= N_{th0} + N_{th1}/\lambda \\
&\equiv \frac{\alpha_{ac} + \left(\frac{1-\Gamma}{\Gamma}\right)\alpha_{wg}}{\partial g/\partial N} + N_{tr} + \frac{1}{\Gamma L} \cdot \ln\frac{1}{R} \, .
\end{aligned}
\tag{18}
$$

From (17) it turns out that the minimum length with respect to threshold current is achieved, if the internal total rate $AN_{th0} + BN_{th0}^2$ without outcoupling contributions in a laser (mainly governed by the active volume) equals the radiative excess rate $B(N_{th} - N_{th0})^2$ (mainly governed by the mirror loss).

In analogy to the description of the threshold of a dot laser in terms of the structural parameter $\lambda$, a similar expression can be obtained in terms of the dot density $\rho_{dot}$

$$
I_{th} = qL_z w (D_1 + D_2\rho_{dot} + D_3/\rho_{dot})
\tag{19}
$$

with constants $D_1$–$D_3$ given by

$$
\begin{aligned}
D_1 &= \frac{\ln(1/R) + L\alpha_{wg}}{(\partial g/\partial N) \cdot \Gamma_{zn}}(A + 2BN_{tr}) \, , \\
D_2 &= N_z L_x L_y L (AN_{tr} + BN_{tr}^2) \, , \\
D_3 &= \left(\frac{\ln(1/R) + L\alpha_{wg}}{(\partial g/\partial N) \cdot \Gamma_{zn}}\right)^2 \frac{1}{L_x L_y N_z L} \cdot B \, .
\end{aligned}
$$

The optimum of the dot density will be achieved for $\rho_{dot,min} = \sqrt{D_3/D_2}$ and the physical meaning of the minimum is the balance between the total rate at density of transparency $AN_{tr} + BN_{tr}^2$ and the radiative excess rate $B(N_{th} - N_{tr})^2$ given by

$$
1 = \frac{B(N_{th} - N_{tr})^2}{AN_{tr} + B \cdot N_{tr}^2} \, .
\tag{20}
$$

The difference $N_{th} - N_{tr}$

$$
N_{th} - N_{tr} = \left(\alpha_{ac} + \frac{1-\Gamma}{\Gamma} \cdot \alpha_{wg} + \frac{\ln(1/R)}{\Gamma L}\right)/(\partial g/\partial N)
\tag{21}
$$

represents the carrier density, which is needed for the resonator gain to cover all internal absorption losses and the mirror loss (outcoupling). In a low dimensional laser, the stationary solutions of (9,11,12) are modified (modification of $G_r$) due to carrier transport processes [210] and carrier relaxation processes [211]. A main result is the approximation of the effective capture time $\tau_{cap}^{eff}$, which is split in a low dimensional laser into a diffusion contribution $\tau_{diff}$ and in a scaled up capture time contribution $\tau_{cap}^{eff}$ [6,7,8,9]

$$
\tau_{cap}^{eff} = \tau_{diff} + \frac{V_b}{V_{ac}} \cdot \tau_{cap}^Q
\tag{22}
$$

which results from the conservation law of particles in the active region/ barrier-scattering processes. This modification of the capture time with the filling factor of the low dimensional structure strongly reduces the carrier capture rate and limits therefore the modulation response as well as the effective gain of the material. The relation (22) can be nicely verified by measuring the modulation response of wire and dot lasers with various ratios of $V_b/V_{ac}$ [8,9] (see also Fig. 24, right). The effect of the capture time (22) on the recombination coefficients $A$, $B$ in the laser threshold equation (12) can be described by a factor $F_B$.

$$F_B = 1 + \frac{\tau_{cap}^{eff}}{\tau_{s,B}^{eff}} \approx 1 + \left[ \tau_{diff} + \frac{\tau_{cap}^Q}{\Gamma} \right] / \tau_{s,B}^{eff}, \tag{23}$$

with $\tau_{cap}^Q$ = LO-phonon based carrier scattering time.

This factor describes the efficiency of carrier capture into the active region of a laser from a reservoir (barrier region) under a certain carrier lifetime $\tau_{s,B}^{eff}$ in the barrier region. The modification of the constants $A$, $B$ is then given by

$$A^* = F_B A + \frac{\tau_{cap}^{eff}}{\tau_{s,B}^{eff} \tau_{esc}}, \tag{24}$$

$$B^* = F_B B. \tag{25}$$

For large filling factors as in QW and bulk lasers, $F_B$ is $\approx 1$. For small filling factors, especially in wire and dot lasers, $F_B$ can become larger than 1. This modifies strongly the threshold current and the quantum efficiency. Compared to lasers with $F_B \approx 1$ the threshold current

$$I_{th}^* = q L_z w \left( C_1^* + C_2^* \lambda + C_3^*/\lambda \right) \tag{26}$$

increases and the quantum efficiency $\eta$

$$\eta^* = \frac{\eta}{F_B} \tag{27}$$

decreases. Therefore, filling factor dependent carrier capture effects described by the factor $F_B$ can counterbalance low dimensional density of state effects as present in wire and dot lasers.

($C_1^*$, $C_2^*$, and $C_3^*$ are calculated according to (14), (15) with coefficients $A^*$, $B^*$.)

A complete picture of $I_{th}$ as a function of the laser length $L$ and the dot density $\rho_{dot}$ is given in Fig. 18. As can be seen, the parameters $L$ and $\rho_{dot}$ must be rather large to obtain low enough thresholds which is due to the small filling factors in the lasers. The respective optimum values for laser length $L_{min}$ (Fig. 19)

$$L_{min}(\rho) = \frac{\lambda_{min}}{N_z L_x L_y} \frac{1}{\rho}$$

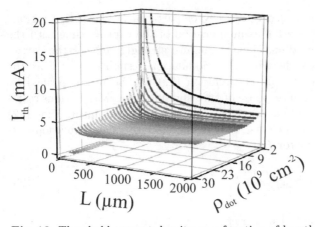

**Fig. 18.** Threshold current density as a function of length and dot density

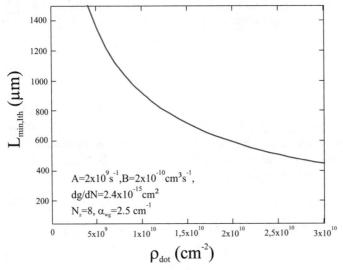

A=2x10$^9$s$^{-1}$,B=2x10$^{-10}$cm$^3$s$^{-1}$,
dg/dN=2.4x10$^{-15}$cm$^2$
N$_z$=8, α$_{wg}$=2.5 cm$^{-1}$

**Fig. 19.** Optimum laser length for minimum threshold current as a function of dot density

$$= \frac{\ln(1/R)}{(\partial g/\partial N)N_z \Gamma_{zn} L_x L_y} \cdot \sqrt{\frac{B}{AN_{th0} + BN_{th0}^2}} \cdot \frac{1}{\rho} \qquad (28)$$

and dot density $\rho_{dot,min}$

$$\rho_{min}(L) = \frac{L\alpha_{wg} + \ln(1/R)}{(\partial g/\partial N)N_z \Gamma_{zn} L_x L_y} \cdot \sqrt{\frac{B}{AN_{tr} + BN_{tr}^2}} \cdot \frac{1}{L} \qquad (29)$$

show a hyperbolic dependence on $\rho_{dot}$ and $L$, respectively.

As the threshold current $I_{th}(L, \rho)$ for fixed length $L$ reaches its minimum for a certain dot density $\rho_{min}(L)$ (see $I_{th}$ v.s. $\rho$ in Fig. 18) the optimization of a dot laser is rather complex. Each dot density $\rho$ possesses its own optimum laser length $L_{min}(\rho)$ and vice versa the same is true for the laser length $L$.

Further important parameters are the ratio between optical losses $\alpha$ and differential gain $(\partial g/\partial N)$ as well as the number of dot stacks $N_z$. From these dependencies it is clear that the number of stacks must be increased and the losses in the resonator must be strongly reduced to obtain low threshold currents at moderate laser lengths and to omit higher level lasing.

### 4.3.2   Modulation Limit

Besides the low threshold of a laser also its ultimate modulation speed is a very important parameter for data communication applications. According to the expected higher gain $g$ and differential gain $\partial g/\partial N$ [5,204] the modulation bandwidth of a dot laser should overcome the modulation bandwidth of a QW laser. The modulation limit in a laser can be given in a compact form by its $K$-factor [212]:

$$K = 4\pi^2 \left(\tau_{ph} + \tau_{cap}^{eff}\right) \tag{30}$$

with

$$\tau_{ph} = \frac{1}{\Gamma v_{gr} \alpha_g} ,$$

which determines the 3 dB-bandwidth of a laser

$$f_{3\,dB} = \frac{2\sqrt{2}\pi}{K} . \tag{31}$$

There are only two time constants which determine $K$. This is the photon lifetime $\tau_{ph}$ (31) and the effective capture time $\tau_{cap}^{eff}$ (22), which an electron–hole pair needs to scatter from the barrier into the quantum dot or wire. The capture time includes transport mechanisms like diffusion, tunneling, and drift from the claddings to the active region of a laser. Furthermore the capture time predominantly determines the gain compression factor. In a three-level system consisting of a barrier reservoir and a two-level active region the damping constant $\epsilon$ and the differential gain of the active region are related to each other by the effective carrier capture time $\tau_{cap}^{eff}$ by

$$\epsilon = \tau_{cap}^{eff} v_{gr} (\partial g/\partial n) \tag{32}$$

which determines the damping in a laser and in addition the ultimate bandwidth (Fig. 20). Especially in low dimensional structures, both time ranges ($\tau_{cap}^{eff}$ and $\tau_{ph}$) can become long due to confinement factor effects and due to transport effects. Very important is the relationship between the gain compression factor $\epsilon$ and differential gain $\partial g/\partial n$ (32). From this relation we see that the gain compression and therefore the maximum modulation speed of

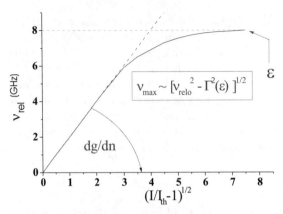

**Fig. 20.** Relaxation frequency as a function of drive current

a low dimensional laser is directly correlated to the differential gain by the capture time $\tau_{cap}^{eff}$: the larger the differential gain the smaller the maximum modulation speed. This relationship explains why low dimensional lasers with high differential gain can show much smaller maximum modulation speed than, e.g. 2D lasers.

### 4.3.3  Key Parameters

Summarizing the above discussion on optimization of low dimensional lasers the following key parameters for laser development can be identified:

- The threshold density $N_{th}$

$$N_{th} = N_{th0} + N_{th1}/\lambda \tag{33}$$

with

$$N_{th1} = \frac{\ln(1/R)}{(\partial g/\partial N)\Gamma_{zn}} \tag{34}$$

$$N_{th0} = [\alpha_{ac} + \alpha_{wg}(1 - \Gamma)/\Gamma]/(\partial g/\partial N) + N_{tr} . \tag{35}$$

- The minimum structural factor $\lambda$

$$\lambda_{min} = \sqrt{C_3/C_2} = \sqrt{\frac{BN_{th1}^2}{AN_{th0} + BN_{th0}^2}} . \tag{36}$$

- The relationship between optimum dot density and optimum laser length

$$\rho_{dot,min} = \frac{\lambda_{min}}{N_z L_x L_y} \cdot \frac{1}{L} , \tag{37}$$

$$L_{min} = \frac{\lambda_{min}}{N_z L_x L_y} \cdot \frac{1}{\rho_{dot}} . \tag{38}$$

- Factor $F_B$ describing carrier capture and its influence on threshold given by

$$F_B = 1 + \frac{\tau_{cap}^{eff}}{\tau_{s,barrier}^{eff}} . \tag{39}$$

- The relation between damping $\epsilon$ and differential gain $\partial g/\partial N$

$$\epsilon = \tau_{cap}^{eff} \cdot v_{gr} \cdot (\partial g/\partial N) \tag{40}$$

with $\Gamma_{zn}$ = confinement factor per dot layer in the growth direction, $\Gamma = \Gamma_x \Gamma_y \Gamma_z \approx N_z L_x L_y \rho_{dot}$.

# 5    Devices and Applications

The advantage of lithography based nanometer fabrication techniques compared to SAD techniques is their high flexibility in defining optical resonators with arbitrary geometry (Figs. 1 and 21). Despite the disadvantage of reduced optical quality of the fabricated dots, the lithography based technique allows the realization of resonators with dots serving as the optical filter and simultaneously as gain media.

## 5.1    Dot- and Wire-DFB-Lasers

Such laser structures are also called complex coupled DFB lasers [14,16,111] [213], as besides a periodic real index function the resonator possesses also a periodic imaginary function. By fine tuning the filling factor (Fig. 14, left and right) keeping the DFB period constant, detailed and systematic dynamic investigations of carrier transport and carrier relaxation can be accomplished which is hardly achieved in SAD structures.

### 5.1.1    Static Characteristics

An impression of the quality of the optical resonators can be obtained by studying sectional views of wire and dot DFB resonators by AFM (Fig. 21, left). From spectra below threshold one can obtain the optical gain in dot DFB lasers as a function of pump current (Fig. 21, right). As expected, the dot laser shows higher differential gain than the 2D laser. The strong barrier–active region interaction of a dot-DFB-laser can be analyzed by measuring the threshold current density as a function of the lateral geometrical resonator filling factor $\Gamma_{geo} = \Gamma_x \Gamma_y$ (Fig. 22), which can be realized by fine tuning the filling factor by lithography based nanometer structuring techniques (Fig. 14). For a given filling factor, the totally injected current splits into a barrier current (carrier recombination in the waveguide and cladding region) and an active region current (radiative and non-radiative carrier recombination in the active region). To simplify the analysis, we assumed in the case of the

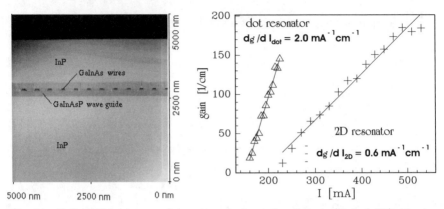

**Fig. 21.** *Left*: AFM picture of a sectional view of a GaInAs/GaInAsP/InP wire DFB laser with a DFB period of $\Lambda = 240$ nm. *Right*: Comparison of differential gain values $dg/dI$ for QW- and dot-laser

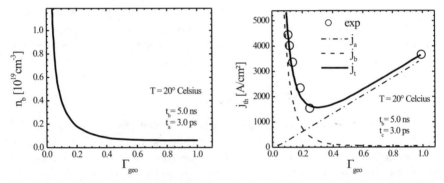

**Fig. 22.** *Left*: Calculated barrier density of a box DFB laser for various dot filling factors. *Right*: Measured total current density (*open circles*) of box DFB lasers for various array filling factors and calculated total current $j_t$ (*solid line*). The barrier current $j_b$ (*dashed line*) and the active current $j_a$ contributions (*dashed dotted line*) are also shown. We assumed a barrier lifetime $t_b \equiv \tau_{s,B}^{\mathrm{eff}}$ of 5.0 ns, and a LO-phonon-carrier scattering time $t_c \equiv \tau_{\mathrm{cap}}^{Q}$ of 3 ps (22)

dot-DFB lasers spontaneous recombination (radiative and non-radiative) in the barrier regions and radiative recombination in the dots. As can be seen from the different contributions of barrier and active region current even at the threshold current minimum ($\Gamma_{\mathrm{geo}} \approx 0.25$ for InGaAs/InGaAsP-dot lasers), the barrier current has roughly the same magnitude as the active region current and increases steeply for non-optimal filling factors due to high barrier carrier density (Fig. 22). For higher filling factors, the dot laser approaches the QW-laser case and the current is mainly determined by the active region current.

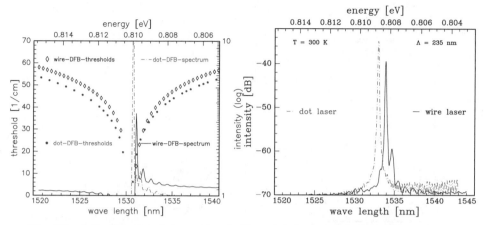

**Fig. 23.** *Left*: Calculated laser threshold and below threshold spectra of wire (*solid line, diamonds*) and dot (*dashed line* and *solid dots*) DFB lasers. Wire: $\kappa_{\text{index}} = 33\,\text{cm}^{-1}$, $\kappa_{\text{gain}} = 80\,\text{cm}^{-1}$ Dot: $\kappa_{\text{index}} = 33\,\text{cm}^{-1}$, $\kappa_{\text{gain}} = 130\,\text{cm}^{-1}$. *Right*: Experimental wire (*solid line*) and dot (*dashed line*) DFB spectra at room temperature

### 5.1.2   Spectra of Dot DFB Lasers

Dot and wire arrays with periodic distances between wires or dots adjusted for Bragg emission wavelength (here 1550 nm) serve as optical filters with complex coupling mechanism between the light field and grating. In Fig. 23, left, the calculated mode threshold (dot filter: solid circles, wire filter: open diamonds) of a dot and a wire DFB laser is depicted. The dot DFB laser shows a stronger asymmetry of the threshold values and therefore also a stronger side mode suppression ratio, which is due to the higher imaginary contribution $\kappa_g$ of the total coupling factor $\kappa_t = \sqrt{\kappa_{\text{index}}^2 + \kappa_{\text{gain}}^2}$ in the dot DFB filter. This theoretical expectation is demonstrated in Fig. 23, right: compared to a conventional QW DFB laser or even to a wire DFB laser, the dot DFB laser shows strongly improved side mode suppression ratio of the order of 40 dB [16] even in the case of a very wide transversal resonator cavity of 64 µm. This makes dot DFB lasers suitable candidates for improved DFB laser emission and for low cost optical single mode laser devices as fabrication tolerances to achieve transversal single mode in these lasers are strongly relaxed due to the two-dimensional DFB-grating.

### 5.2   Modulation Dynamics

From the analysis of the rate equations [7,8,9,209] one finds the well known modulation band width, which depends on the $K$-factor (30) by (31). The modulation band width depends on the one hand on the differential gain $\partial g / \partial n$, which determines the relaxation oscillation frequency of a laser and

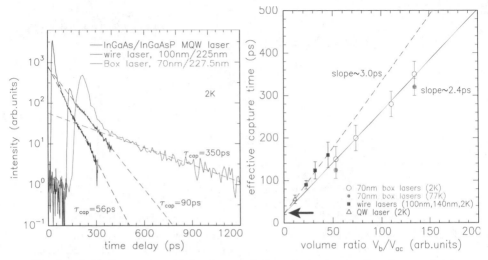

**Fig. 24.** *Left*: Comparison of the time evolution of quantum well-, wire-, and dot-DFB-lasers as function of time. *Right*: Effective capture time vs $V_B/V_{ac} \approx \Gamma$ for wire (*solid circles*) and box lasers (*open circles* 2 K, *solid circles* 77 K) [8,9]. The *arrow* gives the QW laser reference

on the other hand on the damping of the relaxation oscillations mainly determined by the gain compression factor $\epsilon$. These dependencies can be summarized by the $K$-factor given in (30). A further analysis of the 2-level model [7] yields a relation between gain compression factor $\epsilon$ and differential gain $\partial g/\partial n$ (32) by the strongly geometry dependent effective capture time $\tau_{cap}^{eff}$ (22). This relation represents a strong restriction of the maximum achievable modulation band width and leads to strong constraints for the optimization of the modulation band width of a dot/wire laser. An attempt to increase the modulation band width by an increase of the differential gain results simultaneously in an increase of the gain compression via the effective capture time (22) in a laser structure. This strong dependence of the laser response on the filling factor can be seen directly in Fig. 24, where we compare QW wire and dot DFB lasers with respect to their modulation response. The lower the dimension the stronger capture time effects on delay-on time and pulse width of the lasers.

## 5.3 Chirp

The current modulation of a laser enforces the modulation of the carrier density, which result in a modulation of the emission wavelength. In Fig. 25, right side, we depict the wavelength chirp (triangles) during the laser emission after a short ps excitation pulse. Due to the phase shift between the intensity pulse and carrier density pulse, one observes mainly a red shift of the emission according to the falling edge of the carrier pulse. The overall wavelength shift

$$\Delta\nu(t) = -\frac{\alpha}{2\pi}\left[\frac{dlnP(t)}{dt} + \kappa P(t)\right] \tag{41}$$

consists of a structure independent shift according to a change in photon density (and therefore carrier density) and a structure dependent shift expressed by the factor $\kappa$ with $P$ = optical power of the laser, $\kappa = \frac{2\Gamma\epsilon}{V_{ac}\cdot\eta\cdot h\nu}$, $\alpha = -\frac{4\pi}{\lambda}\cdot\frac{\partial n/\partial N}{\partial g/\partial N}$, $\Delta\nu$ = dynamical frequency shift. The overall wavelength shift in low dimensional lasers not only depends on the carrier density change in the active region; also the carrier density change in the passive waveguide region contributes to a refractive index change and therefore to a wavelength change. In Fig. 25, left we compare the wavelength shift of QW-, wire- and dot DFB lasers. The QW and the wire DFB lasers show comparable shifts as expected from the carrier density changes in the active region and in the barrier. The wavelength change of the dot DFB laser, however, is nearly zero. The strong chirp reduction in dot DFB lasers remains, even if we change the pump power by a factor of 4. As the barrier filling factor is larger in dot DFB lasers than in wire or QW DFB lasers, also in dot DFB lasers a chirp effect should occur. However, a more detailed analysis reveals that the barrier carriers experience an increase in carrier temperature with increased pump level (Fig. 25, right). We attribute this to Auger heating [214]. Taking into account the measured increase in carrier temperature in the total refractive index change

$$\Delta n_{\mathrm{eff}} = \frac{\partial n}{\partial N}\Delta N + \frac{\partial n}{\partial T}\Delta T \tag{42}$$

we find nearly a balance between carrier induced refractive index change and temperature induced refractive index change leading to the low observed chirp values in dot DFB lasers. An additional effect comes up in DFB lasers with complex coupling, where the linewidth enhancement factor can be controlled to a certain degree by the different amounts of the real and the imaginary part of the grating coupling factor [111].

## 5.4    High Power Lasers

Output power is another very important characteristic defining the usefulness of a laser diode in modern applications. Semiconductor high brightness lasers (lasers with high power and high beam quality and small emission band width) require a low filamentation tendency and a low saturation intensity of the active gain medium. Filamentation in a semiconductor laser occurs under high field intensities. High light field intensities induce, by carrier-photon interaction, waveguide effects (generation of filaments), which can enhance the light field intensity locally up to the destruction of the laser structure. If the

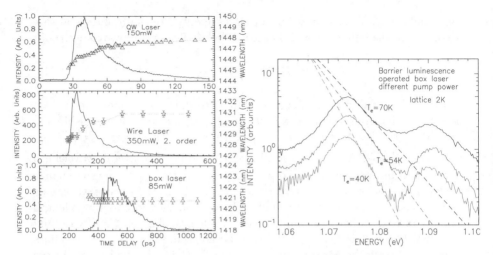

**Fig. 25.** *Left*: Comparison of emission intensity and wavelength chirp of QW, wire, and box lasers as a function of delay time. *Right*: Comparison of the barrier emission of a box laser for different pump powers

average gain in a resonator is low, as in a dot laser, these filamentation effects are more efficiently suppressed. Furthermore to obtain high output power densities on the outcoupling mirror, low confinement factors are wanted. The field intensity on the mirror and therefore the mirror temperature depends on the effective modewidth $d_{\mathrm{mode}}$ given approximately by

$$d_{\mathrm{mode}} = d_{ac}/\Gamma \quad \text{with} \quad d_{ac} = \text{width of active region}. \tag{43}$$

This leads to a relationship between the output power $P_{\mathrm{COD,out}}$ ($P_{\mathrm{COD,out}}$ represents the output power where catastrophic optical damage (COD) of the mirror occurs) and the internal power $P_{\mathrm{int}}$ given by [215,216,217]

$$P_{\mathrm{COD,out}} = \frac{d_{ac} \cdot w}{\Gamma_{ac}} \cdot \frac{1-R}{1+R} \cdot P_{\mathrm{int}}, \tag{44}$$

with $d_{ac}$ = thickness of the active region, $w$ = waveguide width, $R$ = intensity reflectivity of the mirror facet.

From this equation we see that besides a low appropriate output reflectivity $R$, a small confinement factor $\Gamma_a$ of a laser (as present in a dot laser) is helpful to suppress oxidation processes. Reduced nonradiative surface recombination may help to decrease facet overheating and thus to increase the power for catastrophic optical mirror damage. A low confinement factor not only improves the COD behaviour but also results in a low average gain, which reduces the tendency to build up higher optical modes in a resonator (tendency for filamentation).

Some advantages (beam quality, COD stability) are expected when using quantum dot structures for high power applications. Especially the more or less optically passive resonator of a dot laser should result in a beam of very low filamentation and good $M^2$-values. Moreover, owing to the lower filling factor, such lasers should possess lowest intrinsic optical losses, which are strongly needed to allow for a long cavity laser design [218] to meet the requirements of sufficient heat spreading in a high power laser. As the low filling factor is also responsible for a gain saturation at quite low current densities, high power can only be expected from lasers with large dot density or a long cavity. Nevertheless, quantum dot lasers emitting several watts of light have been reported recently. *Heinrichsdorff* et al. [219] have achieved output powers of 3.7 W on 2 mm long devices at a wavelength of 1100 nm and even 4.5 W at 1068 nm after having optimized their laser structures with regard to quantum efficiency and dot density. Impressively high lateral dot densities of more than $10^{11}$ cm$^{-2}$ could be measured on their MOVPE grown material, and the filling factor was further increased by using stacks of 3 or 6 such dot layers. Later, the same group could improve their laser fabrication process thus achieving the same output power at 1140 nm on shorter devices (1.3 mm) [220], which exhibited extremely low internal losses around 2 cm$^{-1}$.

On the other hand, demonstrations of good beam quality and high COD-power are not so clear. This might be due to the fact that resonator losses strongly depend on waveguide design (doping level, waveguide geometry) which was possibly not on the focus of present dot laser structures. Another reason may be the relatively large effective capture time $\tau_{cap}^{eff}$, which results in a carrier accumulation in the barrier of the laser leading to a refractive index change and therefore possibly to a degradation of the beam quality.

# 6   Conclusions

In this contribution, we discussed physical aspects, fabrication aspects, and some applications of dot lasers. From the size reduction we found not only density of states related phenomena, but also transport related phenomena influencing dot laser operation. From this analysis we saw that the filling factor of a dot laser is most prominently characterized by the effective carrier capture time of dots, which strongly determines laser threshold, quantum efficiency, and modulation band width in a counterbalancing scheme with respect to the density of states. For obtaining high power or high modulation band width, the filling factor in dot lasers must be increased showing in a certain respect an opposite optimization behaviour compared to QW lasers. SAD fabrication methods are found most promising in obtaining dots of high quality. Lithography based fabrication methods are interesting fabrication techniques despite the disadvantage of reduced dot quality due to their higher flexibility in defining resonator structures. Resonators with very high side mode suppression ratios can be realized in dot arrays as 2D-DFB gratings. The combination of SAD growth with lateral patterning techniques possibly combines the flexibility of lithography based methods of resonator design with the high quality growth of SADs.

The dynamical analysis reveals a severe limitation in maximum modulation response, which is principally based on the relationship between damping and gain in low dimensional lasers. The dynamic emission analysis shows very low chirp behaviour of dot DFB lasers making these lasers suitable for data communication at longer wavelengths.

## Acknowledgements

The authors would like to thank M. H. Pilkuhn for steady support. For technical assistance we would like to thank J. Porsche, M. Geiger, T. Riedl, and C. Geng for assistance in MOVPE growth, and J. Wang, U. Griesinger, P. Burkard, and H. Gräbeldinger for their support in device processing and characterization. This work was financially supported by the Deutsche Forschungsgemeinschaft and the European Community Esprit Project NANOPT.

# References

1. J. H. C. Casey, M. Panish: *Heterostructure Lasers, Part A: Fundamental Principles* (Academic, New York 1984)
2. J. H. C. Casey, M. Panish: *Heterostructure Lasers, Part B: Materials and Operating Characteristics* (Academic, New York 1984)
3. D. Bimberg, M. Grundmann, N. N. Ledentsov: *Quantum Dot Heterostructures* (Wiley, Chichester 1998)
4. Y. Arakawa, H. Sakaki: Appl. Phys. Lett. **40**, 939 (1982)
5. M. Asada, Y. Miyamoto, Y. Suematsu: IEEE J. Quant. Electron. **QE-22**, 1915 (1986)
6. J. Wang, U. Griesinger, F. Adler, H. Schweizer, V. Härle, F. Scholz: Appl. Phys. Lett. **69**, 287 (1996)
7. J. Wang, U. Griesinger, H. Schweizer: Appl. Phys. Lett. **69**, 1585 (1996)
8. J. Wang, U. Griesinger, H. Schweizer: Appl. Phys. Lett. **70**, 1152 (1997)
9. J. Wang, U. Griesinger, M. Geiger, F. Scholz, H. Schweizer: IEEE J. Sel. Topics Quant. Electron. **3**, 223 (1997)
10. H. Benisty, C. Sotomayor-Torres, C. Weisbuch: Phys. Rev. B **44**, 10945 (1991)
11. M. Asada, Y. Miyamoto, Y. Suematsu: Jap. J. Appl. Phys. **24**, L95 (1985)
12. Y. Miyamoto, Y. Miyake, M. Asada, Y. Suematsu: IEEE J. Quant. Electron. **QE-25**, 2001 (1989)
13. Y. Miyake, H. Hirayama, S. Arai, Y. Miyamoto, Y. Suematsu: IEEE Phot. Technol. Lett. **3**, 191 (1991)
14. U. A. Griesinger, H. Schweizer, V. Härle, J. Hommel, F. Adler, F. Barth, B. Höhing, B. Klepser, F. Scholz: Photon. Technol. Lett. **7**, 953 (1996)
15. H. Hirayama, K. Matsunaga, M. Asada, Y. Suematsu: Electron. Lett. **30**, 142 (1994)
16. U. A. Griesinger, H. Schweizer, S. Kronmüller, M. Geiger, D. Ottenwälder, F. Scholz, M. H. Pilkuhn: IEEE Phot. Technol. Lett. **8**, 587 (1996)

17. J. Marzin, J. Gerard, A. Izraël, D. Barrier, G. Bastard: Phys. Rev. Lett. **73**, 716 (1994)
18. D. S. L. Mui, D. Leonard, L. A. Coldren, P. M. Petroff: Appl. Phys. Lett. **66**, 1620 (1995)
19. D. Bimberg, N. Ledentsov, M. Grundmann, N. Kirstaedter, O. G. Schmidt, M. H. Mao, V. M. Ustinov, A. Y. Egorov, A. E. Zhukov, P. S. Kop'ev, Z. I. Alferov, S. S. Ruvimov, U. Gösele, J. Heydenreich: Jap. J. Appl. Phys. **35**, 1311 (1996)
20. M. Grundmann, R. Heitz, N. Ledentsov, O. Stier, D. Bimberg, V. M. Ustinov, P. S. Kop'ev, Z. I. Alferov, S. S. Ruvimov, P. Werner, U. Gösele, J. Heydenreich: Superlatt. Microstruct. **19**, 81 (1996)
21. K. Nishi, R. Mirin, D. Leonard, G. Medeiros-Ribeiro, P. M. Petroff, A. C. Gossard: J. Appl. Phys. **80**, 3466 (1996)
22. L. Brusaferri, S. Sanguinetti, E. Grilli, M. Guzzi, A. Bignazzi, F. Bogani, L. Carraresi, M. Colocci, A. Bosacchi, P. Frigeri, S. Franchi: Appl. Phys. Lett. **69**, 3354 (1996)
23. A. Sasaki: J. Cryst. Growth **160**, 27 (1996)
24. S. Jeppesen, M. S. Miller, D. Hessman, B. Kowalski, I. Maximov, L. Samuelson: Appl. Phys. Lett. **68**, 2228 (1996)
25. F. Adler, M. Geiger, A. Bauknecht, F. Scholz, H. Schweizer, M. H. Pilkuhn, B. Ohnesorge, A. Forchel: J. Appl. Phys. **80**, 4019 (1996)
26. F. Heinrichsdorff, M. Mao, N. Kirstaedter, A. Krost, D. Bimberg, A. O. Kosogov, P. Werner: Appl. Phys. Lett. **71**, 22 (1997)
27. H. Marchand, P. Desjardins, S. Guillon, J. Paultre, Z. Bougrioua, R. Y. Yip, R. A. Masut: Appl. Phys. Lett. **71**, 527 (1997)
28. R. Tsui, R. Zhang, K. Shiralagi, H. Goronkin: Appl. Phys. Lett. **71**, 3254 (1997)
29. M. Geiger, A. Bauknecht, F. Adler, H. Schweizer, F. Scholz: J. Cryst. Growth **170**, 558 (1997)
30. K. Nishi, M. Yamada, T. Anan, A. Gomyo, S. Sugou: Appl. Phys. Lett. **73**, 526 (1998)
31. S. Rouvimov, Z. Liliental-Weber, W. Swider, J. Washburn, E. R. Weber, A. Sasaki, A. Wakahara, Y. Furkawa, T. Abe, S. Noda: J. Electron. Mater. **27**, 427 (1998)
32. Hao Lee, R. Lowe-Webb, Weidong Yang, P. C. Sercel: Appl. Phys. Lett. **72**, 812 (1998)
33. B. Legrand, B. Grandidier, J. P. Nys, D. Stievenard, J. M. Gerard, V. Thierry-Mieg: Appl. Phys. Lett. **73**, 96 (1998)
34. Y. Tang, D. H. Rich, I. Mukhametzhanov, P. Chen, A. Madhukar: J. Appl. Phys. **84**, 3342 (1998)
35. H. Ishikawa, H. Shoji, Y. Nakata, K. Mukai, M. Sugawara, M. Egawa, N. Otsuka, Y. Sugiyama, T. Futsatsugi, N. Yokoyama: J. Vac. Sci. Technol. A **16**, 794 (1998)
36. N. Carlsson, W. Seifert, A. Petersson, P. Castrillo, M. E. Pistol, L. Samuelson: Appl. Phys. Lett. **65**, 3093 (1994)
37. C. M. Reaves, R. I. Pelzel, G. C. Hsueh, W. H. Weinberg, S. P. DenBaars: Appl. Phys. Lett. **69**, 3878 (1996)
38. W. Seifert, N. Carlsson, A. Petersson, L. Wernersson, L. Samuelson: Appl. Phys. Lett. **68**, 1684 (1996)

39. Jong-Won Lee, A. T. Schremer, D. Fekete, J. R. Shealy, J. M. Ballantyne: J. Electron. Mater. **26**, 1199 (1997)
40. C. Pryor, M. Pistol, L. Samuelson: Phys. Rev. B **56**, 10404 (1997)
41. M. K. Zundel, N. Y. Jin-Phillipp, F. Phillipp, K. Eberl, T. Riedl, E. Fehrenbacher, A. Hangleiter: Appl. Phys. Lett. **73**, 1784 (1998)
42. J. Porsche, A. Ruf, M. Geiger, F. Scholz: J. Cryst. Growth **195**, 591 (1998)
43. J. Porsche, F. Scholz: J. Cryst. Growth **221**, 571 (2000)
44. J. Porsche, M. Ost, T. Riedl, A. Hangleiter, F. Scholz: Mater. Sci. Engin. B **74**, 263 (2000)
45. Y. M. Manz, O. G. Schmidt, K. Eberl: Appl. Phys. Lett. **76**, 3343 (2000)
46. J. Oshinowo, M. Nishioka, S. Ishida, Y. Arakawa: Appl. Phys. Lett. **65**, 1421 (1994)
47. R. Nötzel, T. Fukui, H. Hasegawa, J. Temmyo, T. Tamamura: Appl. Phys. Lett. **65**, 1365 (2854)
48. Y. Sugiyama, Y. Sakuma, S. Muto, N. Yokoyama: Appl. Phys. Lett. **67**, 256 (1995)
49. M. Kitamura, M. Nishioka, J. Oshinowo, Y. Arakawa: Appl. Phys. Lett. **66**, 3663 (1995)
50. K. Mukai, N. Ohtsuka, H. Shoji, M. Sugawara: Appl. Phys. Lett. **68**, 3013 (1996)
51. H. Shoji, Y. Nakata, K. Mukai, Y. Sugiyama, M. Sugawara, N. Yokoyama, H. Ishikawa: Jap. J. Appl. Phys. **35**, L903 (1996)
52. R. Mirin, A. Gossard, J. Bowers: Electron. Lett. **32**, 1733 (1996)
53. K. Kamath, P. Bhattacharya, T. Sosnowski, T. Norris, J. Phillips: Electron. Lett. **32**, 1374 (1996)
54. K. Mukai, N. Ohtsuka, M. Sugawara: Appl. Phys. Lett. **70**, 2416 (1997)
55. Dong Pan, Y. P. Zeng, J. Wu, H. M. Wang, C. H. Chang, J. M. Li, M. Y. Kong: Appl. Phys. Lett. **70**, 2440 (1997)
56. A. A. Darhuber, V. Holy, J. Stangl, G. Bauer, A. Krost, F. Heinrichsdorff, M. Grundmann, D. Bimberg, V. M. Ustinov, P. S. Kop'ev, A. O. Kosogov, P. Werner: Appl. Phys. Lett. **70**, 955 (1997)
57. E. Kuramochi, J. Temmyo, T. Tamamura: Appl. Phys. Lett. **71**, 1655 (1997)
58. R. Leon, Yong Kim, C. Jagadish, M. Gal, J. Zou, D. J. H. Cockayne: Appl. Phys. Lett. **69**, 1888 (1996)
59. C. Lobo, R. Leon: J. Appl. Phys. **83**, 4168 (1998)
60. G. Park, O. B. Shchekin, D. L. Huffaker, D. G. Deppe: Appl. Phys. Lett. **73**, 3351 (1998)
61. D. Lüerssen, R. Bleher, H. Kalt, H. Richter, T. Schimmel, A. Rosenauer, D. Litvinov, A. Kamilli, D. Gerthsen, B. Jobst, K. Ohkawa, D. Hommel: Phys. Stat. Solidi B **178**, 189 (2000)
62. T. Passow, H. Heinke, J. Falta, K. Leonardi, D. Hommel: Appl. Phys. Lett. **77**, 3544 (2000)
63. M. Strassburg, I. Krestnikov, A. Göldner, V. Kutzer, A. Hoffmann, N. Ledentsov, Z. Alferov, D. Litvinov, A. Rosenauer, D. Gerthsen: Proc. ICPS **1**, 70 (1998)
64. D. Hommel: 2002, presentation at ISBLLED March 2002 in Cordoba, Spain
65. M. Klude, T. Passow, R. Kröger, D. Hommel: Electron. Lett. **37**, 1119 (2001)
66. T. Kim, D. Choo, D. Lee, M. Jung, J. Cho, K. Yoo, S. Lee, K. Seo, J. Furdyna: Solid State Commun. **122**, 229 (2002)

67. S. Zaiztsev, N. Gordeev, Y. Shernyakov, V. Ustinov, A. Zhukov, A. Egorov, M. Maximov, P. Kop'ev, Z. Alferov, N. Ledentsov, N. Kirstaedter, D. Bimberg: Superlatt. Microstruct. **21**, 559 (1997)
68. V. Ustinov, A. Egorov, A. Kovsh, A. Zhukov, M. Maximov, A. Tsatsul'nikov, N. Gordeev, S. Zaiztsev, Y. Shernyakov, N. Bert, P. Kop'ev, Z. Alferov, N. Ledentsov, J. Böhrer, D. Bimberg, A. Kosogov, P. Werner, U. Gösele: J. Cryst. Growth **175**, 689 (1997)
69. D. Huffaker, G. Park, Z. Zou, O. Shchekin, D. Deppe: Appl. Phys. Lett. **73**, 2564 (1998)
70. M. Z. K. Eberl, N. Phillipp, T. Riedel, E. Fehrenbacher, A. Hangleiter: Appl. Phys. Lett. **73**, 1784 (1998)
71. T. Riedel, E. Fehrenbacher, A. Hangleiter, M. Z. K. Eberl: Appl. Phys. Lett. **73**, 3730 (1998)
72. J. Porsche, A. Ruf, M. Geiger, F. Scholz: J. Cryst. Growth **195**, 591 (1998)
73. N. Chand, E. Becker, J. van der Ziel, S. Chu, N. Dutta: Appl. Phys. Lett. **58**, 1704 (1991)
74. D. Huffaker, O. Baklenov, L. Graham, B. Streetmann, D. Deppe: Appl. Phys. Lett. **70**, 2356 (1997)
75. S. Fafard, K. Hinzer, S. Raymond, M. Dion, J. McCaffrey, Y. Feng, S. Charbonneau: Science **274**, 1350 (1996)
76. K. Hinzler, S. Fafard, A. Springthorpe, J. Arlett, E. Griswold, Y. Feng, S. Charbonneau: J. Phys. E MSS **8**, 68 (1997)
77. F. Heinrichsdorff, M. Mao, N. Kirstaedter, A. Krost, D. Bimberg, A. Kosogov, P. Werner: Appl. Phys. Lett. **71**, 22 (1997)
78. G. Liu, A. Stintz, H. Li, K. Malloy, L. Lexter: Electron. Lett. **35**, 1163 (1999)
79. H. Saito, K. Nishi, Y. Sugimoto, S. Sugou: Electron. Lett. **35**, 1561 (1999)
80. Z. Xu, M. Wassermeier, Y. Li, P. Petroff: Appl. Phys. Lett. **60**, 586 (1992)
81. E. Kapon, K. Kash, E. M. Clausen Jr., D. Wang, E. Colas: Appl. Phys. Lett. **60**, 477 (1992)
82. M. Walther, E. Kapon, D. Wang, E. Colas, L. Nunes: Phys. Rev. B **45**, 6333 (1992)
83. For $III_A$-planes the group III atoms built up the surface and for $III_B$-planes the group V atoms form the surface. Therefore the growth rate of III-V-crystals is enhanced on $III_A$-planes compared to $III_B$-planes
84. E. Kapon: *Quantum Well Lasers – Quantum Wire Semiconductor Lasers*, S. Zory (Ed.) (Academic, New York 1993)
85. B. Maile, A. Forchel, R. Germann, J. Straka, L. Korte, C. Thanner: Appl. Phys. Lett. **57**, 807 (1990)
86. G. Lehr, R. Bergmann, R. Rudeloff, F. Scholz, H. Schweizer: Appl. Phys. Lett. **61**, 517 (1992)
87. U. A. Griesinger, S. Kronmüller, M. Geiger, D. Ottenwälder, F. Scholz, H. Schweizer: J. Vac. Sci. Technol. B **14** (**6**), 4058 (1996)
88. A. Forchel, P. Ils, K. Wang, O. Schilling, R. Steffen, J. Oshinowo: Microelectron. Engin. **32**, 317 (1996)
89. B. Maile: Dissertation, Universität Stuttgart 1990
90. R. Germann: Dissertation, Universität Stuttgart 1990
91. M. Burkard, C. Geng, A. Mühe, F. Scholz, H. Schweizer, F. Phillipp: Appl. Phys. Lett. **70**, 1290 (1997)
92. J. Cibert, P. Petroff, G. Dolan, S. Pearton, J. English: Appl. Phys. Lett. **49**, 1275 (1986)

230     Heinz Schweizer et al.

93. H. Leier, B. Maile, A. Forchel, G. Weimann, W. Schlapp: Microelectron. Engin. **11**, 43 (1990)
94. H. Leier, A. Forchel, G. Hörcher, J. Hommel, S. Bayer, H. Rothfritz, G. Weimann, W. Schlapp: J. Appl. Phys. **67**, 1805 (1990)
95. H. Leier, A. Forchel, B. Maile, G. Mayer, J. Hommel, G. Weimann, W. Schlapp: Appl. Phys. Lett. **56**, 48 (1990)
96. C. Vieu, M. Schneider, G. Benassayag, R. Planel, L. Birotheau, J. Marzin, B. Descouts: J. Appl. Phys. **71**, 5012 (1992)
97. F. E. Prins, G. Lehr, H. Schweizer, G. W. Smith: Appl. Phys. Lett. **63**, 1402 (1993)
98. F. D. Prins, G. Lehr, M. Burkard, S. Y. Nikitin, H. Schweizer, G. W. Smith: Jap. J. Appl. Phys. **32**, 6228 (1993)
99. M. Burkard, U. A. Griesinger, A. Menschig, H. Schweizer, H. Klein, G. Böhm, G. Tränkle, G. Weimann: J. Vac. Sci. Technol. B **12**, 3677 (1994)
100. W. Beinstingl, Y. Li, H. Weman, J. Merz, P. Petroff: J. Vac. Sci. Technol. B **9**, 3479 (1991)
101. T. Fukui, H. Saito, Y. Tokura: Appl. Phys. Lett. **55**, 1958 (1989)
102. J. Gaines, P. Petroff, H. Kroemer, R. Simes, R. Geels, J. English: J. Vac. Sci. Technol. B **6**, 1378 (1988)
103. M. Miller, C. Pryor, H. Weman, L. Samoska, H. Kroemer, P. Petroff: J. Cryst. Growth **111**, 323 (1991)
104. M. Miller, H. Weman, C. Pryor, M. Krishnamurthy, P. Petroff, H. Kroemer, J. Merz: Phys. Rev. Lett. **68**, 3464 (1992)
105. H. Shoji, Y. Nakata, K. Mukai, Y. Sugiyama, M. Sugawara, N. Yokoyama, H. Ishikawa: Jap. J. Appl. Phys. **35**, L903 (1996)
106. N. Kirstaedter, N. Ledenstov, M. Grundmann, D. Bimberg, V. Ustinov, S. Rivimov, M. Maximov, P. Kop'ev, Z. Alferov, U. Richter, P. Werner, U. Goesele, J. Heydenreich: Electron. Lett. **30**, 1416 (1994)
107. N. Kirstaedter, D. Bimberg, N. Ledentsov, Z. Alferov, P. Kop'ev, V. Ustinov: Appl. Phys. Lett. **69**, 1226 (1996)
108. H. Shoji, Y. Nakata, K. Mukai, Y. Sugiyama, M. Sugawara, N. Yokoyama, H. Ishikawa: IEEE J. Sel. Topics Quant. Electron. **3**, 188 (1997)
109. D. Bimberg, N. Kirstaedter, N. Ledentsov, Z. Alferov, P. Kop'ev, V. Ustinov: IEEE J. Sel. Topics Quant. Electron. **3**, 196 (1997)
110. Y. Nakamura, O. Schmidt, N. Jin-Phillip, S. Kiravittaya, C. Müller, K. Eberl, H. Gräbeldinger, H. Schweizer: J. Cryst. Growth **242**, 339 (2002)
111. R. Schreiner, J. Wiedmann, W. Coenning, J. Porsche, J. Gentner, M. Berroth, F. Scholz, H. Schweizer: Electron. Lett. **35**, 146 (1999)
112. F. C. Frank, J. H. van der Merwe: Proc. Soc. London A **198**, 205 (1949)
113. M. Volmer, A. Weber: Z. Phys. Chem. **119**, 277 (1926)
114. I. N. Stranski, L. Krastanow: Sitzungsber. Kais. Akad. Wiss. Wien/Math.-Naturwiss. Kl. 2b **146**, 797 (1938)
115. S. Lee, I. Daruka, C. S. Kim, A. L. Barabasi, J. L. Merz, J. K. Furdyna: Phys. Rev. Lett. **81**, 3479 (1998)
116. R. Leon, C. Lobo, A. Clark, R. Bozek, A. Wysmolek, A. Kurpiewski, M. Kaminska: J. Appl. Phys. **84**, 248 (1998)
117. P. W. Fry, L. Harris, S. R. Parnell, J. J. Finley, A. D. Ashmore, D. J. Mowbray, M. S. Skolnick, M. Hopkinson, G. Hill, J. C. Clark: J. Appl. Phys. **87**, 615 (2000)

118. A. E. Zhukov, A. R. Kovsh, V. M. Ustinov, A. Y. Egorov, N. N. Ledentsov, A. F. Tatsulnikov, V. M. Maximov, S. V. Zaitsev, Y. M. Shernyakov, A. V. Lunev, P. S. Kopev, Zh. I. Alferov: Semicond. **33**, 1013 (1999)
119. N. N. Ledentsov, V. A. Shchukin, M. Grundmann, N. Kirstaedter, J. Böhrer, O. Schmidt, D. Bimberg, V. M. Ustinov, A. Y. Egorov, A. E. Zhukov, P. S. Kopev, S. V. Zaitsev, N. Y. Gordeev, Z. I. Alferov, A. I. Borovkov, A. O. Kosogov, S. S. Ruvimov, P. Werner, U. Gösele, J. Heydenreich: Phys. Rev. B **54**, 8743 (1996)
120. H. Shoji, Y. Nakata, K. Mukai, Y. Sugiyama, M. Sugawara, N. Yokoyama, H. Ishikawa: Electron. Lett. **32**, 2023 (1996)
121. P. M. Smowton, E. Herrmann, Y. Ning, H. D. Summers, P. Blood, M. Hopkinson: Appl. Phys. Lett. **78**, 2629 (2001)
122. G. S. Solomon, J. A. Trezza, A. F. Marshall, J. S. Harris, Jr.: Phys. Rev. Lett. **76**, 952 (1996)
123. M. K. Zundel, P. Specht, K. Eberl, N. Y. Jin-Phillipp, F. Phillipp: Appl. Phys. Lett. **71**, 2972 (1997)
124. J. Porsche, M. Ost, F. Scholz, A. Fantini, F. Phillipp, T. Riedl, A. Hangleiter: IEEE J. Sel. Topics Quant. Electron. **6**, 482 (2000)
125. Qianghua Xie, J. L. Brown, R. L. Jones, J. E. Van Nostrand, K. D. Leedy: Appl. Phys. Lett. **76**, 3082 (2000)
126. N. Kirstaedter, O. G. Schmidt, N. N. Ledentsov, D. Bimberg, V. M. Ustinov, A. Y. Egorov, A. E. Zhukov, M. V. Maximov, P. S. Kopev, Zh. I. Alferov: Appl. Phys. Lett. **69**, 1226 (1996)
127. M. Sugawara (Ed.): *Self-assembled InGaAs/GaAs Quantum Dots*, Semicond. Semimet. **60** (Academic, San Diego 1999)
128. M. Sugawara, K. Mukai, Y. Nakata, H. Ishikawa, A. Sakamoto: Phys. Rev. B **61**, 7595 (2000)
129. A. Patanè, A. Polimeni, M. Henini, L. Eaves, P. C. Main, G. Hill: J. Appl. Phys. **85**, 1999 (625)
130. K. M. Groom, A. I. Tartakovskii, D. J. Mowbray, M. S. Skolnick, P. M. Smowton, M. Hopkinson, G. Hill: Appl. Phys. Lett. **81**, 1 (2002)
131. Xin-Qi Li, Y. Arakawa: Phys. Rev. B **60**, 1915 (1999)
132. J. Newey: Compound Semicond. **8**, 14 (2002)
133. N. Kirstaedter, N. N. Ledentsov, M. Grundmann, D. Bimberg, V. M. Ustinov, S. Ruvimov, M. V. Maximov, P. S. Kopev, Zh. I. Alferov, U. Richter, P. Werner, U. Gösele, J. Heydenreich: Electron. Lett. **30**, 1416 (1994)
134. K. Sebald, P. Michler, J. Gutowski, R. Kröger, T. Passow, M. Klude, D. Hommel: Phys. Stat. Solidi B **190**, 593 (2002)
135. J. W. Matthews, A. E. Blakeslee: J. Cryst. Growth **27**, 118 (1974)
136. V. M. Ustinov, A. E. Zhukov: Semicond. Sci. Technol. **15**, R41 (2000)
137. K. Nishi, H. Saito, S. Sugou, J. S. Lee: Appl. Phys. Lett. **74**, 1111 (1999)
138. V. M. Ustinov, N. A. Maleev, A. E. Zhukov, A. R. Kovsh, A. Y. Egorov, A. V. Lunev, B. V. Volovik, I. L. Krestnikov, Y. G. Musikhin, N. A. Bert, P. S. Kopev, Zh. I. Alferov, N. N. Ledentsov, D. Bimberg: Appl. Phys. Lett. **74**, 2815 (1999)
139. L. F. Lester, A. Stintz, H. Li, T. C. Newell, E. A. Pease, B. A. Fuchs, K. J. Malloy: IEEE Photon. Technol. Lett. **11**, 931 (1999)
140. G. T. Liu, A. Stintz, H. Li, K. J. Malloy, L. F. Lester: Electron. Lett. **35**, 1163 (1999)

141. K. Mukai, Y. Nakata, K. Otsubo, M. Sugawara, N. Yokoyama, H. Ishikawa: IEEE Photon. Technol. Lett. **11**, 1205 (1999)
142. M. V. Maximov, A. F. Tsatsulnikov, B. V. Volovik, D. A. Bedarev, A. Y. Egorov, A. E. Zhukov, A. R. Kovsh, N. A. Bert, V. M. Ustinov, P. S. Kopev, Zh. I. Alferov, N. N. Ledentsov, D. Bimberg, I. P. Soshnikov, P. Werner: Appl. Phys. Lett. **75**, 2347 (1999)
143. J. Tatebayashi, M. Nishioka, Y. Arakawa: Appl. Phys. Lett. **78**, 3469 (2001)
144. Z. Zou, O. B. Shchekin, G. Park, D. L. Huffaker, D. G. Deppe: IEEE Photon. Technol. Lett. **10**, 1673 (1998)
145. D. L. Huffaker, G. Park, Z. Zou, O. B. Shchekin, D. G. Deppe: Appl. Phys. Lett. **73**, 2564 (1998)
146. G. Park, O. B. Shchekin, S. Csutak, D. L. Huffaker, D. G. Deppe: Appl. Phys. Lett. **75**, 3267 (1999)
147. G. Park, O. B. Shchekin, D. L. Huffaker, D. G. Deppe: Electron. Lett. **36**, 1283 (2000)
148. O. B. Shchekin, D. G. Deppe: Appl. Phys. Lett. **80**, 3277 (2002)
149. O. B. Shchekin, D. G. Deppe: Appl. Phys. Lett. **80**, 2758 (2002)
150. A. E. Zhukov, V. M. Ustinov, A. Y. Egorov, A. R. Kovsh, A. F. Tatsulnikov, V. M. Maximov, N. N. Ledentsov, S. V. Zaitsev, N. Y. Gordeev, V. I. Kopchatov, Y. M. Shernyakov, P. S. Kopev, D. Bimberg, Zh. I. Alferov: J. Electron. Mater. **27**, 106 (1998)
151. F. Schäfer, J. P. Reithmaier, A. Forchel: Appl. Phys. Lett. **74**, 2915 (1999)
152. H. Saito, K. Nishi, I. Ogura, S. Sugou, Y. Sugimoto: Appl. Phys. Lett. **69**, 3140 (1996)
153. D. L. Huffaker, O. Baklenov, L. A. Graham, B. G. Streetman, D. G. Deppe: Appl. Phys. Lett. **70**, 2356 (1997)
154. J. A. Lott, N. N. Ledentsov, V. M. Ustinov, A. Yu. Egorov, A. E. Zhukov, P. S. Kopev, Zh. I. Alferov, D. Bimberg: Electron. Lett. **33**, 1150 (1997)
155. I. L. Krestnikov, N. A. Maleev, A. V. Sakharov, A. R. Kovsh, A. E. Zhukov, A. F. Tsatsulnikov, V. M. Ustinov, Zh. I. Alferov, N. N. Ledentsov, D. Bimberg, J. A. Lott: Semicond. Sci. Technol. **16**, 844 (2001)
156. A. Ramakrishnan, G. Steinle, D. Supper, C. Degen, G. Ebbinghaus: Electron. Lett. **38**, 322 (2002)
157. V. M. Ustinov, A. E. Zhukov, A. Y. Egorov, A. R. Kovsh, S. V. Zaitsev, N. Y. Gordeev, V. I. Kopchatov, N. N. Ledentsov, A. F. Tsatsulnikov, B. V. Volovik, P. S. Kopev, Z. I. Alferov, S. S. Ruvimov, Z. Liliental-Weber, D. Bimberg: Electron. Lett. **34**, 670 (1998)
158. S. DenBaars, C. Reaves, V. Bressler-Hill, S. Varma, W. H. Weinberg, P. M. Petroff: J. Cryst. Growth **145**, 721 (1994)
159. N. Carlsson, K. Georgsson, L. Montelius, L. Samuelson, W. Seifert, R. Wallenberg: J. Cryst. Growth **156**, 23 (1995)
160. J. H. Ryou, R. D. Dupuis, G. Walter, N. Holonyak, Jr., D. T. Mathes, R. Hull, C. V. Reddy, V. Narayanamurti: J. Appl. Phys. **91**, 5313 (2002)
161. M. Jetter, G. Beirne, M. Rossi, J. Porsche, F. Scholz, H. Schweizer: In Proc. of the 14th Indium Phosphide and Related Materials Conf., Venghaus (Ed.) (IEEE, Washington 2002) p. 569
162. T. Riedl, A. Hangleiter, J. Porsche, F. Scholz: Appl. Phys. Lett. **80**, 4015 (2002)
163. G. Walter, N. Holonyak, Jr., J. H. Ryou, R. D. Dupuis: Appl. Phys. Lett. **79**, 3215 (2001)

164. A. J. Williamson, A. Franceschetti, H. Fu, L. W. Wang, A. Zunger: J. Electron. Mater. **28**, 414 (1999)
165. K. Hinzer, S. Fafard, A. J. SpringThorpe, J. Arlett, E. M. Griswold, Y. Feng, S. Charbonneau: Physica E **2**, 729 (1998)
166. K. Hinzer, J. Lapointe, Y. Feng, A. Delage, S. Fafard, A. J. SpringThorpe, E. M. Griswold: J. Appl. Phys. **87**, 1496 (2000)
167. S. Nakamura, M. Senoh, S. Nagahama, N. Iwasa, T. Yamada, T. Matsushita, H. Kiyoku, Y. Sugimoto: Jap. J. Appl. Phys. **35**, L74 (1996)
168. Y. Narukawa, Y. Kawakami, M. Funato, S. Fujita, S. Nakamura: Appl. Phys. Lett. **70**, 981 (1997)
169. F. Widmann, B. Daudin, G. Feuillet, Y. Samson, J. L. Rouviere, N. Pelekanos: J. Appl. Phys. **83**, 7618 (1998)
170. K. Tachibana, T. Someya, Y. Arakawa: Appl. Phys. Lett. **74**, 383 (1999)
171. Y. Kobayashi, V. Perez-Solorzano, J. Off, B. Kuhn, H. Gräbeldinger, H. Schweizer, F. Scholz: J. Cryst. Growth **243**, 103 (2002)
172. K. Tachibana, T. Someya, R. Werner, A. Forchel, Y. Arakawa: Physica E **7**, 944 (2000)
173. B. Damilano, N. Grandjean, S. Dalmasso, J. Massies: Appl. Phys. Lett. **75**, 3751 (1999)
174. S. Nagahama, T. Yanamoto, M. Sano, T. Mukai: Appl. Phys. Lett. **79**, 1948 (2001)
175. K. Tachibana, T. Someya, Y. Arakawa: IEEE J. Sel. Topics Quant. Electron. **6**, 475 (2000)
176. G. Owen, R. Pease, D. Markle, A. Grenville, R. Hsieh, R. von Bünau, N. Maluf: J. Vac. Sci. Technol. B **10**, 3032 (1992)
177. T. Ohfuji, O. Nalamasu, D. Stone: J. Vac. Sci. Technol. B **11**, 2714 (1993)
178. D. Fleming, J. Maldonado, M. Neisser: J. Vac. Sci. Technol. B **10**, 2511 (1992)
179. R. Viswanathan, D. Seeger, A. Bright, T. Bucelot, A. Pomerene, K. Petrillo, P. Blauner, P. Agnello, J. Warlaumont, J. Conwaym, D. Patel: J. Vac. Sci. Technol. B **11**, 2910 (1993)
180. Y. Takahashi, A. Yamada, Y. Oae, H. Yasuda, K. Kawashima: J. Vac. Sci. Technol. B **10**, 2794 (1992)
181. H. Pfeiffer, D. Davis, W. Enichen, M. Gordon, T. Groves, J. Hartley, R. Quickle, J. Rockrohr, W. Stickel, E. Weber: J. Vac. Sci. Technol. B **11**, 2332 (1993)
182. T. Sakata, K. Kumagai, M. Naitou, I. Watanabe, Y. Ohhashi, O. Hosoda, Y. Kokubo, K. Tanaka: J. Vac. Sci. Technol. B **10**, 2842 (1992)
183. H. Lüschner, G. Stengl, A. Chalupka, J. Fegerl, R. Fischer, E. Hammel, G. Lammer, L. Malek, R. Nowak, C. Traher, H. Vonach, P. Wolf: J. Vac. Sci. Technol. B **11**, 2409 (1993)
184. αM-CMS: Cloromethylated Poly-α-Methylstyrol [221], [222]
185. AZPN114: Trade name, Höchst Co., Frankfurt, Germany
186. PMMA: PolyMethylMethAcrylat
187. S. Rishton, D. Kern: J. Vac. Sci. Technol. B **5**, 135 (1987)
188. D. Tennant, G. Doran, R. Howard, J. Denker: J. Vac. Sci. Technol. B **6**, 426 (1988)
189. R. Fastenau, K. Monahan, D. Kyser, S. Phelps: J. Vac. Sci. Technol. B **7**, 1933 (1989)
190. C. Kaden, U. Griesinger, H. Schweizer, M. Pilkuhn: J. Vac. Sci. Technol. B **10**, 2970 (1992)

191. C. Kaden, H. Gräbeldinger, H.-P. Gauggel, V. Hofsäss, A. Hase, A. Menschig, H. Schweizer, R. Zengerle, H. Brückner: Microelectron. Engin. **23**, 469 (1994)
192. M. H. Pilkuhn, A. Forchel, R. Germann, H. Leier, B. E. Maile, G. Mayer, A. Menschig, H. Schweizer: *Proc. 1990 Int. MicroProcess Conference*, JJAP Ser. **4**, 281 (1990)
193. H. Schweizer, U. Griesinger, H. Gauggel, R. Hofmann, J. Wang, N. Lichtenstein, M. Burkard, J. Kuhn, C. Geng, M. Geiger, J. Off, F. Scholz: Electrochem. Soc. Proc. **98-2**, 210 (1998)
194. J. Hommel, F. Schneider, M. Moser, C. Geng, F. Scholz, H. Schweizer: Microelectron. Engin. **23**, 349 (1994)
195. H. Schweizer, G. Lehr, F. E. Prins, G. Mayer, E. Lach, R. Krüger, E. M. Fröhlich, M. H. Pilkuhn, G. W. Smith: Superlat. Microstruct. **12**, 419 (1992)
196. H. Schweizer, G. Lehr, F. E. Prins, E. Lach, E. M. Fröhlich, M. H. Pilkuhn, J. Straka, A. Forchel: Phys. Stat. Solidi B **173**, 331 (1992)
197. G. Lehr, R. Bergmann, F. Scholz, H. Schweizer: Appl. Phys. Lett. **61**, 517 (1992)
198. G. Lehr, R. Bergmann, R. Rudeloff, J. Hommel, F. Scholz, H. Schweizer: Superlat. Microstruct. **11**, 329 (1992)
199. R. Bergmann, A. Menschig, G. Lehr, P. Kübler, J. Hommel, R. Rudeloff, B. Henle, F. Scholz, H. Schweizer: J. Vac. Sci. Technol. B **10**, 2893 (1992)
200. A. Menschig, F. E. Prins, G. Lehr, R. Bergmann, J. Hommel, U. A. Griesinger, V. Härle, F. Scholz, H. Schweizer: in *Nanolithography: A Borderland between STM, EB, IB and X-Ray Lithographies*, M. Gentili et al. (Eds.) (Kluwer Academic, Dordrecht 1994) p. 81
201. H. Schweizer, J. Wang, U. Griesinger, M. Burkard, J. Porsche, M. Geiger, F. Scholz: *Frontiers of Nano-Optoelectronic Systems*, NATO Sci. Ser. II, **6**, 65 (2000)
202. The chemical potential of dot systems cannot be approximated by the integral over states. It must be summed up over discrete energy levels.
203. H. Schweizer, G. Lehr, F. Prins, E. Lach, E. Fröhlich, M. Pilkuhn, J. Straka, A. Forchel: Phys. Stat. Solidi B **173**, 331 (1992)
204. M. Willatzen, T. Tanaka, Y. Arakawa: IEEE J. Quant. Electron. **30**, 640 (1994)
205. U. Bockelmann, G. Bastard: Phys. Rev. B **45**, 1688 (1992)
206. E. Kane: *The kp-method, Semicond. and Semimet.* **1** (Academic, New York 1966)
207. E. Zielinski, H. Schweizer, K. Streubel, H. Eisele, G. Weimann: J. Appl. Phys. **59**, 2196 (1986)
208. J. Bowers: Solid State Electron. **30**, 1 (1987)
209. J. Wang, H. Schweizer: IEEE J. Quant. Electron. **33**, 1 (1997)
210. N. Tessler, G. Eisenstein: IEEE J. Quant. Electron. **29**, 1586 (1993)
211. L. Davis, Y. Lam, Y. Chen, J. Singh, P. Bhattacharya: IEEE J. Quant. Electron. **30**, 2560 (1994)
212. K. Petermann: *Advances in Optoelectronics (ADOP). Laserdiode Modulation and Noise*, T. Okoshi, T. Kamiya (Eds.) (Kluwer Academic, Dordrecht 1988)
213. U. A. Griesinger, H. Schweizer, V. Härle, J. Hommel, F. Barth, B. Höhing, B. Klepser, F. Scholz: IEEE Phot. Technol. Lett. **7**, 953 (1995)
214. A. V. Uskov, J. McInerney, F. Adler, H. Schweizer, M. H. Pilkuhn: Appl. Phys. Lett. **72**, 58 (1998)

215. D. Botez, L. Mawst, A. Bhattacharya, J. Lopez, J. Li, T. Kuech, V. Iakovlev, G. Suruceanu, A. Caliman, A. Syrbu, J. Morris: Electron. Lett. **33**, 2037 (1997)
216. L. Mawst, A. Bhattacharya, J. Lopez, D. Botez, D. Garbuzov, L. DeMarco, J. Connolly, M. Jansen, F. Fang, R. Nabiev: Appl. Phys. Lett. **69**, 1532 (1996)
217. M. Emanuel, N. Carlson, J. Skidmore: IEEE Phot. Technol. Lett. **8**, 1291 (1996)
218. N. Lichtenstein, R. Winterhoff, F. Scholz, H. Schweizer: IEEE J. Sel. Topics Quant. Eletctron. **16**, 564 (2000)
219. F. Heinrichsdorff, C. Ribbat, M. Grundmann, D. Bimberg: Appl. Phys. Lett. **76**, 556 (2000)
220. R. L. Sellin, Ch. Ribbat, M. Grundmann, N. N. Ledentsov, D. Bimberg: Appl. Phys. Lett. **78**, 1207 (2001)
221. S. Imamura: J. Electrochem. Soc. **126(9)**, 1628 (1979)
222. K. Sukegawa, S. Sugawara: Jap. J. Appl. Phys. **20(8)**, L583 (1981)

# Dephasing Processes and Carrier Dynamics in (In,Ga)As Quantum Dots

Paola Borri and Wolfgang Langbein

Fachbereich Physik, Universität Dortmund,
Otto-Hahn Str. 4, 44221 Dortmund, Germany
borri@fred.physik.uni-dortmund.de

**Abstract.** In this contribution the dynamics of the optically induced interband polarization in (In,Ga)As/(Ga,Al)As quantum dots is reviewed, following the advances in the literature of the last ten years. Radiative recombination, exciton–phonon interaction and Coulomb interaction of excitons with other carriers are discussed as mechanisms responsible of the excitonic dephasing in relation to their influence in the decay of the excitonic population. Following the recent progress in this area, the coherent non-linear light–matter interaction under strong light fields that results in optical Rabi oscillations of the excitonic population in quantum dots is also reviewed.

## 1 Introduction

The dynamics of electrons and holes in semiconductors and semiconductor heterostructures have been extensively discussed in the literature, and are governed by their mutual Coulomb interaction, by their interaction with lattice vibrations (phonons) and by their radiative recombination. A powerful technique to study the carrier dynamics in semiconductors utilizes short optical pulses to photoexcite electrons from the valence to the conduction band (i.e. to create electron–hole pairs) that subsequently relax until the photon energy stored in the medium is dissipated and the system has recovered thermal equilibrium. The dynamics of the carriers photoexcited by the short optical pulse can be followed by monitoring two main optical observables: the interband population difference and the induced polarization. Both quantities decay with time after the photoexcitation due to the carrier dynamics, however the induced polarization (which is an ensemble average over many photoexcited dipole moments) is sensitive to scattering events that break the phase relationship among the individual dipoles and/or with the light field creating them (dephasing processes), while the population difference is sensitive to events that change the energy of the carriers. Transient coherent optical spectroscopy addresses the time-evolution of the induced polarization in the time-regime during and immediately following the photoexcitation by a short pulse laser, in which the photoexcited dipoles have retained a defined phase relationship among each other and with the exciting coherent radiation.

P. Michler (Ed.): Single Quantum Dots, Topics Appl. Phys. **90**, 237–268 (2003)
© Springer-Verlag Berlin Heidelberg 2003

In semiconductor structures with at least one direction of translational invariance, electrons and holes can span over a continuum range of energies and wavevectors. The description of light–matter interaction in these systems has to account for many-body Coulomb interactions among the carriers and is generally rather complicated. In semiconductor quantum dots (QDs), the translational invariance is destroyed in all three directions, and electrons and holes have discrete energy levels (see the Chapter by Hawrylak and Korkusiński in this volume). The description of light–matter interaction in QDs can thus recover most of the simplified approaches known from atomic physics for a long time.

This article concentrates on the carrier dynamics in semiconductor QDs epitaxially grown with III–V materials such as (In,Ga,Al)As, with particular emphasis on the dephasing processes. In the following section the theoretical background for the description of the coherent light–matter interaction in QDs is given (Sect. 2.1), together with an overview of the optical experiments used in the literature for coherent optical spectroscopy in QDs (Sect. 2.2). A discussion of the different dephasing processes in (In,Ga)As/(Ga,Al)As QDs will follow in Sect. 3, such as radiative recombination (Sect. 3.1), phonon interactions (Sect. 3.2) and Coulomb interactions (Sect. 3.3). Sect. 4 is dedicated to coherent non-linear spectroscopy of QDs in the presence of strong light fields resulting in optical Rabi oscillations. Finally, a summary of the work is given in Sect. 5 together with an outlook toward the importance of QD dephasing in applications.

# 2   Coherent Spectroscopy of Semiconductor Quantum Dots

As mentioned before, with respect to their optical properties QDs can be in a good approximation described as atom-like objects. Many transient coherent effects in atoms are analyzed describing the atoms as two-level systems. This model assumes that the interacting light has a narrow frequency spectrum such that photons are resonant with an optically allowed transition between two energy levels, and off resonant with respect to all other optical transitions. We begin our discussion on the coherent spectroscopy of QDs with the basic treatment of light–matter interaction in the two-level model.

## 2.1   Theoretical Background

### 2.1.1   Optical Bloch Equations

Let us consider an atom-like two-level system, i.e. a charged particle bound to move in a confining potential with two discrete energy levels $E_a, E_b$ (see Fig. 1a), in the presence of an electromagnetic field. The two-level system is treated quantum-mechanically, i.e. its wavefunction $\Psi$ follows the Schrödinger equation:

$$i\hbar\frac{\partial\Psi}{\partial t} = H\Psi ,\tag{1}$$

where $H$ is the Hamiltonian operator, in general time-dependent, while the electromagnetic field is described classically and obeys the Maxwell equations. This semiclassical treatment of the light–matter interaction is bound to be incomplete, but offers a simpler framework than using the quantum theory of radiation. With the semiclassical treatment it is possible to give a correct description of the influence of an external coherent radiation field to the system (absorption and stimulated emission) but not of the spontaneous emission of radiation from the system which is accounted for only phenomenologically. The treatment of the interaction of atoms with a quantized field can be found in [1,2].

The Hamiltonian $H_0$ of the isolated two-level system in the absence of the electromagnetic field is time-independent and (1) has solutions of the form [1]:

$$\Psi_j(\boldsymbol{r},t) = \varphi_j(\boldsymbol{r})e^{-iE_jt/\hbar} , \qquad H_0\varphi_j = E_j\varphi_j ,\tag{2}$$

with $j = a, b$ and $\varphi_a, \varphi_b$ are orthogonal functions normalized to one; $\boldsymbol{r}$ denotes the spatial coordinate of the charged particle. In the presence of a time-dependent field, the solution of (1) with $H = H_0 + H'(t)$ is sought in the form:

$$\Psi(\boldsymbol{r},t) = c_a(t)\Psi_a(\boldsymbol{r},t) + c_b(t)\Psi_b(\boldsymbol{r},t) ,\tag{3}$$

with the coefficient $c_a$ ($c_b$) such that $|c_a|^2$ ($|c_b|^2$) is the probability of finding the system in the level $a$ ($b$). We require the normalization condition:

$$|c_a(t)|^2 + |c_b(t)|^2 = 1 ,\tag{4}$$

so that the wavefunction in (3) is normalized at all times, i.e. the probability is conserved. For an electric-dipole allowed transition the main contribution to the interaction Hamiltonian $H'$ is:

$$H' = -\boldsymbol{p}\cdot\boldsymbol{E} ,\tag{5}$$

where $\boldsymbol{E}$ is the incident electric field and $\boldsymbol{p}$ is the electric dipole moment of the charged particle. In the so-called dipole approximation, the size of the wavefunction that describes the motion of the particle is negligible compared to the wavelength of the field, and $\boldsymbol{E}$ is evaluated at a fixed spatial position (e.g. at the nucleus of the atom). By substituting (3) into (1), taking into account of (2) and (5), one gets:

$$i\hbar\frac{dc_a}{dt} = c_aH'_{aa} + c_bH'_{ab} , \qquad i\hbar\frac{dc_b}{dt} = c_aH'_{ba} + c_bH'_{bb} ,\tag{6}$$

with

$$H'_{ij} = e^{-i(E_j - E_i)t/\hbar}(-\boldsymbol{p}_{ij} \cdot \boldsymbol{E}), \qquad \boldsymbol{p}_{ij} = \int d^3r \varphi_i^* \boldsymbol{p} \varphi_j. \tag{7}$$

The four elements $H'_{ij}$ ($i, j = a, b$) of the interaction Hamiltonian can be simplified when considering that $H'$ is a self-adjoint operator, since it is the representation of a physical observable, and thus $H'_{ij} = H'^*_{ji}$. Moreover, if the eigenfunctions $\varphi_i$ have a given parity, which is the case for an inversion symmetric Hamiltonian $H_0(\boldsymbol{r}) = H_0(-\boldsymbol{r})$ (like for example in the hydrogen atom with the origin of the coordinates taken in the nucleus), then:

$$\boldsymbol{p}_{ii} = \int d\boldsymbol{r} \varphi_i^*(-e\boldsymbol{r})\varphi_i = 0, \tag{8}$$

where the electric dipole moment of a particle of charge $-e$ has been written. Let us now introduce the following substitutions:

$$\hbar\omega_0 = E_b - E_a, \qquad a = c_a e^{+i\omega_0 t/2}, \qquad b = c_b e^{-i\omega_0 t/2}, \tag{9}$$

and

$$\boldsymbol{s} = (s_1, s_2, s_3) = (ab^* + ba^*, i(ab^* - ba^*), bb^* - aa^*). \tag{10}$$

Then, (6) can be rewritten in the vectorial form:

$$\frac{d\boldsymbol{s}}{dt} = \boldsymbol{\Omega} \times \boldsymbol{s}, \qquad \boldsymbol{\Omega} = \left( \frac{(\boldsymbol{p}_{ab} + \boldsymbol{p}_{ab}^*) \cdot \boldsymbol{E}}{\hbar}, i\frac{(\boldsymbol{p}_{ab} - \boldsymbol{p}_{ab}^*) \cdot \boldsymbol{E}}{\hbar}, -\omega_0 \right), \tag{11}$$

which describes the precession of the vector $\boldsymbol{s}$ (of unitary modulo as follows from (4)) with the angular frequency vector $\boldsymbol{\Omega}$. Equations (11) represent the general form of the interaction of a two-level system with a light field in the semiclassical theory and are called optical Bloch equations, due to their analogy with the equations introduced by Bloch for magnetic resonances describing the behavior of a spin 1/2 particle in an oscillatory magnetic field [3]. The solution of the precession in (11) is in general complicated since $\boldsymbol{\Omega}$ is time dependent. Let us for the moment consider an electric field in the form of a linearly polarized monochromatic wave $\boldsymbol{E} = \boldsymbol{E}_0 \cos \omega t$ and let us suppose that $\boldsymbol{p}_{ab} = \boldsymbol{p}_{ab}^* = \boldsymbol{\mu}$, i.e. a linearly polarized transition (the case of circularly polarized dipole transitions can be found in [4]).

If (11) is rewritten in the rotating system with the angular frequency $\boldsymbol{\omega} = (0, 0, -\omega)$ one gets:

$$\frac{d\boldsymbol{s}'}{dt} = \left( \frac{\boldsymbol{\mu} \cdot \boldsymbol{E}_0(1 + \cos 2\omega t)}{\hbar}, \frac{\boldsymbol{\mu} \cdot \boldsymbol{E}_0 \sin 2\omega t}{\hbar}, \omega - \omega_0 \right) \times \boldsymbol{s}', \tag{12}$$

where $\boldsymbol{s}'$ is the transformed vector in the rotating system. Therefore, in the rotating system the precession is described by a constant angular frequency, if one neglects the terms rapidly oscillating at the frequency $2\omega$. This is the so-called rotating wave approximation [3], which is valid if $\boldsymbol{\mu} \cdot \boldsymbol{E}_0/\hbar \ll \omega$ and one considers averaged effects over observation times much longer than the optical period $2\pi/\omega$ (note that for visible light $2\pi/\omega \approx 10^{-15}\,\mathrm{s}$). The vectors $\boldsymbol{s}$ and $\boldsymbol{s}'$ are related by the matrix of the rotation and one finds:

$$
\begin{aligned}
s_1 &= s_1' \cos\omega t + s_2' \sin\omega t\,, \\
s_2 &= -s_1' \sin\omega t + s_2' \cos\omega t\,, \\
s_3 &= s_3'\,.
\end{aligned}
\tag{13}
$$

The precession of $\boldsymbol{s}'$ with constant angular frequency is finally given by:

$$
\frac{d\boldsymbol{s}'}{dt} = \boldsymbol{\Omega}_\mathrm{e} \times \boldsymbol{s}'\,, \qquad \boldsymbol{\Omega}_\mathrm{e} = (\omega_\mathrm{R}, 0, \omega - \omega_0)\,, \qquad \omega_\mathrm{R} = \frac{\boldsymbol{\mu} \cdot \boldsymbol{E}_0}{\hbar}\,,
\tag{14}
$$

and can be easily solved. The frequency $\omega_\mathrm{R}$ is called the Rabi frequency. It is interesting to point out the physical meaning of the components of the vector $\boldsymbol{s}'$. The component $s_3' = s_3$ gives the difference between the excited-state and the ground-state occupation probabilities (10), and it is also called population difference or inversion. To understand the physical meaning of $s_1'$ and $s_2'$ let us write the expectation value of the dipole moment operator for the state $\Psi$:

$$
\langle \boldsymbol{p} \rangle = \int d\boldsymbol{r}\Psi^* \boldsymbol{p}\Psi = \boldsymbol{\mu}(ab^* + ba^*) = \boldsymbol{\mu}s_1\,.
\tag{15}
$$

From (13) it follows that $s_1'$ and $s_2'$ are the amplitudes of the induced dipole moment in phase and in quadrature with the driving electric field, respectively [from (8) it follows that the dipole moment of the system at equilibrium without field is zero].

With the initial conditions $\boldsymbol{s}'(t = 0) = (0, 0, -1)$ one finds as solution of (14):

$$
|b|^2 = \frac{\omega_\mathrm{R}^2 \sin^2\left(\sqrt{(\omega - \omega_0)^2 + \omega_\mathrm{R}^2}\,t/2\right)}{(\omega - \omega_0)^2 + \omega_\mathrm{R}^2}\,.
\tag{16}
$$

Two examples of this solution are shown in Fig. 1b.

For the resonant case $\omega = \omega_0$, the occupation probability oscillates in time with the Rabi frequency and reaches a maximum value of 1. Out of resonance, the probability oscillates faster and never reaches its maximum value of 1. Note that using (4) the population inversion is given by $s_3' = 2|b|^2 - 1$. This oscillatory behavior of the occupation probability versus time (Rabi oscillations) is the physical origin of the coherent optical effect called transient nutation [3,4] in which the transmission of light, immediately following its switching-on, exhibits a modulation due to the interaction with the medium which is absorbing and re-emitting light sinusoidally versus time, according to (16).

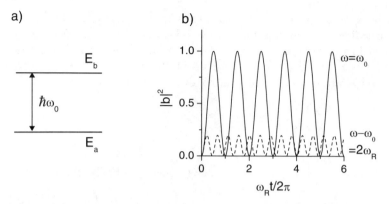

**Fig. 1.** (a) Energy scheme of a two-level system. (b) Time-evolution of the excited state occupation probability according to (16) in the resonant case $\omega = \omega_0$ (*solid line*), and for a detuning $\omega - \omega_0$ given by twice the Rabi frequency (*dashed line*)

### 2.1.2   Ensemble of Identical Two-level Systems: Density Matrix

In real systems, the transient nutation showed a damped oscillation toward a steady-state value [3,4]. There are various physical mechanisms as the origin of this damping. First, the spontaneous emission from level $b$ to level $a$ results in a finite radiative lifetime of level $b$. Additionally, one might have to take into account interactions with the environment such as collisions with other atoms for gases, or lattice vibrations and Coulomb interactions with other carriers for solids. It is in general complicated to account for these effects since the dynamics of a large number of other particles have to be considered. In a simple approach, one considers the environment as a thermal bath without memory and the time evolution of the system is described by an exponential decay to the equilibrium value. This simple approach introduces phenomenological damping constants for the time evolution of the vector $s'$, with the result that $s'$ no longer maintains its unitary modulo [3]. This is, however, not compatible with the description followed up to now where the unitarity of $s'$ is a direct consequence of the conservation probability in (4).

A consistent description of the damping mechanisms has to abandon the possibility of describing a two-level system with the wavefunction $\Psi$ (called a pure state in quantum mechanics) and has to introduce the density matrix formalism [4]. In this case, the system is not in a pure state but in a "mixture" of states, and it is described by the probabilities $P_k$ of being in the state $\Psi_k = c_{a,k}\Psi_a + c_{b,k}\Psi_b$. The density matrix of the system is defined as:

$$\rho = \sum_k P_k \begin{pmatrix} |c_{a,k}|^2 & c_{a,k}c_{b,k}^* \\ c_{b,k}c_{a,k}^* & |c_{b,k}|^2 \end{pmatrix}. \tag{17}$$

We use this description for systems consisting of a large ensemble of identical systems which are in different pure states or for a single system where physical quantities measured over long times can be equivalently treated as an average over a large ensemble of statistically independent realizations. The density matrix describes statistical properties of the system. Let us introduce two physically measurable quantities that are important in optical experiments: the population difference $\Delta N$ and the polarization $\boldsymbol{P}$ of the system. These are *macroscopic* properties of the system, i.e. ensemble averages, and are given by:

$$\Delta N = n \sum_k P_k(|c_{b,k}|^2 - |c_{a,k}|^2) = n \sum_k P_k s_{3,k} \,,$$

$$\boldsymbol{P} = n \sum_k P_k \langle \boldsymbol{p} \rangle_k = n\boldsymbol{\mu} \sum_k P_k s_{1,k} \,, \tag{18}$$

with $n$ number of systems per unit volume in the ensemble. Therefore the diagonal terms in the density matrix are associated to the population of the energy eigenstates, while the off-diagonal terms are associated to the phase–sensitive superposition of the two-level dipoles in the ensemble and, thus, to their mutual coherence. We now introduce a new vector $\boldsymbol{S'}$ which describes the macroscopic inversion and polarization:

$$S_1' = \sum_k P_k s_{1,k}' \,, \qquad S_2' = \sum_k P_k s_{2,k}' \,, \qquad S_3' = \sum_k P_k s_{3,k}' \,, \tag{19}$$

and obeys the optical Bloch equations with damping:

$$\frac{\mathrm{d}S_3'}{\mathrm{d}t} = \omega_{\mathrm{R}} S_2' - \frac{S_3' + 1}{T_1} \,,$$

$$\frac{\mathrm{d}S_2'}{\mathrm{d}t} = (\omega - \omega_0) S_1' - \omega_{\mathrm{R}} S_3' - \frac{S_2'}{T_2} \,,$$

$$\frac{\mathrm{d}S_1'}{\mathrm{d}t} = -(\omega - \omega_0) S_2' - \frac{S_1'}{T_2} \,. \tag{20}$$

We have introduced $T_1 > 0$ as the decay time of the population difference, such that the system returns to its ground state at equilibrium without field $[\boldsymbol{S'} = (0, 0, -1)]$ due to the spontaneous emission from the level $b$. $T_2 > 0$ is the damping constant of the polarization, also called the dephasing time. $T_1$ and $T_2$ are also known as the longitudinal and transverse relaxation time, respectively, in analogy with the theory of magnetic resonance. In general, they are related by the following equation:

$$\frac{1}{T_2} = \frac{1}{2T_1} + \frac{1}{T_2'} \,. \tag{21}$$

$T_2' > 0$ is often called the "pure" or "elastic" dephasing time. For example, the collision broadening in atoms is often of this type [2], i.e. collisions can leave the atom in the same energy level (elastic collisions) and only change the phases of $c_{a,k}$ and $c_{b,k}$ without changing their amplitudes. The macroscopic polarization is sensitive to the phase relations of the terms $c_{a,k} c_{b,k}{}^*$ in

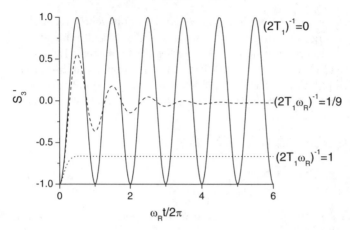

**Fig. 2.** Solution of the population difference from the optical Bloch equations with radiative damping $T_1$ and without pure dephasing ($T_2' = 0$) in the resonant case, for different ratios of the Rabi frequency to the damping rate

the superposition $\sum_k P_k s_{1,k}$, and thus random phase changes caused by the elastic collisions result in a decay of the macroscopic polarization additional to the one given by the population damping. The factor of two between the elastic dephasing rate and the radiative one in (21) should be noted. This can be understood since the polarization is proportional to the *amplitude* of excited-state occupation probability $c_b$ while the population inversion is proportional to $|c_b|^2$.

In the presence of damping, the components of $\mathbf{S}'$ decay with time and tend to steady-state values. Equation (20) can be solved analytically [3] in the resonant case $\omega = \omega_0$. In Fig. 2 this solution is shown for the population difference $S_3'$ with the initial condition $S_3'(0) = -1$ in the case of pure population damping, i.e. $1/T_2' = 0$ for different ratios of the Rabi frequency to the damping rate $1/T_2'$. Significant oscillations occur for large Rabi frequency and long dephasing time.

The steady state solution of (20) results in constant amplitudes of the polarization in phase and in quadrature with the incident electric field. These terms allow one to define a susceptibility with its absorptive and dispersive part and link the damping constants to the spectral linewidth of the absorption [2]. In the limit of linear susceptibility ($S_3' + 1 \ll 1$), one finds a Lorentzian spectral lineshape of the absorption $\alpha_{\omega_0}(\omega)$ centered on $\omega_0$ with a full width at half maximum (FWHM) given by $\gamma/\hbar = 2/T_2$.

Let us briefly comment that the introduction of the damping constants $T_1$, $T_2$ is the simplest approximation to describe the relaxation of the system to its equilibrium, which is also called the relaxation-time approximation. The general description of the processes that bring the system to thermal equilibrium via interactions with its environment, such as collisions, radiative and non-radiative recombination, interactions with other charges, considers dephasing processes that are non-instantaneous functions of the environment (non Markovian behavior) [5].

### 2.1.3   Inhomogeneously Broadened Transitions

In most of the real cases the system under observation is an inhomogeneous ensemble, i.e. it is formed by non-identical two-level systems. Each two-level system can be described by (20) but parameters such as the transition frequency, the dipole moment and the damping constants can have different values for the different systems in the ensemble. The formal description of an inhomogeneous ensemble of a large number of systems with density $n$ defines how a given parameter is statistically distributed in the ensemble. The distribution of transition frequencies is often the most important inhomogeneous effect in the optical properties of an ensemble. In many cases, an inhomogeneous ensemble is approximated assuming that only a distribution of resonance frequencies $\omega_0'$ is present, represented by a function $G(\omega_0')$ such that $nG(\omega_0')d\omega_0'$ is the number of systems per unit volume with transition frequencies between $\omega_0'$ and $\omega_0' + d\omega_0'$. A well known example is the Doppler broadening for atomic gases which leads to a Gaussian distribution of resonance frequencies [2]. The macroscopic population difference and polarization are given by the statistical average over the inhomogeneous distribution:

$$\Delta N = n \int_{-\infty}^{+\infty} S_3(\omega_0')G(\omega_0')d\omega_0' ,$$

$$\boldsymbol{P} = n\boldsymbol{\mu} \int_{0}^{\infty} S_1(\omega_0')G(\omega_0')d\omega_0' . \tag{22}$$

From (22) the absorption spectral lineshape $\alpha(\omega)$ is given by the convolution of the Lorentzian lineshape of the individual two-level system $\alpha_{\omega_0'}(\omega)$, centered in $\omega_0'$ and homogeneously broadened by $\gamma/\hbar$, with the distribution [2] $G(\omega_0')$:

$$\alpha(\omega) = \int_{-\infty}^{+\infty} \alpha_{\omega_0'}(\omega)G(\omega_0')d\omega_0' . \tag{23}$$

### 2.1.4   Beyond the Two-level Approximation: Many-level Systems

Even for discrete energy level systems, in many real situations the interaction of light cannot be described in the two-level approximation. This is the case if several transitions of one system are near resonance with the incident field. For example, in InAs semiconductor quantum dots the ground-state excitonic transition and the exciton–biexciton transition are separated by only a few meV which is below the spectral width of a short optical pulse [6]. Even if the other levels do not lead to optical transitions close in energy, one might have to take into account that the two levels under optical study interact with the other levels, for example via phonon-mediated transitions. In this case the radiative lifetime of level $b$ toward level $a$ is not a realistic description of the relaxation of the population difference of the considered two-level transition but several radiative and non-radiative population relaxation channels are present.

A convenient way to describe the light–matter interaction in the presence of more than two levels is to generalize the density matrix in (17) by considering the wavefunction $\Psi_k = \Sigma_j a_{j,k} \Psi_j$ with $j$ that runs over the number of levels under consideration. The generic element of the matrix is thus defined as $\rho_{ij} = \Sigma_k P_k a_{i,k} a_{j,k}^*$. The time-dependent equation for $\rho_{ij}$ is given by:

$$\frac{d\rho_{ij}}{dt} = -\frac{i}{\hbar} \sum_k (\rho_{kj} H'_{ik} - \rho_{ki}^* H'^*_{jk}) + \left(\frac{d\rho_{ij}}{dt}\right)_{rel}. \tag{24}$$

The last term in the equation represents the relaxation of the system back to thermal equilibrium due to the damping mechanisms. In the relaxation-time approximation, phenomenological damping constants can be introduced both for the diagonal and the off-diagonal elements of the density matrix, to account for the relaxation of the populations and of the mutual coherences.

## 2.2   Optical Experiments

The optical study of the dephasing time can be addressed experimentally in many ways. As mentioned in the previous section, the homogeneous width of the absorption spectral lineshape is inversely proportional to the dephasing time. Thus, experiments can be performed either in the time domain to directly address the transient decay of the polarization induced by a pulsed coherent light field or in the spectral domain by measuring the steady-state optical absorption lineshape. Generally, the response of the medium to the incident field depends on the field intensity. For example, only in the linear response limit (i.e. in the first order to the incident field amplitude) is the absorption lineshape Lorentzian with the FWHM given by $\gamma/\hbar = 2/T_2$. At higher orders, effects such as the power broadening [2], the quadratic Stark shift [4], and the optical Stark splitting (the signature of Rabi oscillations in the spectral domain) become important.

To separate effects at the different orders in the incident field amplitude, the population and the polarization (or the diagonal and off-diagonal elements of the density matrix) can be expanded in a Taylor series. Consequently, the optical Bloch equations separate into a series of equations which can be truncated to a desired order. This perturbation approach allows one to describe nonlinear signals associated with the third-order polarization, which for example in media with inversion symmetry is the lowest nonlinear polarization allowed in the electric-dipole approximation. Third-order nonlinearities were

shown to be very powerful in the measurement of the dephasing time in transient coherent experiments, or of the homogeneous broadening in continuous-
wave spectrally resolved experiments, overcoming many limitation of linear
spectroscopy.

One obvious limitation of linear spectroscopy in determining the homogeneous lineshape is the inhomogeneous broadening, which has to be de-
convoluted from the total line broadening expressed in (23). In the time domain, this translates into an additional decay rate (sometimes also indicated
as $1/T_2^*$ [3]) of the macroscopic first-order polarization inversely proportional
to the inhomogeneous spectral width $\sigma$ of $G(\omega_0')$. In strongly inhomogeneously
broadened systems, linear spectroscopy usually fails in measuring the dephasing time since $\gamma \ll \hbar\sigma$. Note that, recently, a novel linear technique based on
a time-resolved speckle-analysis of the resonant Rayleigh scattering (RRS)
in systems with static disorder allowed to measure the $T_2$ time also in the
presence of large inhomogeneous broadening [7].

Even when isolating one system from the inhomogeneous ensemble, linear
spectroscopy might be severely affected by backgrounds. This is for example the case when light is detected in the transmitted or reflected directions
(which are the directions of propagation of the first order polarization in
planar samples) and only a small part of it contains the effect of the investigated resonance. The analysis of the secondary emission such as RRS or
photoluminescence might, once again, overcome the problem. Indeed, recent
achievements in performing photoluminescence spectroscopy with high spatial and spectral resolutions allowed to infer the homogeneous broadening
from isolated excitons weakly confined by the lateral disorder in thin GaAs
quantum wells (QWs) [8] and in strongly confined single InGaAs QDs [9] (see
also the Chapter by Bayer in this volume).

Experiments based on third-order signals, such as four-wave mixing (FWM)
in the transient coherent domain after pulsed excitation [5] or spectral hole-
burning in the frequency domain with continuous wave excitation [10] allow
one to overcome the presence of an inhomogeneous distribution and can be
detected background-free with appropriate selection in the direction and/or
in the frequency domain. The formal treatment of third-order non-linearities
can be found in [4] while the application of transient FWM to measure the
dephasing time in semiconductors is reviewed in [5,11]. A comprehensive review of laser spectroscopy including the subjects of nonlinear, coherent and
time-resolved spectroscopies can be found in [10].

Measurements of third-order nonlinearities have been successfully applied
to (In,Ga)As quantum dots. Transient FWM spectroscopy was performed on
localized excitons in thin GaAs QWs [12,13] and on strongly confined InGaAs
QDs [14,15,16]. In the frequency domain, the third-order polarization induced
by two narrow-bandwidth continuous wave lasers was investigated on isolated
excitons localized by the lateral disorder in thin GaAs QWs [17,18].

# 3    Dephasing of Excitons in (In,Ga)As Quantum Dots

The optically induced interband polarization in a semiconductor quantum dot under a coherent light field involves the coherent superposition of Coulomb-correlated valence and conduction band states (excitonic polarization). Beside its fundamental interest, the study of the dephasing time $T_2$ of the excitonic polarization in QDs has recently received renewed attention for its implication in the potential applications of quantum dots for quantum information processing. We will here review the physical mechanisms that are responsible for the dephasing of the excitonic polarization in (In,Ga)As QDs. We subdivided these mechanisms into three categories: radiative processes, exciton–phonon interactions, and Coulomb interactions of excitons with other carriers.

## 3.1    Radiative Processes

The radiative lifetime of excitons is a $T_1$ time (see Sect. 2.1) that influences the dephasing time $T_2$ as given in (21). In the absence of interactions with phonons and other carriers and of non-radiative recombination channels, the dephasing time in QDs is limited by the radiative decay. The radiative lifetime of the optically active ground-state Wannier excitons, going from quantum dots in the strong confinement regime to free excitons in quantum wells, has been discussed theoretically in [19,6]. This treatment considers a quantum disk of radius $R_0$, with a height smaller than twice the exciton Bohr radius $a_B$, i.e. in strong vertical confinement. Three size regimes are distinguished: (i) the strongly confined quantum-dot case when also $R_0 < a_B$, (ii) the mesoscopic case when $a_B < R_0 < \lambda_{ex}/n$ with $\lambda_{ex}$ the wavelength of the exciton resonance and $n$ the refractive index, and (iii) the quantum well case when $\lambda_{ex}/n \ll R_0$.

Case (i) is reached in (In,Ga)As QDs for $R_0$ smaller than about 5 nm and results in a radiative lifetime in the nanosecond range ($\sim 1$–$2$ ns), independent of the dot size, essentially given by the band-to-band recombination rate of an electron–hole pair in the bulk crystal. Case (ii) is an intermediate size range of weak confinement, which allows the separation of variables into relative and center–of–mass motion in the exciton wavefunction. In this case, a radiative lifetime that decreases with increasing size of the quantum disks is found, according to an inverse proportionality to the so-called coherence volume [20]. This is the regime for excitons weakly confined by the lateral disorder in narrow quantum wells which results in radiative lifetimes in the range of 100 ps [21]. Case (iii) is practically reached for radii larger than about 150 nm (which corresponds to $\sim \lambda_{ex}/2n$) in (In,Ga)As quantum wells and results in the free-exciton radiative lifetime of 10–20 ps [22].

Measured photoluminescence decay times agree with the predicted radiative lifetimes in the nanosecond range in strongly confined InGaAs QDs [23,24,25]. For $T_1 = 1$ ns one gets from (21) $T_2 \leq 2$ ns (i.e. a FWHM homogeneous broadening $\gamma \geq 0.66$ μeV). Therefore, the radiative broadening in strongly confined (In,Ga)As QDs is in the sub-μeV range. Recently, $T_2$ times

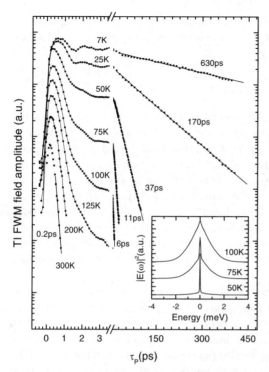

**Fig. 3.** Time-integrated four-wave mixing on $In_{0.7}Ga_{0.3}As/GaAs$ self-assembled quantum dots [15]. The dephasing time inferred from the exponential decays are indicated together with the exponential fits. *Inset*: Power spectrum of the time-integrated four-wave mixing field (after [15])

of 630 ps at 7 K and 370 ps at 5 K were measured with transient FWM in $In_{0.7}Ga_{0.3}As/GaAs$ QDs [15] and $In_{0.5}Ga_{0.04}Al_{0.46}As/Ga_{0.92}Al_{0.08}As$ QDs [16], respectively, approaching the radiative lifetime limit (see Fig. 3). Similarly, a FWHM homogeneous broadening of less than 2 μeV at 2 K was measured on a single $In_{0.6}Ga_{0.4}As$ QD [9]. We should comment that previous experiments on the photoluminescence linewidth of the ground-state exciton on single InGaAs dots showed low-temperature values in the 50 μeV range [26,27]. These higher values are now understood as a "dynamical" broadening from the spectral jitter of the single-dot luminescence emission frequency $\omega_0$ occurring over time scales much longer than the $T_1$ exciton decay time [28,29].

For excitons weakly confined by lateral disorder in thin GaAs QWs the FWHM homogeneous broadening at low temperature ($\sim 5$ K) is measured to be in the range of 20–50 μeV [8,12,17,18] and it is found to be mainly given by a $T_1$ time, with no substantial pure dephasing (21). However, it appears questionable whether such a population lifetime $T_1$ (which is then in the range of 15–30 ps) is given by the radiative lifetime which is expected to be more in the range of $\sim 100$ ps for localized excitons in disordered QWs, as mentioned before (see also [30]).

## 3.2    Exciton–Phonon Interaction

The interaction of excitons with phonons in III–V semiconductor quantum dots is a topic intensively discussed in the literature since the early 1990s and is still under debate. Two types of dynamics influenced by the exciton–phonon interaction can be distinguished: the population *relaxation* from excited states into the exciton ground state and the *dephasing* (i.e. the homogeneous broadening) of the transition from the crystal ground state to the ground-state exciton (which we label 0–$X$). The first aspect has been largely debated since the first claims of the so-called "phonon bottleneck" effect. The second aspect has recently received renewed attention after the measurement of very long dephasing times of the 0–$X$ transition in strongly confined InGaAs QDs.

Let us first discuss the phonon bottleneck effect in the exciton relaxation in QDs. When Fermi's golden rule is used, the exciton relaxation rate via spontaneous emission of longitudinal acoustic (LA) phonons is estimated to be very inefficient (i.e. the relaxation time is larger than the radiative lifetime) in dots of small sizes with energy level separations more than a few meV. This is due to the energy conservation, requiring an acoustic phonon energy equal to the level spacing, and the energy cut-off in the exciton–phonon matrix element proportional to the inverse of the dot size. Moreover, due to the small energy dispersion of the longitudinal optical (LO) phonons, a simple consideration based on energy conservation predicts a strongly reduced exciton relaxation via emission of LO phonons unless the energy spacing equals the LO-phonon energy ([31] and references therein).

The experimental observation of the phonon bottleneck, i.e. the reduced exciton relaxation rate from the inefficient phonon emission at low temperature in QDs, has been controversial. A large number of publications indicated that the phonon bottleneck does not exist even in strongly confined InGaAs QDs. Measurements of efficient low-temperature exciton relaxation were explained at high carrier densities with Auger-like carrier–carrier scattering and at low carrier densities with multi-phonon processes [32,33] or relaxation mediated by an efficient hole–acoustic-phonon scattering in the presence of closely spaced hole levels [34,35]. Recently, the presence of a continuum of states extending at lower energies of the absorption edge of the wetting layer, which was proposed to mediate fast intradot relaxation, was reported in several photoluminescence excitation spectra of single QDs [36,35,37].

Alternatively, it was recently pointed out that, since the coupling strength of excitons with LO phonons depends on the local charge density and thus on the difference between the electron and hole wavefunctions, pyramidal InGaAs QDs with high aspect ratio can be strongly coupled to the LO phonons

beyond the prediction of Fermi's golden rule [38]. The strong coupling of excitonic states with LO phonons is solved non-perturbatively and leads to
the formation of new eigenstates: the excitonic polarons [39]. The problem
of an excitonic excited-state to ground-state transition in the presence of
$n$ LO phonons in a given mode can be solved in strict analogy with the problem of an atomic excited-state to ground-state transition in the presence of
$n$ photons of a quantized photon mode in cavity quantum-electrodynamics
(similar considerations for the electron–LO phonon coupled system can be
found [40,41,42,43]). In this picture, the exciton occupation probability oscil-
lates between the excited and the ground state with a Rabi frequency related
to the exciton–phonon coupling strength (in analogy with the optical Rabi
frequency, which in the limit of $n \gg 1$ tends to the semiclassical solution
discussed in Sect. 2.1 [3]). In other words, the coupled exciton–LO phonon
system conserves its energy which is exchanged repeatedly at the Rabi frequency between the exciton and the LO phonons. However, when the finite
lifetime of the LO phonons is included, the excited-state occupation probability undergoes a damped Rabi oscillation with a decay time which defines
the relaxation time. In a similar form, it is for example calculated that the
LO–phonon mediated electron relaxation in QDs is quite efficient both at $0\,\mathrm{K}$
and at $300\,\mathrm{K}$ [40,41,44]. This occurs for a wide energy detuning between the
electronic energy level splitting and the LO phonon energy (tens of meV, in
the order of the Rabi splitting $\hbar\omega_R$), in agreement with experiments [43,45],
thus relaxing the strict energy conservation of Fermi's golden rule. Recent
experiments indeed indicate that in flat, truncated InGaAs/GaAs quantum
dots a suppressed exciton relaxation occurs in qualitative agreement with the
expectation from Fermi's golden rule [46]. This was explained by a more similar shape of electron and hole wavefunctions and thus by a decreased strength
of the exciton–LO phonon coupling in flat, truncated dots as compared to
pyramidal dots with high aspect ratio.

When considering the dephasing time of the $0$–$X$ transition from the crystal ground state to the optically allowed ground-state exciton, the phonon-
mediated transition from the ground to excited-excitonic state (i.e. the opposite of the exciton relaxation previously discussed) introduces a $T_1$ damping
constant given by the inverse of the transition rate. While the relaxation rate
is non-zero even for zero temperature due to spontaneous phonon emission,
the activation rate which is a stimulated process in the scattering picture
tends to zero as the phonon occupation number tends to zero. This holds
also in the exciton–polaron description [39], as for an atom in its ground state
in the vacuum field which does not exhibit any Rabi oscillation [3]. Therefore, the $T_1$ phonon contribution to the dephasing of the $0$–$X$ transition at
low temperature is expected to be small. On the other hand, as it is stated
in (21), dephasing can occur without a change of the occupation probability
when pure dephasing times are important. The theoretical treatment of pure
dephasing of the $0$–$X$ transition due to exciton–phonon interaction in semi-

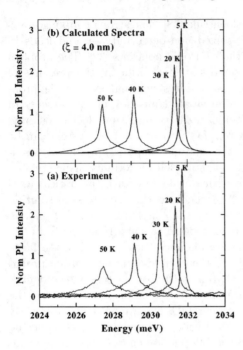

**Fig. 4.** (after [47]) (**a**) Experimental photoluminescence spectra of a single CdTe/ZnTe quantum-dot exciton. (**b**) Calculated spectra for different temperature using an exciton localization length of 4 nm

conductor QDs has been discussed in recent literature [47,48,49,50,51,52]. As pointed out in [47,50] the homogeneous lineshape of the $0-X$ transition for excitons strongly localized in QDs can be described similarly to electrons localized on lattice defects in bulk semiconductors [53]. The starting point of this treatment considers that the crystal ground state and the state where the localized exciton is optically created correspond to different adiabatic potentials for the nuclear motion. The modification of the adiabatic potential for the nuclear motion due to the electronic excitation can be expressed in different orders of the coupling of the nuclear displacement with the created electronic density. A shift of the potential minima is a first-order contribution while a change of the curvature is a second-order contribution. A relative shift of the potential minima implies that optical transitions can occur with absorption and emission of phonons, like the electronic transitions of molecules with roto-vibrational bands governed by the Franck–Condon principle.

The mathematical treatment of the exciton–phonon interaction to the first-order contribution and without involving excitonic excites states is shown in [47,51,52]. In this case the form of the exciton–phonon Hamiltonian allows for an analytical solution giving polaron eigenstates. The exciton–phonon interaction in this form does not lead to a change of the excitonic population and thus account only for pure dephasing. When LO phonons are considered, the absorption (emission) spectra calculated for (In,Ga)As QDs exhibit several side lines, separated by the LO phonon frequencies, above (below) the so-called zero (optical) phonon line (ZOPL), in agreement with experiments [38]. In the first-order contribution, the ZOPL is unbroadened [49,50,51]. When acoustic phonons are taken into account as a continuum of bulk-like acoustic phonon modes, a good approximation for dots embedded in a matrix with similar elastic properties such as (In,Ga)As QDs on (Ga,Al)As barriers, the ZOPL decomposes into a sharp zero-phonon line (ZPL) and a broad band caused by the multiple acoustic-phonon assisted transitions, which is asymmetric toward high (low) energies for the absorption (emission) spectrum at low temperature. To first-order contribution the ZPL is unbroadened, while the width of the broad band increases with decreasing the size of the dot [47]. The weight of the ZPL allows one to define the Huang–Rhys factor of the exciton–acoustic-phonon coupling, which increases with decreasing size of the dot and with increasing temperature [47,52]. Therefore, this composite non-Lorentzian lineshape of the ZOPL is expected to be more visible in strongly confined dots and was experimentally observed in II–VI epitaxially grown QDs (Fig. 4) but not in localized excitons in narrow GaAs QWs [8,18].

In the time domain, a strongly non-Lorentzian lineshape translates into a non-exponential decay of the induced polarization. This was observed in InGaAs QDs at temperatures below $100\,\mathrm{K}$ [15], showing a long exponential dephasing corresponding to the sharp ZPL and a fast initial dephasing corresponding to the broad acoustic-phonon band (Fig. 3). The experimental temperature dependence [15] of the width of the ZPL and of the broad band is shown in Fig. 5. The ZPL broadening shows a linear and an activated temperature dependence as also measured in [9], i.e. can be expressed as

$$\gamma = \gamma_0 + aT + b\frac{1}{e^{E_A/k_B T} - 1}. \tag{25}$$

The values of the parameters $\gamma_0, a, b, E_A$ ($k_B$ is Boltzmann's constant) to fit the experimental data in [15] are indicated in Fig. 5. As already mentioned, in the first-order contribution the ZPL is not broadened by pure dephasing. The measured broadening was interpreted as a $T_1$ inelastic broadening given by the radiative decay ($\gamma_0$) and by phonon-assisted transitions into other exciton states. The linear temperature dependence is *not* expected to be observed in QDs [8], contrary to bulk and QWs [14], unless transitions occur into exciton states separated by less than $k_B T$ even at low temperatures. This linear dependence was therefore attributed to transitions among the fine-structure states of the exciton ground state which are separated by only $\sim 100\,\mu\mathrm{eV}$ in InGaAs QDs [54]. However, this interpretation appears in contradiction with recent measurements of the exciton spin-flip time much longer than the radiative lifetime [55]. The activated temperature dependence was attributed to the phonon-assisted transition toward the excited exciton state separated by $E_A$ involving the first excited hole level, which is consistent with typical energy level spacings calculated in InGaAs QDs [56].

Theoretically, a broadening of the ZPL from pure dephasing occurs when second-order contributions are considered. This was shown in [49] for the

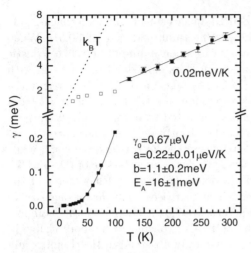

**Fig. 5.** Full width at half maximum homogeneous linewidth estimated from the dephasing time of the $0$–$X$ transition in $In_{0.7}Ga_{0.3}As/GaAs$ self-assembled quantum dots (after [15]). Below 100 K, the *open squares* are the linewidths of the broadband while the *closed squares* are the linewidths of the ZPL and the *solid line* is a fit according to (25). The *dotted line* is the thermal energy $k_BT$. Above 100 K the *solid line* is a linear fit of the homogeneous linewidth with the indicated slope

exciton–LO phonon interaction, with the inclusion of exciton excited states. The resulting ZPL broadening is estimated to have an activated behavior at low temperature and vanishes at $T = 0$. A ZPL broadening from pure dephasing also occurs in the first-order contribution when introducing phonon damping from a finite phonon lifetime [57]. This has been recently proposed for the exciton–LA phonon interaction [52] as an alternative explanation of the ZPL broadening below 100 K measured in [15].

The weak temperature dependence of the width of the broad band caused by the multiple acoustic-phonon assisted transitions is shown in Fig. 5. If $k_BT$ is bigger than the cutoff energy in the exciton–acoustic-phonon coupling, roughly given by the inverse of the localization length, all the phonon modes that contribute to the coupling are occupied and the width of the broad band is given by the cutoff energy [15]. However, the experimental data are still awaiting a clear comparison with theoretical expectations.

At higher temperatures from 100 K to 300 K the homogeneous lineshape of InGaAs QDs is dominated by a broadening in the few-meV range which increases approximately linearly with temperature [9,15,58] (Fig. 5) and is comparable in value with the broadening of excitons in $(In,Ga)As/(Al,Ga)$ QWs [14,59]. Its interpretation its more complex since the role of the exciton excited states in the modelling of the dephasing (inelastic and pure) cannot be neglected [49]. Moreover, both optical and acoustic phonons have to be taken into account, which still awaits a theoretical treatment in the coupled exciton–phonon model including the excited states.

## 3.3    Coulomb Interactions

Of particular importance are the Coulomb interactions occurring in a QD beyond the one-exciton limit. Pauli blocking allows no more than two carriers to occupy the lowest lying two-fold spin degenerate states of electrons ($e_0$) and holes ($h_0$) in a QD, which form the ground-state biexciton. Additional carriers have to occupy higher states. At high ($> 2$) exciton occupancies, single-dot photoluminescence spectra with a large number of emission lines are observed [60,61,62,63]. In fact, the transition energy of the optical recombination of one electron–hole pair depends on the number of additional electrons and holes present in the dot as "spectators" due to the Coulomb interaction between carriers [64,65] (see also the Chapter by Hawrylak and Korkusiński in this volume). Most of the work reported so far on biexcitons and multiexcitons in (In,Ga)As QDs focused on the energy position and on the strength of the optical transitions, while their homogeneous broadening has been addressed only very recently.

It is generally found that with increasing number of carriers, both inside the dots and in the barrier/wetting layer material, the homogeneous broadening of excitonic ground-state transitions, i.e. transitions that involve the recombination of one $e_0 h_0$ pair, increases [66,67,68,69,70,71,72,73]. Two mechanisms of homogeneous broadening due to Coulomb interactions were found to be important. (i) Pure dephasing due to Coulomb interaction of the carriers in the QD with delocalized carriers in the barrier/wetting layer material. This broadening has been investigated in self-assembled InGaAs QDs theoretically [67,70] and experimentally [69,72,73]. (ii) Final state damping given by the rapid population relaxation of the excited multiexciton with only one $e_0 h_0$ pair into its ground state with two $e_0 h_0$ pairs, mediated by Auger-like intradot carrier–carrier scattering and phonon emission in the presence of several carriers in the excited states. The second type of broadening has been investigated experimentally only recently in InGaAs QDs [71,72,73].

The dephasing time of excitonic ground-state transitions is measured in [72,73] with FWM in electrically pumped InGaAs QDs. At low temperature the FWM decay of the GS transitions is dominated by the zero-phonon line, and the dephasing due to Coulomb interactions with the injected carriers is clearly distinguished [72]. With increasing injection current, the $0–X$ transition from the crystal ground state to the ground-state exciton is found to be broadened dominantly by pure dephasing from Coulomb interactions with carriers in the barrier/wetting layer material. At high injection currents such that the majority of dots is occupied with two $e_0 h_0$ pairs, the dephasing of the biexciton to exciton transition (labelled $XX–X$) is measured. At low temperature, also the $XX–X$ transition is dominantly broadened by pure dephasing from Coulomb interactions with carriers in the barrier/wetting

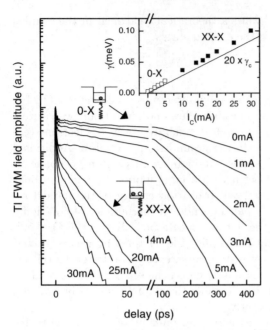

**Fig. 6.** Time-integrated four-wave mixing at 10 K and different injection currents in electrically pumped $In_{0.7}Ga_{0.3}As/GaAs$ self-assembled quantum dots (after [72]). The dependence of the homogeneous broadening on injection current deduced from the signal decay at long delays is shown in the *inset* (*squares*), together with twenty times the estimated carrier capture rate per dot (*line*). Sketches of the $0–X$ and $XX–X$ transitions probed in the experiment are shown (*curly arrows* are the interacting photons)

layer material with the same dependence on injection current as for the $0–X$ transition (Fig. 6). The temperature dependence of the $XX–X$ homogeneous broadening extrapolated to zero injection current is also inferred and it is found to obey to (25) with the following parameters: $\gamma_0 = 2\,\mu eV$, $a = 0.37 \pm 0.04\,\mu eV/K$, $b = 1.1 \pm 0.2\,meV$, $E_A = 16 \pm 1\,meV$. The identical thermally activated part of the $0–X$ and $XX–X$ transitions indicates correlated exciton and biexciton scattering via phonon absorption, similar to observations in InGaAs/GaAs and GaAs/AlGaAs quantum wells [14,74] and in CdSe/ZnSe quantum dots [75]. The linear temperature dependence with a coefficient approximately twice the one of the $0–X$ transition is not consistent with the interpretation of this trend as being due to transitions among the fine-structure states of the exciton ground state which are separated by less than $k_B T$. In fact, within this picture the linear coefficients for the $0–X$ and $XX–X$ transitions should be equal, since $XX$ is a singlet spin state as the 0 state and thus does not have a fine-structure splitting. $\gamma_0 = 2\,\mu eV$ corresponds to three times the exciton radiative broadening, indicating an

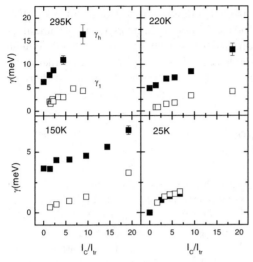

**Fig. 7.** Homogeneous broadening $\gamma_h$ (*solid squares*) of the $X^n$–$(X^{n-1})^*$ transitions deduced from the dephasing time measured with FWM in electrically-pumped $In_{0.7}Ga_{0.3}As/GaAs$ self-assembled quantum dots (after [73]). The homogeneous broadening of the $0$–$X$ transition without electrical injection is shown for comparison. $\gamma_h$ is compared with the broadening $\gamma_l$ (*open squares*) deduced from the population relaxation of the excited multiexciton $(X^{n-1})^*$ measured with differential transmission spectroscopy. $I_C/I_{tr}$ is the injection current normalized to the value at transparency

uncorrelated radiative decay of the three involved excitons in the $XX$–$X$ transition.

The dephasing of the optical transition from the ground-state multiexciton with two $e_0h_0$ pairs and additional carriers in the excited states to the excited multiexciton with one less $e_0h_0$ pair (labelled $X^n$–$(X^{n-1})^*$, $n > 2$) is measured in electrically pumped InGaAs QDs in [72,73]. At low temperature ($< 30\,K$), the dephasing is found to be dominated by the rapid relaxation of the excited multiexciton to its ground state which was measured with differential transmission spectroscopy (Fig. 7). With increasing temperature, pure dephasing mechanisms due to phonon interactions become important in the homogeneous broadening of the $X^n$–$(X^{n-1})^*$ transition. With the occurrence of thermal equilibrium among the dots in the ensemble and with the wetting layer ($T > 150\,K$), pure dephasing by Coulomb interactions with carriers in the wetting layer becomes important and increases with increasing injection current and temperature. At room temperature and with high injection current, a large homogeneous broadening above $10\,meV$ is measured. This value of the homogeneous broadening was found to depend on the carrier confinement energy in the dots, mainly via the change of the thermal population of carriers in the wetting layer for different confinement energies and thus of the pure dephasing from Coulomb interactions. InGaAs QDs with confinement energies of $\sim 100\,\text{meV}$ showed homogeneous broadenings $> 25\,\text{meV}$ [69], while strongly confined dots with $\sim 300\,\text{meV}$ confinement energy showed a homogeneous broadening of $\sim 9\,\text{meV}$ at room temperature and comparably high injection current [76], consistent with a reduced thermal carrier population in the wetting layer for large confinement energies.

## 4    Optical Rabi Oscillations in (In,Ga)As Quantum Dots

In Sect. 2 we have calculated the time evolution of the population difference for a two-level system interacting with a strong coherent light field by solving the optical Bloch equations (Figs. 1 and 2). Significant oscillations in the population difference occur at resonance for large Rabi frequencies (proportional to the electric field amplitude and the transition dipole moment) and long dephasing times. This oscillatory behavior of the population difference versus time was observed in early times in molecules [77] and recently also in semiconductor quantum wells [78], where however many-body Coulomb interactions strongly modify the phenomenon compared to the two-level solution.

Recently the optical Rabi oscillations in semiconductor quantum dots have received a lot of attention, besides their fundamental interest, for their potential application in quantum computing. It is proposed that a resonant Rabi rotation using a pulsed excitation with $\pi$ pulse area, where the pulse area is the time integrated Rabi frequency (see Sect. 2): $\theta = \int_{-\infty}^{\infty} \omega_R(t)\mathrm{d}t = (\hbar)^{-1} \int_{-\infty}^{\infty} \boldsymbol{\mu} \cdot \boldsymbol{E}_0(t)\mathrm{d}t$, can coherently flop the occupation probabilities of the crystal ground state 0 with the one exciton state $X$ in the QD, or the one exciton state $X$ with the biexciton state $XX$. Using the polarization selection rules for the excitonic transitions and considering that the $X$–$XX$ transition is frequency shifted from the 0–$X$ transition by the biexciton binding energy, logic operations between two quantum bits are feasible in QDs if the pulse duration is much shorter than the dephasing time of the optical excitonic transitions [79].

In analogy with the oscillations versus time, optical Rabi oscillations of the population inversion which is left *after* a pulsed excitation occur *versus pulse area*. In Fig. 8 the solution of the population difference $s_3'(t \gg t_0)$ from (14) is shown versus pulse area and versus the energy detuning $\hbar(\omega - \omega_0)$, considering a Gaussian pulse $\boldsymbol{E}_0(t) = \boldsymbol{E}_{00} e^{-(t/t_0)^2}$ with $t_0 = 1\,\text{ps}$ and with the initial condition $s_3'(t \ll t_0) = -1$.

Rabi oscillations versus pulse area were measured first in Rb atoms [80]. The first observation in single semiconductor quantum dots has been reported at low temperature for localized excitons in narrow GaAs QWs [81]. The transition dipole moment of the excitonic ground-state transition was inferred to be between 50 and 100 Debye [82], i.e. more than one order of magnitude

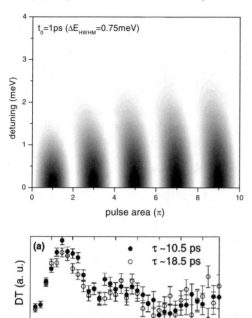

**Fig. 8.** Contour plot (*black* = +1, *white* = −1) of the population difference $s'_3(t \gg t_0)$ versus detuning and pulse area, after solving numerically (14) for a Gaussian driving field $\boldsymbol{E}_0(t) = \boldsymbol{E}_{00}e^{-(t/t_0)^2}$ with $t_0 = 1\,\mathrm{ps}$ and the initial condition $s'_3(t \ll t_0) = -1$. The energy half-width at half maximum of the pulse intensity in the spectral domain is also indicated

**Fig. 9.** Measured probe differential transmission versus pump field amplitude, after the excitation of isolated excitons localized by the lateral disorder in a thin GaAs quantum well at 6 K with a 6 ps pump pulse (after [81]). Two data sets are shown for different pump-probe delays and show Rabi oscillations of the exciton occupation probability

larger than in atoms, while the dephasing time is of a few tens of picoseconds at low temperature (see Sect. 3.1). Strongly damped Rabi oscillations versus pulse area were measured (Fig. 9). The damping was interpreted as due to Coulomb interactions with nearly degenerate delocalized excitons present in these weakly confined systems, which are excited non-resonantly, resulting in an exciton relaxation rate that increases with increasing excitation intensity.

Self-assembled InGaAs quantum dots are good candidates for the observation of Rabi oscillations, due to the long dephasing time (∼1 ns) of the excitonic ground-state transition at low temperature [9,15,16], the strong confinement energies in the range of 100–300 meV and the transition dipole moments expected to be of several tens of Debye [84,85]. First experiments on InGaAs QDs showed indications of optical Rabi oscillations versus pulse area for the optical transition from the crystal ground state to an *excited* exciton state [86], which has a dephasing time 5–10 times shorter than the excitonic ground-state transition [35]. In the experiment only the first maximum and minimum of the oscillation was observed. The measured amplitude of the oscillation was reduced compared to the ideal case shown in Fig. 8, proba-

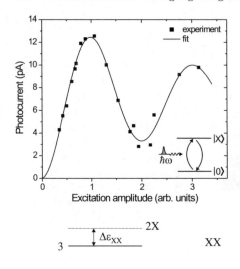

**Fig. 10.** Rabi oscillations in the photocurrent measured on a single InGaAs quantum-dot photodiode, under picosecond optical excitation resonant to the excitonic ground-state transition, versus the excitation field amplitude (after [83])

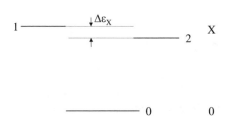

**Fig. 11.** Sketch of the four-level system describing the crystal ground-state (0), the exciton ($X$) and biexciton ($XX$) resonances in a single quantum dot. The biexciton binding energy and the exciton fine-structure splitting are indicated as $\Delta\varepsilon_{XX}$ and $\Delta\varepsilon_X$, respectively

bly due to the dephasing. An optical Stark splitting (the signature of Rabi oscillations in the spectral domain) in the photoluminescence spectrum of the excitonic ground state was also observed under continuous wave resonant excitation of an excitonic excited state in InGaAs QDs [87].

Experiments on optical Rabi oscillations of the excitonic *ground-state transition* versus pulse area in self-assembled InGaAs QDs have been reported recently [83,88]. In [83] coherent photocurrent spectroscopy was performed on a single InGaAs QD photodiode by resonant picosecond excitation of the excitonic ground state at low temperature. The amplitude of the photocurrent versus pulse area reflects the Rabi oscillation of the exciton occupancy. A clear first maximum and minimum in the amplitude of the photocurrent versus pulse area are reported (Fig. 10). The amplitude of the measured oscillation is still reduced as compared to the full modulation shown in Fig. 8 indicating that the system deviates from the ideal two-level model without damping.

A systematic study of the effect of the dephasing and of the biexciton resonance on the Rabi oscillations versus pulse area of the excitonic ground-state transition in an InGaAs QD *ensemble* was performed in [88,89]. The biexciton resonance is included by considering a four-level model as sketched

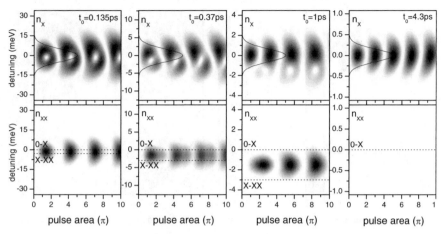

**Fig. 12.** Contour plot (*black* = 1, *white* = 0) of the excitonic ($n_X$) and biexcitonic ($n_{XX}$) population versus detuning and pulse area, determined by solving numerically (24) for the four-level system in Fig. 9 with a Gaussian electric field $E_0(t) = E_{00}e^{-(t/t_0)^2}$ for different values of $t_0$ as indicated. An infinitely long dephasing time is considered. The pulse intensity $|E_0(\omega)|^2$ in the spectral domain is indicated (*solid line*). *Dotted lines* indicate the energy position of the ground-state-to-exciton and exciton–biexciton transitions

in Fig. 11 and solving (24) for the corresponding $4 \times 4$ density matrix. In Fig. 12 the solution for the exciton ($n_X$) and biexciton ($n_{XX}$) occupation probability is shown versus detuning and pulse area as for Fig. 8 in the simplified case of zero exciton fine-structure splitting ($\Delta \varepsilon_X = 0$), linearly polarized field along the 0-1-3 transition, same $0$–$X$ and $X$–$XX$ transition dipole moment, $\Delta \varepsilon_{XX} = 3\,\text{meV}$ biexciton binding energy and without damping mechanisms, i.e. infinitely long dephasing times. Different pulse durations $t_0$ of Gaussian electric fields are considered, corresponding to FWHM of the intensity spectrum ranging from 11.5 meV to 0.36 meV, i.e. larger or smaller than the biexciton binding energy.

The inclusion of the biexciton resonance in the calculations creates a biexciton density which has the maximum oscillation amplitude for a pulse carrier frequency in resonance with *half* the $0$–$XX$ transition energy. When the pulse spectral width is larger than the biexciton binding energy a complex interplay in the oscillations of the excitonic and biexcitonic occupation probabilities occurs. When the pulse spectral width is smaller than the biexciton binding energy, the oscillation of $n_X$ and $n_{XX}$ are distinct, with $n_{XX}$ oscillating with a longer period than $n_X$ (see Fig. 12 for $t_0 = 1\,\text{ps}$). When the pulse spectral width is smaller than half of the biexciton binding energy, a pulse in resonance with the $0$–$X$ transition is completely out of resonance for the $XX$ occupation, and its interaction with the $0$–$X$ transition recovers the form of the solution of a two-level system shown in Fig. 8.

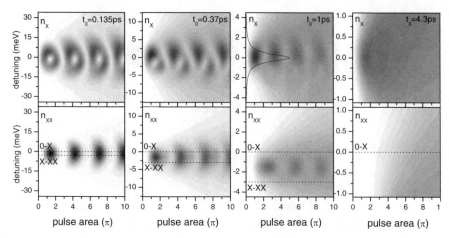

**Fig. 13.** As Fig. 12, but for a finite dephasing time of 1.5 ps for both the $0$–$X$ and $X$–$XX$ transitions. The Lorentzian absorption cross-section of the $0$–$X$ transition is shown (*solid line*)

The effect of a finite dephasing time on the Rabi oscillations versus pulse area was also studied in [88,89]. In Fig. 13, the solution for the exciton and biexciton occupation probability is shown versus detuning and pulse area as for Fig. 12 with the inclusion of a dephasing time $T_2 = 1.5$ ps for both the $0$–$X$ and $X$–$XX$ transitions. The inclusion of a finite dephasing reduces the *amplitude* of the Rabi oscillations and adds a broadening versus the detuning, as compared to the case without damping shown in Fig. 12. When the dephasing time is much shorter than the pulse duration, the incoherent regime is reached and Rabi oscillations do not occur (see the case for $t_0 = 4.3$ ps in Fig. 13). It is important to point out that when the dephasing time is longer than the pulse duration and the oscillations are visible, several Rabi oscillations versus pulse area occur. Thus a finite dephasing quenches the visibility but does not strongly damp the oscillations versus pulse area since the interaction time with the pulse is constant.

In [88,89] the differential transmission ($\Delta T/T$) of a probe pulse spectrally degenerate to a strong pump pulse exciting the ground-state excitonic transition of an InGaAs QD ensemble embedded in an optical waveguide was measured at 10 K for different pump pulse durations. The experimental data are shown in Fig. 14. The indicated pulse durations are the FWHM of the field intensity and correspond to the values of $t_0$ used in the calculations shown in Figs. 12 and 13. Rabi oscillations strongly damped versus pulse area were measured in the ensemble. The pulse area at the first minimum in the oscillations slightly changed for different pulse durations (see Fig. 14). A sizeable two-photon absorption (TPA) occurred in the waveguide at high peak pump–pulse intensities, i.e. for short pulse duration, which partly covered the observation of Rabi oscillations (see dotted lines in Fig. 14a). With the use

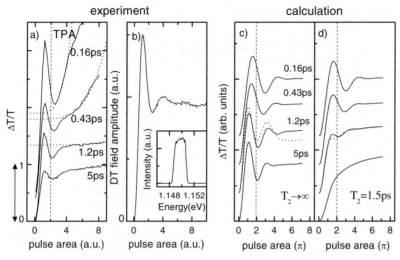

**Fig. 14.** Measured and calculated differential transmission intensity for a degenerate pump-probe experiment in resonance with the excitonic ground-state transition of an InGaAs QD ensemble embedded in an optical waveguide (after [89]). (**a**) Time-integrated differential transmission probe intensity versus pump pulse area measured for different pump pulse durations (intensity FWHM) as indicated. *Curves* are vertically displaced for clarity. The *vertical dashed line* is a mark for the eyes. *Dotted lines* show the effect of two-photon absorption (TPA). (**b**) Differential transmission probe field amplitude measured for a spectrally square–shaped pulse. The pulse spectrum is shown in the *inset*. (**c**) Differential transmission intensity calculated for different pulse durations and for an infinitely long dephasing time, vertically displaced for clarity. *Dotted lines* are calculations for a spectrally square–shaped pulse of $\sim 1\,\mathrm{ps}$ duration. The $2\pi$ pulse area is marked with a *dashed line*, for reference. (**d**) As (**c**) for 1.5 ps dephasing time

of a spectrally square-shaped pulse of $\sim 1\,\mathrm{ps}$ duration, a second oscillation maximum could be distinguished in the experiment (Fig. 14b). Simulations (see Fig. 14c,d) of $\Delta T/T$ including the integration over the inhomogeneous broadening of the transition energies and the spatial averaging over a hyperbolic secant field profile of the transverse electric waveguide mode in the QD plane indicated that the measured strong damping versus pulse area is due to a *distribution of transition dipole moments in the ensemble*, while the change of the pulse area at the oscillation minimum with pulse duration is due to the biexciton resonance. Calculations performed for a spectrally square-shaped pulse also indicated an improved visibility of the second oscillation maximum in this case (dotted lines in Fig. 14c,d). These findings pointed out how nonuniformity in a dot ensemble is critical for the observation of Rabi oscillations, while Rabi oscillations in a single InGaAs quantum dot were calculated to be quite "robust" as a function of pulse area even when dephasing processes and biexcitonic effects were included.

# 5   Summary and Outlook

In summary, the understanding of the mechanisms responsible of the excitonic dephasing in quantum dots is undergoing a rapid and very active evolution after the recent experimental progress in addressing the homogeneous broadening by isolating excitonic emission lines of single quantum dots or directly measuring the exciton dephasing time with transient coherent spectroscopy in quantum-dot ensembles. The achieved results have stimulated several theoretical discussions, in major part still active and open. The present stage of study in the field of excitonic dephasing in III–V epitaxially grown quantum dots can be summarized as follows. (i) At low temperature (∼5 K) the dephasing time of the ground-state exciton in strongly confined InGaAs QDs is in the nanosecond range as compared to a dephasing time in the few tens of picoseconds for localized excitons in thin GaAs quantum wells. This is only partly explained by the change in the exciton radiative lifetime from the mesoscopic to the strong confinement (see Sect. 3.1) since processes additional to the radiative decay are present, the origin of which is still under discussion. (ii) The influence of the exciton–phonon coupling is presently under intense discussion, in particular after the measurement of a temperature-dependent dephasing which is non-exponential, corresponding a non-Lorentzian homogeneous lineshape, in strongly confined InGaAs dots. Current theoretical reports show the need to solve the coupled exciton–phonon system non-perturbatively in these dots. In such a model, the optical transition that involves the creation or emission of a ground-state exciton might also involve the absorption or emission of phonons, with the consequent formation of optical phonon satellites and an acoustic–phonon broadband superimposed on a zero-phonon line. Quantitative comparisons between experiments and theory of the temperature-dependent width of the ZPL, of the broad band and of the ZPL weight are in progress (see Sect. 3.2). (iii) The experimental study of Coulomb interactions on the ground-state exciton dephasing time has revealed two important mechanisms: (a) a pure dephasing due to Coulomb interaction of the carriers in the QD with delocalized carriers in the barrier/wetting layer material and (b) a final state damping given by the rapid population relaxation of multiexcitons with one less exciton in the ground state. Moreover first experimental studies on the temperature-dependent dephasing of the biexciton to exciton transition have revealed a correlated scattering of the exciton and the biexciton with phonons (see Sect. 3.3).

In the application of quantum dots in opto-electronics devices such as lasers/amplifiers, the homogeneous broadening is an important parameter (see also [90]). It sets the intrinsic limit to a delta-function-like density of states which is the key property for expected superior performance of quantum-dot compared to bulk and quantum-well devices (see also the Chapter by Schweizer et al. in this volume). The study in Sect. 3.3 shows that barrier engineering can reduce the pure dephasing from Coulomb interactions with carrier in the barrier/wetting layer in electrically pumped InGaAs QD lasers/amplifiers. However, a homogeneous broadening of at least a few meV is measured at room temperature even in strongly confined InGaAs QDs, simply due to phonon interaction. Moreover, the rapid relaxation of multi-excitons occurring in a QD in the gain case under strong carrier injection (Sect. 3.3) adds a further contribution to the homogeneous broadening which can also be up to a few meV. These values should be taken into account in future design of QD devices. Indication of a lasing behavior dominated by the homogeneous broadening has already been reported in the recent literature for InGaAs QD lasers [91].

Let us also briefly remark that the measured long dephasing times in InGaAs at low temperature give promising perspectives for the application of excitonic transitions in quantum information processing. In this respect, the demonstration of a coherent manipulation of the excitonic population which has been achieved recently in single localized excitons in thin GaAs QWs, in single InGaAs QDs and in InGaAs QD ensembles (Sect. 4) appears very attractive. In fact, strong coherent light–matter coupling resulting in a $\pi$ rotation of the excitonic population is feasible in InGaAs QDs with optical pulses of $\sim 1$ ps duration, more than two orders of magnitude shorter than the dephasing time at low temperature.

## Acknowledgements

We would like to acknowledge the support and collaboration with Prof. U. Woggon, Prof. D. Bimberg, Prof. J. M. Hvam, Dr. S. Schneider and fruitful discussions with Prof. R. Zimmermann and Prof. A. Knorr. Paola Borri is supported by the European Union with the Marie Curie Individual Fellowship contract nr. HPMF-CT-2000-00843.

# References

1. L. I. Schiff: *Quantum Mechanics* (McGraw-Hill, Singapore 1968)
2. R. Loudon: *The Quantum Theory of Light* (Oxford Science Publications, Oxford 1983)
3. L. Allen, J. H. Eberly: *Optical Resonance and Two-Level Atoms* (Wiley, New York 1975)
4. Y. R. Shen: *The Principles of Nonlinear Optics* (Wiley, New York 1984)
5. J. Shah: *Ultrafast Spectroscopy of Semiconductors and Semiconductor Nanostructures* (Springer, Berlin, Heidelberg 1996)
6. D. Bimberg, M. Grundmann, N. N. Ledentsov: *Quantum Dot Heterostructures* (Wiley, Chichester 1999)
7. W. Langbein, J. M. Hvam, R. Zimmermann: Phys. Rev. Lett. **82**, 1040 (1999)
8. D. Gammon, E. S. Snow, B. V. Shanabrook, D. S. Katzer, D. Park: Science **273**, 87 (1996)
9. M. Bayer, A. Forchel: Phys. Rev. B **65**, 041308 (2002)

10. W. Demtröder: *Laser Spectroscopy* (Springer, Berlin, Heidelberg 1998)
11. C. F. Klingshirn: *Semiconductor Optics* (Springer, Berlin, Heidelberg 1995)
12. X. Fan, T. Takagahara, J. E. Cunningham, H. Wang: Solid State Commun. **108**, 857 (1998)
13. J. Erland, J. C. Kim, N. H. Bonadeo, D. G. Steel, D. Gammon, D. S. Katzer: Phys. Rev. B **60**, R8497 (1999)
14. P. Borri, W. Langbein, J. M. Hvam, F. Martelli: Phys. Rev. B **60**, 4505 (1999)
15. P. Borri, W. Langbein, S. Schneider, U. Woggon, R. Sellin, D. Ouyang, D. Bimberg: Phys. Rev. Lett. **87**, 157401 (2001)
16. D. Birkedal, K. Leosson, J. M. Hvam: Phys. Rev. Lett. **87**, 227401 1 (2001)
17. N. H. Bonadeo, G. Chen, D. Gammon, D. S. Katzer, D. Park, D. G. Steel: Phys. Rev. Lett. **81**, 2759 (1998)
18. J. Guest, T. H. Stievater, G. Chen, E. A. Tabak, B. G. Orr, D. G. Steel, D. Gammon, D. S. Katzer: Science **293**, 2224 (2001)
19. M. Sugawara: Phys. Rev. B **51**, 10743 (1995)
20. J. Bellessa, V. Voliotis, R. Grousson, X. L. Wang, M. Ogura, H. Matsuhata: Phys. Rev. B **58**, 9933 (1998)
21. D. S. Citrin: Phys. Rev. B **48**, 2535 (1993)
22. L. C. Andreani: in E. Burstein, C. Weisbuch (Eds.): *Confined Electrons and Photons: New Physics and Applications*, NATO Sci. Ser. **340** (Plenum, New York 1995) pp. 57–112
23. A. Fiore, P. Borri, W. Langbein, J. M. Hvam, U. Oesterle, R. Houdré, R. P. Stanley, M. Ilegems: Appl. Phys. Lett. **76**, 3430 (2000)
24. R. Heitz, H. Born, T. Lüttgert, A. Hoffmann, D. Bimberg: Phys. Stat. Solidi (b) **221**, 65 (2000)
25. E. Dekel, D. V. Regelman, D. Gershoni, E. Ehrenfreund, W. V. Schoenfeld, P. M. Petroff: Solid State Comm. **117**, 395 (2001)
26. K. Leosson, J. R. Jensen, J. M. Hvam, W. Langbein: Phys. Stat. Solidi (b) **221**, 49 (2000)
27. K. Ota, N. Usami, Y. Shiraki: Physica E **2**, 573 (1998)
28. V. Türck, S. Rodt, O. Stier, R. Heitz, R. Engelhardt, U. W. Pohl, D. Bimberg, R. Steingrüber: Phys. Rev. B **61**, 9944 (2000)
29. H. D. Robinson, B. B. Goldberg: Phys. Rev. B **61**, R5086 (2000)
30. W. Langbein, R. Zimmermann, E. Runge, J. M. Hvam: Phys. Stat. Solidi (b) **221**, 349 (2000)
31. U. Bockelmann: Phys. Rev. B **48**, 17637 (1993)
32. B. Ohnesorge, M. Albrecht, J. Oshinowo, A. Forchel, Y. Arakawa: Phys. Rev. B **54**, 11532 (1996)
33. R. Heitz, M. Veit, N. N. Ledentsov, A. Hoffmann, D. Bimberg, V. M. Ustinov, P. S. Kop'ev, Z. I. Alferov: Phys. Rev. B **56**, 10435 (1997)
34. T. S. Sosnowski, T. B. Norris, H. Jiang, J. Singh, K. Kamath, P. Bhattacharya: Phys. Rev. B **57**, R9423 (1998)
35. H. Htoon, D. Kulik, O. Baklenov, A. L. Holmes Jr., T. Takagahara, C. K. Shih: Phys. Rev. B **63**, 241303 (2001)
36. Y. Toda, O. Moriwaki, M. Nishioka, Y. Arakawa: Phys. Rev. Lett. **82**, 4114 (1999)
37. C. Kammerer, G. Cassabois, C. Voisin, C. Delalande, P. Roussignol, A. Lemaitre, J. M. Gérard: Phys. Rev. B **65**, 033313 (2001)
38. R. Heitz, I. Mukhametzhanov, O. Stier, A. Madhukar, D. Bimberg: Phys. Rev. Lett. **83**, 4654 (1999)

39. O. Verzelen, R. Ferreira, G. Bastard: Phys. Rev. Lett. **88**, 146803 (2002)
40. O. Verzelen, R. Ferreira, G. Bastard: Phys. Rev. B **62**, R4809 (2000)
41. X.-Q. Li, H. Nakayama, Y. Arakawa: Phys. Rev. B **59**, 5069 (1999)
42. S. Hameau, Y. Guldner, O. Verzelen, R. Ferreira, G. Bastard, J. Zeman, A. Lemaître, J. M. Gérard: Phys. Rev. Lett. **83**, 4152 (1999)
43. S. Sauvage, P. Boucaud, R. P. S. M. Lobo, F. Bras, G. Fishman, R. Prazeres, F. Glotin, J. M. Ortega, J. M. Gérard: Phys. Rev. Lett. **88**, 177402 (2002)
44. L. Jacak, J. Krasnyj, D. Jacak, P. Machinkowski: Phys. Rev. B **65**, 113305 (2002)
45. S. Marcinkevičius, A. Gaarder, R. Leon: Phys. Rev. B **64**, 115307 1 (2001)
46. R. Heitz, H. Born, F. Guffarth, O. Stier, A. Schliwa, A. Hoffmann, D. Bimberg: Phys. Rev. B **64**, 241305(R) 1 (2001)
47. L. Besombes, K. Kheng, L. Marsal, H. Mariette: Phys. Rev. B **63**, 155307 (2001)
48. T. Takagahara: Phys. Rev. B **60**, 2638 (1999)
49. A. V. Uskov, A.-P. Jauho, B. Tromborg, J. Mørk, R. Lang: Phys. Rev. Lett. **85**, 1516 (2000)
50. S. V. Goupalov, R. A. Suris, P. Lavallard, D. S. Citrin: Nanotechnol. **12**, 518 (2001)
51. B. Krummheuer, V. M. Axt, T. Kuhn: Phys. Rev. B **65**, 195313 (2002)
52. R. Zimmermann, E. Runge: in J. H. Davies, A. R. Long (Eds.): Proc. 26th Int. Conf. on the Physics of Semiconductors (Institute of Physics Publishing, Bristol 2002)
53. G. Mahan: *Many-Particle Physics* (Plenum, New York 1990)
54. M. Bayer, A. Kuther, A. Forchel, A. Gorbunoy, V. B. Timofeev, F. Schäfer, J. P. Reithmaier, T. L. Reinecke, S. N. Walck: Phys. Rev. Lett. **82**, 1748 (1999)
55. M. Paillard, X. Marie, R. Renucci, T. Amand, A. Jbeli, J. M. Gérard: Phys. Rev. Lett. **86**, 1634 (2001)
56. O. Stier, M. Grundmann, D. Bimberg: Phys. Rev. B **59**, 5688 (1999)
57. S. V. Goupalov, R. A. Suris, P. Lavallard, D. S. Citrin: in Zh. Alferov, L. Esaki (Eds.): *10th Int. Symp. on Nanostructures: Physics and Technology*, SPIE Proc. **5023** (SPIE Bellingham WA, 2003) p. 235
58. K. Matsuda, K. Ikeda, T. Saiki, H. Tsuchiya, H. Saito, K. Nishi: Phys. Rev. B **63**, 121304 (2001)
59. P. Borri, W. Langbein, J. Mørk, J. M. Hvam, F. Heinrichsdorff, M.-H. Mao, D. Bimberg: Phys. Rev. B **60**, 7784 (1999)
60. E. Dekel, D. Gershoni, E. Ehrenfreund, D. Spektor, J. M. Garcia, P. M. Petroff: Phys. Rev. Lett. **80**, 4991 (1998)
61. A. Zrenner: J. Chem. Phys. **112**, 7790 (2000)
62. A. Hartmann, Y. Ducommun, E. Kapon, U. Hohenester, E. Molinari: Phys. Rev. Lett. **84**, 5648 (2000)
63. E. Dekel, D. V. Regelman, D. Gershoni, E. Ehrenfreund, W. V. Schoenfeld, P. M. Petroff: Phys. Rev. B **62**, 11038 (2000)
64. F. Troiani, U. Hohenester, E. Molinari: Phys. Rev. B **62**, R2263 (2000)
65. A. J. Williamson, A. Franceschetti, A. Zunger: Europhys. Lett. **53**, 59 (2001)
66. H. Kamada, J. Temmyo, M. Notomi, T. Furuta, T. Tamamura: Jpn. J. Appl. Phys. **37**, 4194 (1997)
67. A. V. Uskov, K. Nishi, R. Lang: Appl. Phys. Lett. **74**, 3081 (1999)
68. K. Matsuda, T. Saiki, H. Saito, K. Nishi: Appl. Phys. Lett. **76**, 73 (2000)

69. P. Borri, W. Langbein, J. M. Hvam, F. Heinrichsdorff, M.-H. Mao, D. Bimberg: Appl. Phys. Lett. **76**, 1380 (2000)
70. A. V. Uskov, I. Magnusdottir, B. Tromborg, J. Mørk, R. Lang: Appl. Phys. Lett. **79**, 1679 (2001)
71. Z. L. Yuan, E. R. D. Foo, J. F. Ryan, D. J. Mowbray, M. S. Skolnick, M. Hopkinson: Phys. Stat. Solidi (b) **224**, 409 (2001)
72. P. Borri, W. Langbein, S. Schneider, U. Woggon, R. L. Sellin, D. Ouyang, D. Bimberg: Phys. Rev. Let. **89**, 187401 (2002)
73. P. Borri, W. Langbein, S. Schneider, U. Woggon, R. L. Sellin, D. Ouyang, D. Bimberg: IEEE J. Sel. Topics Quant. Electron **8**, 984 (2002)
74. W. Langbein, J. M. Hvam: Phys. Rev. B **61**, 1692 (2000)
75. F. Gindele, K. Hild, W. Langbein, U. Woggon: Phys. Rev. B **60**, R2157 (1999)
76. P. Borri, S. Schneider, W. Langbein, U. Woggon, A. E. Zhukov, V. M. Ustinov, N. N. Ledentsov, Z. I. Alferov, D. Ouyang, D. Bimberg: Appl. Phys. Lett. **79**, 2633 (2001)
77. R. G. Brewer, R. L. Shoemaker: Phys. Rev. Lett. **27**, 631 (1971)
78. R. Schülzgen, R. Binder, M. E. Donovan, M. Lindberg, K. Wundke, H. M. Gibbs, G. Khitrova, N. Peyghambarian: Phys. Rev. Lett. **82**, 2346 (1999)
79. P. Chen, C. Piermaroccchi, L. J. Sham: Phys. Rev. Lett. **87**, 067401 (2001)
80. H. M. Gibbs: Phys. Rev. A **8**, 446 (1973)
81. T. H. Stievater, X. Li, D. G. Steel, D. S. Katzer, D. Park, C. Piermarocchi, L. Sham: Phys. Rev. Lett. **87**, 133603 (2001)
82. J. R. Guest, T. H. Stievater, X. Li, J. Cheng, D. G. Steel, D. Gammon, D. S. Katzer, D. Park, C. Ell, A. Thränhardt, G. Khitrova, H. M. Gibbs: Phys. Rev. B **65**, 241310(R) (2002)
83. A. Zrenner, E. Beham, S. Stufler, F. Findeis, M. Bichler, G. Abstreiter: Nature **418**, 612 (2002)
84. L. C. Andreani, G. Panzarini, J.-M. Gérard: Phys. Rev. B **60**, 13276 (1999)
85. G. Panzarini, U. Hohenester, E. Molinari: Phys. Rev. B **65**, 165322 (2002)
86. H. Htoon, T. Tagakahara, D. Kulik, O. Baklenov, A. L. H. Jr., C. K. Shih: Phys. Rev. Lett. **88**, 087401 (2002)
87. H. Kamada, H. Gotoh, J. Temmyo, T. Tagakahara, H. Ando: Phys. Rev. Lett. **87**, 246401 (2001)
88. P. Borri, W. Langbein, S. Schneider, U. Woggon, R. L. Sellin, D. Ouyang, D. Bimberg: Phys. Rev. B **66**, 081306(R) (2002)
89. P. Borri, W. Langbein, S. Schneider, U. Woggon, R. L. Sellin, D. Ouyang, D. Bimberg: Phys. Stat. Solidi (b) **233**, 391 (2002)
90. P. Borri: in M. Grundmann (Ed.): *Nano-Optoelectronics, Concepts, Physics and Devices* (Springer, Berlin, Heidelberg 2002), p. 411
91. M. Sugawara, K. Mukai, Y. Nakata, H. Ishikawa, A. Sakamoto: Phys. Rev. B **61**, 7595 (2000)

# Solid-State Cavity-Quantum Electrodynamics with Self-Assembled Quantum Dots

Jean-Michel Gérard

Nanophysics and Semiconductors Laboratory
CEA/DRFMC/SP2M, 17 rue des Martyrs, 38054 Grenoble Cedex 9, France
jmgerard@cea.fr

**Abstract.** In this contribution, the recent development of cavity-quantum electro-dynamics experiments in all semiconductor microcavities using self-assembled quantum dots as "artificial atoms" is reviewed. In the weak coupling regime, a strong enhancement of the spontaneous emission rate (Purcell effect) can be observed for collections of dots as well as single dots. This effect permits us to achieve a regime of "nearly" single-mode spontaneous emission and is the key for the efficient operation of the first solid-state single-mode single-photon source, which is based on a single quantum dot in a pillar microcavity. Several major issues for future work, such as the quest for a strong coupling regime for single quantum dots in cavities and the feasibility and performance of single QD lasers are also discussed.

## 1 Introduction

Cavity quantum electro dynamics (CQED) has become since 1990 a major source of inspiration for basic research in optoelectronics [1,2,3,4]. In the 1980s, a beautiful series of experiments on atoms in microwave and optical cavities had demonstrated that optical processes including spontaneous emission (SE) can be deeply modified by using a cavity to tailor the emitter–field coupling [5]. Among other effects observable in the low-$Q$ CQED regime, the modification of the emission diagram, the enhancement or inhibition of the SE rate, the funneling of SE photons into a single mode and the control of the SE process on the single photon level are particularly attractive for applications in optoelectronics. For very high-$Q$ (i.e. weakly damped) cavities SE can even become a reversible process, in the so-called strong-coupling regime.

The high quality of some solid-state microcavities available since the early 1990s has permitted some major achievements, such as the strong coupling for quantum wells in planar cavities [6], the fabrication of high-efficiency microcavity LEDs which exploit SE angular redistribution [7], and low-threshold vertical-cavity surface emitting lasers [8] or microsphere lasers [9]. However, the usual broad spectra of solid-state emitters (bulk semiconductors or quantum wells, rare earth atoms, etc.) has been for long a major hindrance to the observation of several important CQED effects, including SE rate enhancement (often called "Purcell effect" [10]). For bulk material or quantum wells, non-radiative recombination at microcavity sidewalls can also play a domi-

P. Michler (Ed.): Single Quantum Dots, Topics Appl. Phys. **90**, 269–314 (2003)
© Springer-Verlag Berlin Heidelberg 2003

nant role in recombination dynamics, and hide intrinsic modifications of the
SE rate induced by the microcavity.

Self-assembled semiconductor quantum dots (QDs) are particularly well
suited for performing solid-state CQED experiments, as shown for instance
by the observation of a strong Purcell effect for QDs in semiconductor microcavities [11,12,13,14,15,16] and by the recent development of a single-mode
solid-state source of single photons based on a single QD in a pillar microcavity [16,17]. This device is the first optoelectronic component whose operation
actually relies on CQED, through the Purcell effect.

In this review paper, we first briefly recall some relevant properties of
QDs in this context (Sect. 2), and some CQED basics (Sect. 3). We present
available 0D semiconductor microcavities in Sect. 4, and compare their respective assets for CQED. Section 5 is devoted to a discussion of the current
status of the quest for a strong coupling regime with single QDs. We review in Sect. 6 experimental data obtained in the weak coupling regime, with
an emphasis on the observation of strong Purcell effects and "nearly"-single
mode SE. Well-established as well as more prospective applications of these
CQED effects in the field of nanooptoelectronics and quantum information
processing are finally discussed in Sect. 7.

## 2   Some Assets of Self-Assembled QDs
## for Quantum Optics

Solid-state CQED experiments have been until now mostly performed on
InAs/GaAs QDs in GaAs/GaAlAs microcavities. We will concentrate our
attention on this very mature system, although other QDs, such as II–VI
self-assembled QDs [18,19], QDs formed by interface fluctuations in quantum wells [20,21] or semiconductor nanocrystals [22,23] are also potentially
interesting in this context.

It has been known since 1985 [24] that strained-layer epitaxy can be used
to build defect-free nanometer scale InAs rich clusters in GaAs, which constitute potential traps for both electrons and holes. These self-assembled QDs
support well-separated discrete electronic states and exhibit a single narrow
emission line ($\ll kT$) under weak excitation conditions [25]. A linewidth close
to the limit imposed by the exciton radiative lifetime ($\sim 1\,\mu eV$) has been observed at 2 K [26,27]. This permits us to fully exploit the potential of high-$Q$
cavities for SE control. When the temperature is raised, the electron–phonon
interaction causes the onset of a broadband line related to elastic acoustic phonon–exciton interactions as well as a broadening of the zero-phonon
line [18,26,27,28,29]. At 300 K, the first of these effects is dominant and the
emission linewidth of single QDs can be as large as several meV [29].

Due to the strong 3D electronic confinement in QDs, few-particle Coulomb
effects dominate single QD emission spectra under high excitation conditions [30,31]. For instance, "exciton" and "biexciton" lines, into which pho-

tons are emitted when the QD contains respectively one or two electron–hole pairs, are separated by typically one to three meV for InAs QDs. This splitting varies from QD to QD, and from one laboratory to another, a fact that highlights the high sensitivity of these effects to the precise shape and size of the QD. Under non-resonant pumping, the QD emission is influenced by the fluctuating charge distribution in its surroundings [18,26], which entails a significant broadening of the exciton and multi-exciton lines. This effect as well as phonon-related ones shows that a single QD cannot in general be considered as an *isolated* artificial atom and puts in practice severe limits on the implementation of CQED effects on QDs in optoelectronic devices as discussed in Sect. 7.

Self-assembled QDs also display a large electric dipole moment $d$ for their fundamental optical transition, providing an efficient coupling to the electromagnetic field. Due to the compressive strain experienced by InAs, this dipole is (randomly) oriented in the $x$–$y$ epilayer plane. Absorption experiments on InAs QD arrays have shown that the oscillator strength per QD is of the order of 10 [32]; this corresponds to values for $d_x$ (or $d_y$) around $9 \times 10^{-19}$ C·m, which is ten times larger than typical values for atoms emitting in the same wavelength range. This feature is reflected by the rather short radiative lifetime ($\sim 1$ ns) of InAs QDs in bulk GaAs.

The peak emission wavelength of InAs QD arrays can be chosen typically in the 0.9–1.2 µm range at 10 K since the average size of the QDs depends reproducibly on the growth conditions in molecular beam epitaxy [33]. Such QDs display a high radiative quantum yield ($\eta \sim 1$), provided that carrier thermoemission is negligible. This corresponds to $T < 100$–200 K depending on their average size.

This excellent yield is due to the trapping of excitons in the defect-free QDs, which prevents further diffusion towards non-radiative recombination centers [34]. This trapping is obviously the key to the observation of intrinsic properties for QDs in mesa structures or in 0D microcavities, which present free surfaces close to the QDs under study [35].

Though moderate, size fluctuations lead to a significant dispersion of QD bandgaps within arrays, revealed by the inhomogeneous broadening of QD array emission lines ($\sim 15$ to 100 meV). This feature constitutes at first sight a major difficulty when one wants to place a single QD in resonance with a discrete cavity mode. This difficulty can however be turned to some extent into an asset. Instead of placing a single QD inside the cavity (which is in practice rather tricky), it is possible to insert a collection of a few hundred QDs, which are statistically all different thanks to size fluctuations. The detuning between the cavity mode and the center of the QD distribution can be easily adjusted so that a single QD on average is resonant with the cavity mode. A fine tuning of the resonance can then be obtained using small temperature changes [15].

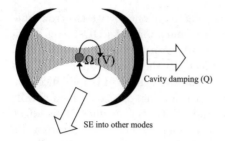

Cavity damping (Q)

SE into other modes

**Fig. 1.** Schematic configuration for single atom CQED

Finally, let us mention a major practical advantage of QDs for applications: unlike atoms, QDs can be pumped in a non-resonant way, by exciting the GaAs barrier. This opens a route toward compact, electrically pumped optoelectronic devices exploiting CQED effects on QDs.

# 3   Weak and Strong Coupling Regimes: Some Basics [36]

We consider in this section the model system formed by a single localized emitter, initially in its excited state and resonantly coupled to a single empty mode of a lossless microcavity (Fig. 1). It is well known that in the ideal case for which cavity losses and emitter's relaxation can be neglected, this coupled system experiences a Rabi oscillation. Spontaneous emission becomes a reversible process. Since the cavity stores the emitted photon in the emitter's neighbourhood, it can be reabsorbed and reemitted again and again. The angular frequency $2\Omega$ of the Rabi oscillation is proportional to the strength of the atom–field coupling, and therefore to the vacuum field amplitude at the emitter's location. As we will see later, $\Omega$ is directly related to the effective volume $V$ of the cavity, which allows us to quantify its ability to concentrate vacuum field fluctuations at a given location.

In practice, however, microcavities are not perfect. The finite reflectivity of the mirrors, either due to residual absorption or transmission, allows photons to leak out of the cavity. For an open cavity, SE into a continuum of non-resonant modes opens also another path for the radiative relaxation of the emitter. As long as such decoherence processes, affecting either the photon or the emitter, are slow enough on the scale of the Rabi period, the coupled system is still in a strong coupling regime and experiences a damped Rabi oscillation. Faster decoherence processes lead to overdamping; in this so-called "weak coupling regime", the emitter relaxes monotonically down to its ground state. When the main decoherence process is due to cavity damping, characterized by the resonance quality factor of the mode, $Q$, it is still possible to tailor to a large extent the SE of this emitter and in particular its SE rate $1/\tau_{\mathrm{cav}}$, by playing both with the strength of the emitter-field coupling (i.e. $V$), and with the cavity losses (i.e. $Q$). After having recalled the expressions of $\Omega$ and $1/\tau_{\text{cav}}$, we will present the $V$s and $Q$s of available solid-state microcavities in order to discuss their potential for solid-state CQED experiments.

## 3.1   The Strong Coupling Regime

The angular frequency $2\Omega$ of the Rabi oscillation depends on the coupling strength, i.e. on the emitter dipole and magnitude of the vacuum field. After introducing some notation, we will recall the expression of $\Omega$.

Field quantization leads to the following expression for the electric field operator for the cavity mode:

$$\hat{E}(r,t) = i\varepsilon_{\text{max}}\, f(r)\hat{a}(t) + \text{h.c.}\,, \tag{1}$$

where h.c. means Hermitian conjugate, $\hat{a}$ is the photon creation operator and $f$ the mode spatial function. $f$ is a complex vector which describes the local field polarization and relative field amplitude; it obeys Maxwell's equations and is normalized so that its norm is unity at the antinode of the electric field. The normalization prefactor $\varepsilon_{\text{max}}$, which is often named, in a somewhat improper way, "maximum field per photon" can be estimated by expressing that the vacuum-field energy for a mode with angular frequency $\omega$ is $\hbar\omega/2$:

$$\varepsilon_{\text{max}} = \sqrt{\hbar\omega/2\varepsilon_0 n^2\, \mathrm{V}}\,, \quad \text{where} \quad V = \frac{1}{n^2}\int\int_r\int n(r)^2\,|f(r)|\,d^3r\,. \tag{2}$$

In this expression, $n$ is the refractive index at the field maximum and $V$ the effective cavity volume, which describes how efficiently the cavity concentrates the electromagnetic field in a restricted space. More precisely, $V$ is the volume of a cavity, defined by Born–von Kármán periodic boundary conditions, which would provide the same maximum field per photon as the cavity under study.

For an electric-dipole transition, the atom–field interaction Hamiltonian can be written as:

$$\hat{H}_{\text{int}} = i\varepsilon_{\text{max}}\, d.f(r_e)\,|g\rangle\,\langle e|\,\hat{a} + \text{h.c.}\,, \tag{3}$$

where $|g\rangle$ and $|e\rangle$ designate the ground and excited states of the emitter, modelled as a two-level system, $r_e$ its location, and $d$ its electric dipole.

In the limit of negligible dissipation, the system is described by the Jaynes–Cummings Hamiltonian:

$$\hat{H} = \hbar\omega\hat{\sigma}_z + \hbar\omega(\hat{a}^+\hat{a} + 1/2) + i\hbar\Omega(\hat{\sigma}_-\hat{a}^+ - \hat{\sigma}_+\hat{a})\,, \tag{4}$$

where $\hat{\sigma}_+$, $\hat{\sigma}_-$, $\hat{\sigma}_z$ are pseudospin operators for the two-level system, and $\Omega$ the Rabi pulsation defined by:

$$\hbar\Omega = |\varepsilon_{\text{max}}\, d.f(r_e)|\,. \tag{5}$$

The interaction Hamiltonian couples only the (degenerate) states $|e, n\rangle$ and $|g, n+1\rangle$ of the uncoupled emitter/cavity system. Therefore, the spectrum of the Jaynes–Cummings Hamiltonian consists in a non-degenerate fundamental level $|g, 0\rangle$, and in a ladder of pairs of entangled states, $|e, n\rangle \pm |g, n+1\rangle$ ($n = 0, 1, \ldots$) split by $2\hbar\Omega\sqrt{n+1}$. If the system is initially prepared in the state $|e, 0\rangle$, it will experience a Rabi oscillation at the pulsation $2\Omega$ as stated earlier.

In practical cases, this coherent evolution is observable only if all decoherence processes, affecting either the emitter or the photon, are slow on the Rabi period timescale. For isolated atoms in cavities, the main decoherence processes are the cavity damping, related e.g. to the finite mirror reflectivity, and SE into additional modes. It is reasonable to consider that these effects arise because of the coupling of the quantum system to reservoirs, which permits a simple treatment of this problem within the density matrix formalism. In the limit of a weak excitation of the system, the eigenenergies of the first excited doublet become [37]:

$$\tilde{\omega}_\pm = \omega \pm \sqrt{\Omega^2 - \left(\frac{\Delta\omega_e - \Delta\omega_c}{4}\right)^2} - \frac{i}{4}(\Delta\omega_e + \Delta\omega_c), \qquad (6)$$

where $\Delta\omega_e$ and $\Delta\omega_c$ are the emitter and cavity linewidths arising from these decoherence channels. The system is in the strong coupling regime if and only if $\Omega > |\Delta\omega_c - \Delta\omega_e|/4$. Its eigenstates are then non-degenerate entangled states. However, dissipation induces a reduction of their energy splitting and limits their lifetime, as reflected by their finite linewidth $(\Delta\omega_e + \Delta\omega_c)/2$ (FWHM).

For single QDs, other phenomena such as electron–phonon or electron–electron scattering are usually the main decoherence processes affecting the emitter. It is not possible to treat in the general case the emitter decoherence as being due to a weak coupling to an appropriate reservoir. Considering theoretically a single QD in the strong coupling regime, *Wilson-Rae* and *Imamoglu* have shown for instance that the strong interaction of the QD exciton with acoustic phonons must be treated in a non-perturbative way [38].

For sake of simplicity, approximate (sufficient) conditions such as $\Omega > \max(\Delta\omega_c, \Delta\omega_e)$ or $\Omega > (\Delta\omega_c + \Delta\omega_e)/2$ are often used. For single QDs at low temperature in semiconductor microcavities, it is reasonable to assume that exciton decoherence is negligible compared to cavity damping. In this case, the calculation leading to (6) is perfectly valid. The relevant condition for strong coupling becomes $\Omega > \Delta\omega_c/4$, and is therefore less severe than the previously mentioned approximate conditions. The eigenenergies of the first excited doubled of the coupled QD/cavity system can be written as [37]:

$$\tilde{\omega}_\pm = \omega \pm \sqrt{\Omega^2 - \left(\frac{\Delta\omega_c}{4}\right)^2} - \frac{i}{4}\Delta\omega_c \quad \text{if } \Omega > \Delta\omega_c/4 \text{ (strong coupling)}, \quad (7)$$

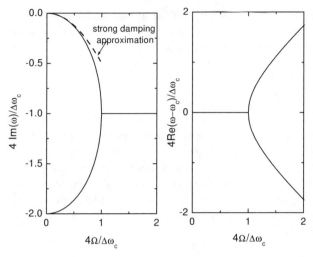

**Fig. 2.** Real and imaginary parts of the angular frequencies of the fundamental doublet of states of the coupled cavity-emitter system, as a function of $\Omega/\Delta\omega_c$

$$\tilde{\omega}_\pm = \omega \pm i\sqrt{\left(\frac{\Delta\omega_c}{4}\right)^2 - \Omega^2} - \frac{i}{4}\Delta\omega_c \text{ if } \Omega < \Delta\omega_c/4 \text{ (weak coupling). (8)}$$

## 3.2   Weak Coupling Regime: The Purcell Effect

In the weak-coupling regime, both eigenenergies have the same real part, so that the Rabi oscillation is lost. SE is an irreversible process; the emitter experiences a monotonic relaxation towards its ground state, as when it is in free space. However, it is still possible to tailor SE to a large extent in this regime, and in particular to control the emitter's SE rate.

Let us first consider the strong damping limit ($\Omega \ll \Delta\omega_c = \omega/Q$); the eigenvalues $\tilde{\omega}_\pm$ become $\omega - i\Delta\omega_c/2$, associated to the cavity mode and $\omega - 2i\Omega^2/\Delta\omega_c$ for the emitter. The emitter relaxation by SE is thus characterized by an exponential decay and its SE rate is given by:

$$\frac{1}{\tau} = \frac{4\Omega^2 Q}{\omega}.$$  (9)

This value is to be compared with the SE rate of this emitter when it is imbedded in a transparent homogeneous medium of refractive index $n$:

$$\frac{1}{\tau_{\text{free}}} = \frac{|\boldsymbol{d}|^2 \omega^3 n}{3\pi\varepsilon_0 \hbar c^3},$$  (10)

where $\boldsymbol{d}$ is the electric dipole of the localized emitter.

Using these expressions, along with (2) and (5) we see that the emitter's SE rate is enhanced (or inhibited) in the cavity with respect to its value for a homogeneous surrounding material by a factor

$$\frac{\tau_{\text{free}}}{\tau_{\text{cav}}} = \frac{3Q\,(\lambda_{\text{c}}/n)^3}{4\pi^2 V} \times \frac{|\boldsymbol{d}.\boldsymbol{f}\,(\boldsymbol{r}_{\text{e}})|^2}{|\boldsymbol{d}|^2}. \tag{11}$$

Whereas the first term is only related to cavity properties $(Q, V)$, the second one (which is always smaller than 1) depends on the relative field amplitude at the emitter's location and on the orientation matching of the transition dipole and electric field. In order to find a figure of merit describing the ability of the cavity for SE enhancement, it is convenient to consider the SE rate of an "ideal" emitter, whose properties allow one to maximize this effect. This ideal emitter should be located at a maximum of the electric field, with its dipole aligned with the local electric field. This cavity figure of merit takes the form proposed by Purcell fifty years ago [10]

$$F_{\text{p}} = \frac{\tau_{\text{free}}}{\tau_{\text{cav}}} = \frac{3Q\,(\lambda_{\text{c}}/n)^3}{4\pi^2 V}. \tag{12}$$

The Purcell factor $F_{\text{p}}$ can become much larger than 1 for sufficiently low volume and high-$Q$ microcavities. In other words, the emitter's SE rate into a single cavity mode can be orders of magnitude larger than its SE rate in free space, as confirmed by spectacular early experiments on atoms in cavities ($\times 500$ SE rate enhancement [5]). This point constitutes a major difference with planar cavities, which provide limited SE rate enhancement factors, $\times 3$ at most for perfect metallic mirrors [39,40] and $\times 5$ at most for realistic metallic mirrors [41].

It should be recalled at this point that the derivation of the Purcell factor is only valid in the strong damping limit ($\Omega \ll \omega/Q$). We see from Fig. 2 that this approximation is very good as soon as $\Omega < \omega/2Q$. It is still not too bad even when the strong damping condition is not satisfied. We see for instance from (8) that at the critical coupling ($\Omega = \Delta\omega_{\text{c}}/4$), the decay rate is typically given by $\Delta\omega_{\text{c}}/2$ which is only a factor of 2 larger than the approximate value $4\Omega^2/\Delta\omega_{\text{c}}$.

It is also interesting to rewrite the strong damping condition as

$$\Delta\omega_{\text{e}} = \frac{1}{\tau_{\text{cav}}} = \frac{4\Omega^2 Q}{\omega} \ll \frac{\omega}{Q} = \Delta\omega_{\text{c}}, \tag{13}$$

which shows that the homogeneous broadening due to the radiative recombination $\Delta\omega_{\text{e}}$ is much smaller than the linewidth of the cavity mode $\Delta\omega_{\text{c}}$. It is then valid to treat the single cavity mode as a continuum and to estimate the SE rate using the Fermi Golden Rule, which was the original approach of Purcell. (In principle, the emitter should be coupled to a continuum of modes, delocalized over the cavity and the outside world. It is possible to show in a rigorous way that this continuum can be replaced by a single "quasi-mode", which has the same spatial dependence than the confined mode of the perfect microcavity and a spectral density of modes which is that of the coupled cavity–reservoir system [42].) We briefly outline this derivation here, in order to highlight more easily the role of an eventual detuning between the emitter and the cavity mode.

For an electric dipole transition, the Fermi golden rule can be written as

$$\frac{1}{\tau} = \frac{2\pi}{\hbar^2}\rho(\omega_e).\langle|\langle \boldsymbol{d} \cdot \hat{\boldsymbol{\varepsilon}}(\boldsymbol{r}_e)\rangle|^2\rangle \ ; \tag{14}$$

where $\rho(\omega_e)$ is the density of photon modes at the emitter's angular frequency $\omega_e$ and where the averaging of the squared dipolar matrix element is performed over the various modes seen by the emitter.

The insertion of the radiating dipole inside the cavity will change its SE rate in three ways: the spectral density of modes, the amplitude of the vacuum field and its orientation with respect to the radiating dipole are indeed all modified. We evaluate in the following the resulting change of the SE rate for a cavity supporting a single mode (angular frequency $\omega_c$, linewidth $\Delta\omega_c$ and quality factor $Q = \omega_c/\Delta\omega_c$). In this case, the mode density seen by the emitter is a normalized Lorentzian function

$$\rho_{\text{cav}}(\omega) = \frac{2}{\pi\Delta\omega_c} \times \frac{\Delta\omega_c^2}{4(\omega - \omega_c)^2 + \Delta\omega_c^2}. \tag{15}$$

The "free-space" mode density for a homogeneous medium of refractive index $n$ is:

$$\rho_{\text{free}}(\omega) = \frac{\omega^2 V' n^3}{\pi^2 c^3}, \tag{16}$$

where $V'$ is a normalization volume. The field operator for each free-space mode is:

$$\hat{\boldsymbol{E}}(\boldsymbol{r},t) = \mathrm{i}\boldsymbol{\varepsilon}\sqrt{\frac{\hbar\omega}{2\varepsilon_0 n^2 V}} \ \mathrm{e}^{\mathrm{i}\boldsymbol{k}\boldsymbol{r}}\hat{a}(t) + \text{h.c.}, \tag{17}$$

where $\boldsymbol{\varepsilon}$ is a unit vector describing the polarization of the mode.

Using the Fermi golden rule for both estimates, we can compare the emitter's SE rate in the single-mode cavity and in a homogeneous surrounding medium:

$$\frac{\tau_{\text{free}}}{\tau_{\text{cav}}} = \frac{3Q\,(\lambda_c/n)^3}{4\pi^2 V} \frac{\Delta\omega_c^2}{4(\omega_e - \omega_c)^2 + \Delta\omega_c^2}\xi^2\,|\boldsymbol{f}\,(\boldsymbol{r}_e)|^2 \ , \tag{18}$$

where $\xi = |\boldsymbol{d} \cdot f(\boldsymbol{r}_e)|\,/\,|\boldsymbol{d}| \cdot |\boldsymbol{f}(\boldsymbol{r}_e)|$ describes the orientation matching of $\boldsymbol{d}$ and $\boldsymbol{f}\,(\boldsymbol{r}_e)$, and where the factor 3 stems from a $1/3$ averaging factor accounting for the random polarization of free-space modes with respect to the dipole. We thus retrieve the enhancement factor given by (14) for an emitter perfectly on resonance with the cavity mode ($\omega_e = \omega_c$). Here again, the *Purcell factor* $F_{\text{p}}$ appears as the largest SE rate enhancement which can be induced by the cavity. In order to observe the Purcell effect in its full magnitude, our "ideal" emitter should be well matched with the frequency, spatial distribution and polarisation of the mode.

Until now, decoherence processes affecting the emitter have been neglected, so that the homogeneous linewidth of the emitter *placed inside the microcavity* is much smaller than the cavity linewidth according to (9). If other decoherence processes are efficient, (14) still holds as long as $\Delta\omega_e \ll \Delta\omega_c$.

Let us consider now that decoherence processes (other than SE) are so strong that the homogeneous linewidth of the emitter is comparable to or larger than the cavity linewidth. In this case, $\rho_{cav}(\omega)$ must be replaced in the expression of the Fermi golden rule by [43]:

$$\int L(\omega)\rho_{cav}(\omega)d\omega, \tag{19}$$

where $L(\omega)$ is a normalized Lorentzian function describing the distribution of optical transitions of the homogeneously broadened emitter. For an emitter on resonance with the mode, it is obvious to see that the magnitude of the SE rate enhancement is reduced compared to the previous case. In (18), one has to replace $1/Q$ by $1/Q_{em} = \Delta\omega_e/\omega_e$ when the homogeneous broadening is very large ($\Delta\omega_e \gg \Delta\omega_c$), by $1/2Q$ when $\Delta\omega_e = \Delta\omega_c$ and by $1/Q + 1/Q_{em}$ (in a good approximation) in the general case, for an emitter in resonance ($\omega_e = \omega_c$).

As a result, the Purcell effect is washed out when the emitter is spectrally much broader than the cavity mode. This explains why the lack of appropriate emitters has long been a major hindrance to the observation of the Purcell effect in solid-state microcavities, until QDs have been used in this context.

# 4    Presentation of Available 0D Semiconductor Microcavities

Unlike the microwave spectral range, where superconducting mirrors are available, lossless metallic mirrors do not exist at optical frequencies. Major efforts have been devoted in the 1990s to the development of dielectric microcavities able to confine light on the wavelength scale in one or several dimensions. Two basic effects can be used, alone or in combination, for that purpose. The first one is total internal reflection, which has been exploited for a long time in optical fibers (2D confinement) and is used to provide 3D confinement in microspheres [9,44,45], microtores [46] and microdisks [47]. The second one is distributed Bragg reflection, which is implemented in 1D dielectric mirrors and more generally in photonic bandgap crystals [48]. Micropillars [49,50] exploit both waveguiding along the pillar axis and reflection by 1D Bragg mirrors to generate 3D confinement, whereas 2D photonic crystal membranes combine waveguiding in the semiconductor slab and lateral confinement by the photonic crystal [51]. It should be noted that these 0D dielectric microcavities do not constitute perfect "photonic boxes". Besides a discrete series of three-dimensionally confined modes, these open cavities also support a continuum of non-resonant modes. Recent progress in the fabrication of 3D photonic crystals at optical frequencies [52] might soon change this state of affairs. Whereas the evanescent field of free-space modes in the photonic bandgap energy range can be made arbitrarily low at the emitter's location for thick crystals, a tailored defect can be used to create one (or a few) resonant cavity modes in the same energy range [48].

We give in this section a brief overview of this quest for low-$V$/high-$Q$ dielectric microcavities. Quite amazingly, the study of QDs in cavities has had important by-products in the field of microcavity characterization. Though some microcavities, such as micropillars, can be studied by reflection spectroscopy [49,53], it is much more convenient to probe their resonant modes in emission, using a QD array as a broadband internal light source [35,50]. This approach, which is widely used today to probe other complex photonic structures such as microdisks or photonic crystal (PC) microcavities, displays several important advantages. Unlike reflectometry, for which only modes having the same symmetry as the probe beam can be studied, all confined modes appear on emission spectra. On a practical side, high-resolution data can be obtained in a simple microphotoluminescence experiment. Let us note finally that the additional absorption loss entailed by the insertion of a QD array is in general negligible, which allows one to probe the optical properties of the empty cavity. The comparison of nominally identical micropillars containing either a QD array or several QWs has been for instance a powerful tool for understanding the formation of confined cavity polaritons in the latter case [54].

## 4.1   Pillar Microcavities

Micropillars have received much attention since the first fabrication of vertical cavity surface emitting lasers [55,56]. Their interest for SE rate enhancement and low-threshold lasing has been acknowledged as early as 1991 [57,58].

We present in Fig. 3 a typical micropillar fabricated through the reactive ion etching (RIE) of a GaAs/AlAs $\lambda$-cavity resonant around 0.9 μm, as well as a microphotoluminescence (μPL) spectrum obtained at low temperature. The series of sharp PL lines corresponding to the resonant cavity modes as well as a background emission of the QD array into the continuum of non-resonant modes are clearly seen on this spectrum. This feature can be used for comparing in a single time-resolved experiment microcavity effects for QDs which are in or out of resonance with a cavity mode (Sect. 6). A systematic study of the mode energies as a function of the pillar radius $r$ highlights the main effects of an increased lateral confinement, which are an overall blueshift of the resonant modes and an increased energy separation between successive modes [50,59].

These experimental trends and the mode energies can be quantitatively understood using a simple model, based on a decoupling of the vertical and lateral confinements. Having in mind the formation of resonant modes from plane waves in planar cavities as a result for vertical confinement, we use each pair of counter-propagating guided modes $I$ of the GaAs cylinder to build a resonant mode $i$. In a micropillar, guided photons are localized by distributed Bragg reflection. Since their velocity is modified by the lateral confinement, the resonance wavelength is modified as well; it shifts with respect to the normal incidence resonance wavelength of the planar cavity $\lambda_{2D}$ and is given by [50]

$$\lambda_i = n_{\text{eff}}^I L = \lambda_{2D} \frac{n_{\text{eff}}^I}{n} , \tag{20}$$

where $n$ is the refractive index of GaAs, $n_{\text{eff}}^I$ the effective index of the guided mode $I$, and $L$ the thickness of the GaAs cavity layer.

As shown in Fig. 4, this simple description accounts amazingly well for the experimental results for pillar radius larger than $\sim \lambda/n$. This is due to the combination of three effects. Firstly, the electromagnetic field distribution is very similar for a given mode in AlAs and GaAs waveguides, due to the strong index contrast with the air; as a result, different guided modes are not significantly mixed at interfaces. Secondly, although the continuity conditions for the normal electric field component are not satisfied at all interfaces, this component is very small compared to in-plane components. Finally, the relative index change $n/n_{\text{eff}}^I$ is the same (within less than 3%) for GaAs and AlAs waveguides. Therefore, the resonance condition (20) remains the same as for the planar cavity, but for the replacement of $n$ by $n_{\text{eff}}^I$ [50].

This model has been used to estimate the field distributions of the resonant modes and their effective volume. For the fundamental mode, the effective height is $\sim 2\lambda_c/n$ and the effective area $\sim \pi r^2/4$. Therefore the 1 µm diameter pillar shown in Fig. 3 is able to confine light within an effective volume as small as $\sim 5(\lambda_c/n)^3$.

According to this model, we also expect the same $Q$ for the planar cavity and for the resonant modes of the micropillar. We show in Fig. 5 typical data obtained for the fundamental mode on a series of micropillars etched in a high-finesse planar microcavity ($Q = 5200$). For large enough pillars, $Q$ takes a constant value and is indeed very similar to the planar cavity $Q$. Below a certain critical diameter, here 3 µm, $Q$ decreases [53,60], which indicates a reduction of the escape time of the photons. This trend reflects the onset of several diffusion/diffraction processes, which can scatter confined photons out of the pillar. In contrast with previous work, the major part of the bottom mirror is etched in order to quench the diffraction by the finite aperture of the pillar foot. We attribute therefore the $Q$ spoiling to the scattering by the residual roughness of the pillar sidewalls. This interpretation is supported by the excellent description of the experimental behaviour for $Q$ by a simple

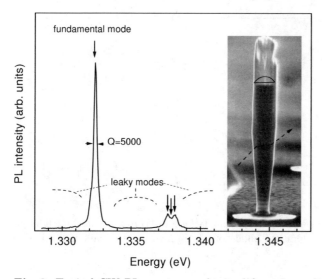

**Fig. 3.** Typical CW PL spectrum, obtained for a 3 μm diameter micropillar containing InAs QDs. The *arrows* indicate the calculated energies for the resonant modes of the pillar

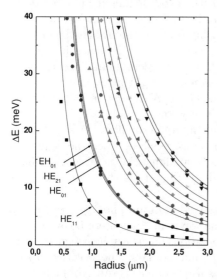

**Fig. 4.** Spectral shift of the resonant modes of GaAs/AlAs micropillars with respect to the normal incidence resonance energy of the planar cavity, as a function of the pillar radius $r$. *Dots*: experimental results. *Lines*: model

model, which assumes that the scattering probability is proportional to the mode intensity at the surface of the pillar [53].

The optimization of the etching step has allowed important progress in this respect. $Q$s as large as 2100 for $r = 0.5$ μm [11] and 630 for $r = 0.3$ μm [61] have been obtained respectively by reactive ion etching (RIE) and chemically assisted ion beam etching (CAIBE). For these very narrow pillars ($r < 2\lambda/n$),

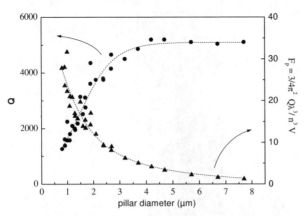

**Fig. 5.** Typical dependence of the cavity quality factor $Q$ (*circles*) and Purcell factor $F_p$ (*triangles*) for the fundamental mode of GaAs/AlAs micropillars as a function of their diameter $d$

the various assumptions sustaining our simple model break down and a more refined theoretical description is compulsory. Recent calculations using the finite-difference time-domain (FDTD) method [62] highlight an *intrinsic Q* spoiling for such small radius micropillars. This effect comes in large part from the different relative index changes resulting from the lateral confinement in GaAs and AlAs waveguides. This effect can be taken into account in the design of the planar cavity which is etched into micropillars. According to FDTD calculations, optimized layer thicknesses might allow one to get $Q$s close to $Q_{2D}$ and as large as 10 000 in the GaAs/AlAs system [62], provided additional *extrinsic* losses are kept sufficiently low.

Considering now the Purcell factor of these micropillars, we see that $F_p$ tends to increase when the pillar diameter is reduced, since the reduction of $V_{eff}$ overcompensates the effect of the $Q$ spoiling. $F_p$s as large as 32 have been estimated (from measured $Q$s and calculated $V$s) for 1 μm diameter micropillars [11].

In practice, the sensitivity of $Q$ on the sidewall morphology is such that reproducibility is still a challenging issue for pillar diameters below 1.5 μm typically. Therefore, novel material combinations have been investigated as an alternate route toward lower $V$/higher $F_p$ micropillars. The selective oxidation of AlAs into low-index $AlO_x$ permits us to replace GaAs/AlAs DBRs by high-contrast GaAs/AlOx ones [63,64]. A three-fold decrease of $V_{eff}$ can thus be obtained thanks to the strong reduction of the spreading of the confined modes into the DBRs. For planar cavities, high-$Q$ values can also be obtained with a small number of DBR periods (up to $Q \sim 3000$ with 4 period-DBRs on both sides [64]). However, the highest reported $F_p$ in this system is only 14 due to the important roughness of the pillar sidewalls [65]. Though some progress can be expected from technological developments, radiation losses (due in particular to the guided mode mismatch at the $AlO_x$/GaAs interface) intrinsically limit $Q$ to values lower than 1000 for $d < 1\,\mu m$ [62]. Minor improvements – if any – are therefore expected compared to the GaAs/AlAs system.

Finally, let us note that $AlO_x$ layers are also commonly used to provide electrical and lateral optical confinement in low-threshold vertical-cavity surface emitting lasers [8]. Typical figures of merit for such microcavities are $Q = 650$ and $V_{eff} \sim 15(\lambda_c/n)^3$ for a $1\,\mu m$ diameter aperture in $AlO_x$ [66].

## 4.2    Microdisks

It is well known that semiconductor microdisks support a series of whispering gallery modes (WGMs), which are guided by total internal reflection – and tightly confined – at the lateral edge of the disk. A semiconductor microdisk, such as the one shown in Fig. 6, is usually fabricated using a combination of RIE, which etches a multilayer into a cylindrical post, and of the partial selective wet etching of a sacrificial layer to form the disk pedestal [47,67,68,69,70,71]. Compared to the pillars, for which the resonant modes are partially delocalized in the DBRs and all over the pillar area, this geometry provides a more efficient 3D confinement. For a small radius GaAs disk ($\lambda_c/n < R < 20\lambda_c/n$), the effective area of the lowest radial quantum number WGM is approximately given by $0.86\lambda_c^2(R/\lambda_c)^{3/2}$, where all lengths are expressed in $\mu m$. The effective vertical thickness $L_{eff}$ of the WGM can be evaluated by considering the field distribution of the guided mode of the air-confined GaAs slab. For the $250\,nm$ thick GaAs disk of Fig. 6, $H_{eff}$ is $175\,nm$ or $\sim 0.6\lambda_c/n$, when $\lambda_c$ is around $1\,\mu m$, so that the effective volume is of the order of $6(\lambda_c/n)^3$.

Figure 6 displays a typical $\mu PL$ spectrum obtained on a single $3\,\mu m$ diameter microdisk containing an InAs QD array [70]. The sharp lines constitute the WGM contribution to the spectrum, while the broad background is due to the emission of the QD array into non-resonant modes. The large number of observable WGMs is quite remarkable, since only one or two WGMs (with TE polarization and radial quantum number $n_r = 1$) are clearly observed for QW emitters, and only above the lasing threshold. An attempt to identify these various WGMs shows that both TE and TM modes are observed. Within the broad spectral range covered by the QD array, all modes with $n_r = 1$ are observed as well as some higher order WGMs ($n_r = 2$ and probably 3).

In this experiment, WGMs display $Q$s around 3000, which are orders of magnitude smaller than the limits set by intrinsic radiation losses (e.g. $Q_{int} > 10^8$ for $d > 2\,\mu m$ [67]) or by the modal absorption of the QD array ($Q_{abs} \sim 20\,000$ here). As is well known, the main losses for empty microdisks are due to scattering by the roughness of the disk edge. By replacing in the process the RIE step by a non-selective wet chemical etching, a very significant improvement of WGM $Q$s has been obtained through a reduction of the sidewall roughness. $Q$s as large as $12\,000$ for $d = 1.8\,\mu m$ [70] and $17\,000$

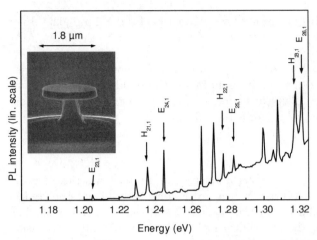

**Fig. 6.** Typical CW PL spectrum, obtained for a 3 μm diameter microdisk containing InAs QDs. The *arrows* indicate the estimated energies of some whispering gallery modes. *Insert*: Scanning electron micrograph displaying a 1.8 μm diameter GaAs microdisk fabricated using a two-step wet-etching process

for $d = 4.5$ μm [71] have been reported. Let us note that the losses introduced by the probe QD array are no longer negligible compared to the total losses; it is necessary for such high-$Q$ structures to work at the transparency threshold to measure accurately the empty cavity $Q$ in the μPL experiment.

These large $Q$s also correspond to a very high Purcell factor, of the order of 150 for $d = 1.8$ μm and $Q = 12\,000$. Microdisks appear therefore as excellent candidates for the study of the Purcell effect in a solid-state microcavity. Since they provide the largest available $Q$s among low-$V$ semiconductor microcavities, they open also interesting opportunities for the observation of other effects such as thresholdless or low-threshold lasing as well as strong coupling for a single QD as discussed later.

### 4.3  Photonic Crystal Microcavities

The ultimate approach for obtaining very small cavities exploits photonic crystals (PCs) as confining material. Thanks to impressive recent progress, 3D PCs are now probably mature enough to build microcavities at optical frequencies [52]. The development of such microcavities is obviously an important challenge since they do not support – unlike other 0D microcavities – a continuum of non-resonant modes.

Until now, only 1D [72,73] or 2D [51,74] PC cavities have been fabricated, using a combination of waveguiding in a semiconductor slab and reflection on the PC to obtain 3D photon confinement. As an example, we present in Fig. 7 a 2D PC-microcavity processed within an air-bridge and designed for operation around 1 μm. In this geometry, defect modes with an effective volume

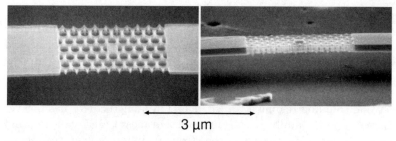

3 µm

**Fig. 7.** Scanning electron micrograph of an air-bridge microcavity based on a 2D PC and designed for an operation around 1 µm

as small as $0.3(\lambda_c/n)^3$ can be formed, thanks to the small field penetration in the surrounding PC, combined with a strong vertical confinement. In PC cavities, $Q$ is limited by optical scattering. Novel designs optimized using FDTD calculations [75] allow one to reduce these losses by more than one order of magnitude. State-of-the-art microcavities in a 2D photonic crystal slab display very impressive figures of merit ($V_{\text{eff}} = 0.43(\lambda_c/n)^3$, $Q = 2800$) and display in particular the highest Purcell factors among solid-state microcavities ($F_p = 490$) [76].

### 4.4 Conclusion

We have shown in this section that many approaches allow one to build high-$F_p$ 0D semiconductor microcavities. Compared to large silica *microspheres* ($V_{\text{eff}} \sim 1000(\lambda_c/n)^3$, $Q \sim 10^9$) [45] or microtores ($Q \sim 10^8$) [46] these structures provide much stronger photon confinement, but exhibit a much smaller $Q$. In the prospect of CQED experiments, these 0D solid-state microcavities thus have different assets. The long-lived whispering gallery modes (WGM) of the microspheres are for instance ideal for fabricating very-low threshold lasers [9]. The small mode volume of semiconductor cavities allows one to get a large Purcell factor for moderate $Q$s (100–1000); the constraint on the emitter's linewidth is then less severe. This widens the choice of usable emitters for observing the Purcell effect. The same considerations hold obviously within the family of the 0D semiconductor microcavities, when comparing, e.g. relatively high-$Q$ microdisks and ultimately small volume PC structures.

## 5    Single QDs and the Strong Coupling Regime

The strong-coupling regime for QWs in planar cavities has received much attention since its discovery in 1992 [6]. Though the need for a description of this regime in terms of the formation of entangled exciton–photon states (the cavity polaritons) is now commonly accepted, it is clear that the analogy with the standard "single atom in a cavity" CQED situation is not very deep. This is mostly related to the fact that polaritons form 2D bands in this planar system, so that polariton scattering plays a central role in their linear and non-linear optical properties. It is thus quite interesting to investigate the system formed by a single QD in a 0D microcavity, which mimics much more closely the standard CQED system.

The strong coupling regime for a single QD in a microcavity has not yet been observed experimentally. Having in mind that most CQED concepts and experiments on atoms are related to the strong coupling regime, this observation is clearly an important milestone for solid-state CQED. It would open in particular novel potential applications in the field of quantum information technology. A quantum computing scheme has been proposed, which uses single QDs as support for qubits, and a coherent coupling to a cavity mode as a way to mediate interactions between distant QDs and realize two-qubit operations [77].

Compared to the atomic case, single QDs in solid-state microcavities display two major assets, the large QD oscillator strength and the availability of very small cavity modes:

- The tight mode confinement leads to very large values of the vacuum field amplitude; $\varepsilon_{max}$ is indeed around $2 \times 10^5$ V/m when $V_{eff} = (\lambda/n)^3$ and $\lambda \sim 1$ μm.
- The electric dipole components, estimated from $d_{x,y} = \sqrt{e^2 \hbar f / 2m\omega}$, are of the order of $9 \times 10^{-29}$ C·m, which is significantly larger than typical values for atomic optical transitions ($2.4 \times 10^{-29}$ C·m for the 5s–5p transition of rubidium at 1.59 eV).

According to (6), the Rabi energy $\hbar\Omega$ of a QD placed at an antinode of the vacuum field is around $\sim 100$ μeV when $V_{eff} = (\lambda/n)^3$. This value is typically two orders of magnitude larger than for a cold atom trapped in an optical microcavity [78].

In order to observe the strong coupling regime, however, the Rabi oscillation should be faster than all decoherence processes affecting the photon or the exciton. The strong homogeneous broadening of QD lines due to electron–phonon interaction rules out this possibility at high temperature ($T > 100$ K) for InAs QDs. On the other hand, the observation of single QD emission linewidths in the few μeV range show that exciton decoherence can be neglected when a low temperature and weak excitation conditions are used [26,27,28].

In such a case, the strong coupling regime is established if cavity damping is small enough ($\Delta E_{cav} < 4\hbar\Omega$) as discussed in Sect. 3. We compare therefore in Fig. 8 the reported mode linewidths for various semiconductor microcavities to $4\hbar\Omega$ estimated from (5). We assume here that the InAs QD is located at an antinode of the mode. This figure shows very clearly that semiconductor micropillars still display too strong cavity losses to reach a strong coupling regime with a single InAs QD.

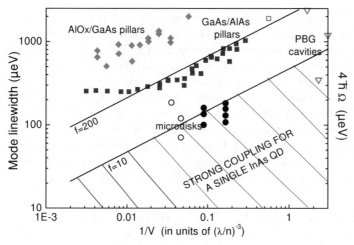

**Fig. 8.** The mode linewidth of some state-of-the-art 0D microcavities (*dots*) is compared to an estimate of $4\hbar\Omega$ for an ideally coupled QD (*lines*), calculated for two different oscillator strengths ($f = 10$ and $f = 200$). Experimental data are from CNRS/LPN [65,70] (*full symbols*), UCSB [71] (*open disks*), MIT [72] and Caltech [76] (*open triangles*), and Stanford (*open square* [61])

In this context, wet-etched microdisks [70,71] and photonic crystal microcavities [72,76] offer the most promising approaches at the moment, taking respectively the benefit of their record $Q$s and $V$s (among semiconductor microcavities). A single InAs QD located at the mode antinode and on perfect resonance would already experience a strong coupling regime in such cavities, although at most two Rabi oscillations would be observed because of the cavity damping. Significant novel breakthroughs concerning the $Q/\sqrt{V}$ ratio of microdisks and PC microcavities are clearly required for InAs QDs in the strong coupling regime to get some practical interest. In the meantime, controlling the QD location within the microcavity will be a key issue. In this context, *Zwiller* et al. have demonstrated the nanofabrication of microdisks at specific locations, defined using a preliminary cartography of the QD sample by μPL [79].

Obviously, an increase of the QD oscillator strength is also quite appealing. Self-assembled QDs made of II–VI semiconductors or cubic GaN could for instance be used, but 0D microcavity technology is still in its infancy for these systems. QDs formed by monolayer fluctuations of the thickness of GaAs QWs are very attractive thanks to their large oscillator strength resulting from exciton delocalization ($f > 300$) on large terraces [37]. Such QDs could in principle allow one to observe the strong coupling regime in microdisks, photonic bandgap cavities and state-of-the art micropillars.

Let us finally note that the coupling of a single QD (either a self-assembled InAs QD or a nanocrystal) to a dielectric microsphere has also been investigated [22,23,80]. Large radius silica microspheres ($r \sim 50\,\mu m$) can display a very high-$Q$ (up to $10^9$), but the estimated Rabi energy for a single QD is smaller than the QD linewidth, as a consequence for their large mode volume and for the non-ideal location of the QD, which is placed outside the sphere, in the evanescent field of the WGM. By contrast, small ($r \sim 4\,\mu m$) glass microspheres provide a much improved confinement [$V_{\mathrm{eff}} \sim 60(\lambda/n)^3$], but until recently, this was obtained at the expense of a strong decrease of $Q$ ($\sim 7000$) [80]. Thanks to the recent development of high-$Q$ ($> 10^6$) small diameter ($r \sim 5\,\mu m$) microspheres, this system is now well suited to the observation of the strong coupling regime for a single nanocrystal, or a single InAs QD in a mesa structure, placed in its close vicinity [81].

# 6    CQED Effects in the Weak Coupling Regime

As shown in the previous section, 0D solid-state microcavities are not yet mature enough to achieve a strong coupling regime for a single InAs QD. Though SE remains an irreversible process, it is still possible to modify emission properties to a large extent in the weak coupling regime. Such effects as SE angular redistribution, SE rate enhancement and inhibition, nearly-single mode emission, can be observed and exploited. This section is devoted to the presentation of experimental results on QD SE tailoring in optical microcavities. Some data obtained in planar cavities will be used to recall their potential and limitations. We will then discuss in detail experimental observations of the Purcell effect for collections of QDs and single QDs in 0D microcavities.

## 6.1    InAs QDs in Planar Cavities

The properties of InAs QD arrays inserted inside GaAs/AlAs planar microcavities have been studied by several groups [14,35]. Such cavities display reflectivity spectra very similar to those of reference empty cavities. The clean insertion of QDs in microcavities by MBE and the weak additional absorption introduced by the few QD arrays permit one to build high-$Q$ cavities. $Q$s as large as 5200 have already been presented (to be compared with 10 000, the highest value reported for MBE grown empty GaAs/AlAs cavities [82]).

Planar microcavities strongly modify the natural radiation of an embedded light emitter as is well known from the early work of *Kastler* [83]. Since the microcavity quantizes the photon wavevector along its axial direction $z$, a monochromatic emitter will emit photons only into specific modes for which the in-plane photon wavevector is also well defined. The far-field radiation diagram then consists in a series of narrow and bright fringes, which are cones of axis $z$. Assuming a constant SE rate, the integrated emission intensity remains the same as in free space; the microcavity redistributes SE angularly through interference effects. This effect, which can be extremely useful in practice to improve light extraction from the high-index semiconductor [84], has been observed experimentally for various emitters such as molecules [85], rare earth atoms [86] and quantum wells [87] as well as QDs [35], using angularly resolved PL. If only normal incidence emission is collected, a spectacular spectral narrowing of the PL peak is observed because of the filtering of the broad emission band of the QD array by the microcavity. The PL emission from QDs in resonance can be enhanced by two to three orders of magnitude compared to a reference QD array in bulk GaAs [35]. On the other hand, PL from off-resonance QDs is drastically reduced. These effects are well understood both qualitatively and quantitatively using simple models [35]. Angular-resolved PL using a QD array as broadband emitter can also be used to map the dispersion relation of the cavity modes over a large spectral range.

By contrast, planar GaAs/AlAs microcavities entail only minor modifications on the QD SE rate. It is well known that the SE rate of an emitter having its electric dipole in the plane of the microcavity is enhanced by at most a factor of 3, if the microcavity is bounded by ideal metallic mirrors [39,40,88,89]. This result is achieved however only for a half-wavelength cavity. Due to the penetration of the confined field inside DBRs, the effective length of semiconductor microcavities is in fact much larger than $\lambda$ ($\sim 4\lambda$ for the GaAs/AlAs system). For such a planar cavity, theory predicts a modification of the SE rate of a collection of independent emitters smaller than $\pm 20\%$, SE being slightly inhibited at energies lower than the normal-incidence resonance, and slightly enhanced above. Time-resolved data on InAs QDs [14,90] are in excellent agreement with these predictions, as well as data on rare-earth atoms in dielectric microcavities [91].

Let us recall that a strongly accelerated PL decay rate ($\times 2$) has been observed in early experiments on quantum wells [92] or organic dyes [88,93], and attributed to SE enhancement by the planar microcavity. A major interest of these experiments on QDs is to show that – unlike QDs or rare-earth atoms – multiple quantum wells or dye molecules (in concentrated solutions) cannot be treated as independent light emitters in these experiments. The analysis of these experimental features is quite complex in practice, due to an important contribution of collective effects, such as superradiance and stimulated emission [88]. Let us finally mention that a more recent experiment on a single GaAs quantum well in a low-$Q$ GaAs/AlAs microcavity led, on the contrary, to moderate experimental lifetime alterations, in agreement with theory [94].

## 6.2 Purcell Effect for QD Arrays in 0D Microcavities

The first experimental evidence of a strong enhancement of the SE rate of QDs came from studies performed on large collections of InAs QDs [11]. Obviously, the QDs contributing to the emission in some cavity mode have different SE rates, which depend on their spatial location and spectral detuning with respect to the cavity mode. This might be considered as a major limitation at first sight, since the average SE rate enhancement for these QDs is much smaller than for a perfectly coupled QD. This drawback turns into an asset when one tries to analyze quantitatively experimental data. Since the spatial location of the QDs in the cavity is in general not known, interpretation of single QD data must be achieved with great care. By contrast, a statistical averaging can be safely performed on the QD locations (and spectral detuning eventually) when numerous QDs are coupled to the cavity mode. Data obtained on QD arrays have therefore been essential to demonstrate the Purcell effect on QDs in semiconductor microcavities.

The first experiment was conducted on the series of GaAs/AlAs micropillars presented in Figs. 3–5 by time-resolved μPL at 8 K, using a set-up based on a pulsed Ti:sapphire laser and a streak-camera. The InAs QDs under study emit around 1.35 eV and display a 1.3 ns radiative lifetime when placed in bulk GaAs. Since some background emission into leaky modes is observed in μPL spectra (Fig. 3), it is possible to compare *in a single experiment* the SE rate of QDs, which are only coupled to leaky modes or in resonance with the fundamental cavity mode. The experiment is conducted under weak excitation conditions (less than one electron-pair per pulse per QD on average) so as to avoid state filling, that would entail a saturation of the QD emission and a quenching of the PL decay at short times. Whereas off-resonance QDs exhibit a 1.05 ns lifetime, quite similar to QDs in bulk GaAs, QDs in resonance experience a clear shortening of the PL decay time, by a factor as large as 5 for a 1 μm diameter micropillar. This selectivity proves the intrinsic nature of this effect, since extrinsic effets – such as increased non-radiative recombination in etched structures – would affect similarly in-resonance and off-resonance QDs.

As shown in Fig. 9, the experimental PL decay time shortening is much smaller than the Purcell figure of merit of the micropillars under study. However, several elements permit us to conclude safely that the observed effect is actually due to the enhancement of the SE rate into the cavity mode. For instance, the variation of the PL decay time correlates well with $F_p$, but neither with the pillar diameter $d$ nor with its $Q$. It is also possible to model as well time-resolved profiles, by performing a statistical averaging of the profiles for random locations and spectral detunings of the QDs [11][1] For high $F_p$s, the radiative lifetime of well-matched QDs can become comparable to the carrier

---

[1] Please note that in [11], a random orientation of the QD dipole in all three dimensions was assumed, which is not correct. For instance, transmission experiments in a guided wave confirmation have confirmed that the heavy-hole–light-hole mixing is negligible [95]. As for compressively strained QWs, the dipole associated to the fundamental optical transition is randomly oriented in the plane of the sample. Figures 8 and 9 present correct fits, obtained for an in-plane random dipole orientation, and taking into account the finite relaxation time of the electron–hole pairs.

relaxation time $\tau_{rel}$, which is of the order of 30–40 ps according to the PL rise time. This effect must be included in the model to reproduce the experimental data. For the model data shown in Fig. 9, we take $\tau_{rel} = 35$ ps in all calculations and use no additional fitting parameter. Finally, highly nonexponential profiles are obtained for high-$F_p$ micropillars ($F_p > 15$) both experimentally and theoretically. We have therefore plotted and compared in Fig. 9b the initial decay rates $1/\tau_{on}$, estimated between 100 and 700 ps after the excitation pulse.

Interestingly, simple hand-waving arguments already give a good first-estimate of $\tau_{free}/\tau_{on}$, as shown now. Considering (18), one notes that the spectral averaging of the Lorentzian factor gives a factor $1/2$. (This averaging is necessary since the mode linewidth ($< 0.5$ meV) is smaller than the spectral resolution of the set-up (1 meV) in this experiment.) Furthermore, the QDs are distributed all over the cross-section of the pillar so that the antinode enhancement for in-plane directions is lost; this introduces typically a factor varying between $1/4$ for large diameter pillars and $1/3$ for the smaller ones. In addition, the random in-plane orientation of the dipole gives $\xi^2 \sim 1/2$. Finally, the fundamental mode is two-fold degenerate, and the SE rate into the leaky modes is of the order of $1/\tau_{free}$. The total SE rate is around $(2F_p/12 + 1)/\tau_{free}$, which leads to $\tau_{free}/\tau_{on} \sim 6$ when $F_p \sim 30$ in reasonable agreement with the experimental data.

It is also important to check in such experiments that QDs can actually be treated as independent light emitters. In principle, power dependent collective effects could modify the PL decay rate. For a weak excitation power, emitted photons can be reabsorbed by the QD array, which would lead to photon recycling and to an apparent *slowing* of the PL decay. On the other hand, amplification by stimulated emission (ASE) could be observed when more than one electron–hole pair per QD is injected on average, which would *enhance* the emission rate. Simple considerations show that both effects can be neglected in this experiment [96].

Several other groups have reported in the last few years the observation of the Purcell effect on collections of QDs in time-resolved experiments. *Graham* et al. [66] have studied $\tau_{free}/\tau_{on}$ as a function of the spectral detuning of the QDs within the cavity mode linewidth, for oxide-apertured vertical microcavities. The SE rate enhancement factor follows the expected Lorentzian spectral dependence and its maximum value (2.9) is well understood when the Purcell factor $F_p = 9$ and the random spatial distribution of the QD are taken into account. *Solomon* et al. have presented similar data, obtained on a pillar microcavity [14].

Thanks to their larger $F_p$s, microdisks provide much stronger effects [12]. $\tau_{free}/\tau_{on}$ as large as 12 have been observed for 1.8 μm diameter microdisks ($Q = 10\,000$, $F_p = 160$). Here again, it is compulsory to take the finite carrier relaxation time into account when modeling experimental data [12].

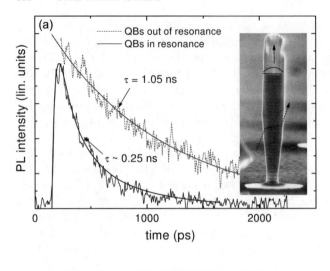

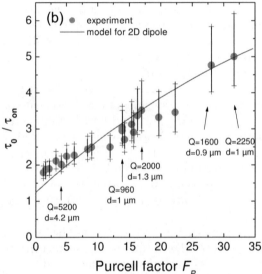

**Fig. 9.** (a) Time-resolved PL for QDs in or out of resonance with the fundamental cavity mode of a circular micropillar ($d = 1\,\mu m$, $Q = 2250$, $F_{\mathrm{p}} = 32$). (b) Experimental dependence of the PL decay time $\tau_{\mathrm{on}}$ as a function of the Purcell factor $F_{\mathrm{p}}$ (*dots*). $d$ and $Q$ are indicated for some of the pillars under study. Error bars correspond to a $\pm 70\,\mathrm{ps}$ uncertainty on $\tau_{\mathrm{on}}$. The *solid lines* in (a) and (b) show the result of a calculation of the PL profile and average decay rate, for in-resonance QDs, assuming a random in-plane dipole and a 35 ps relaxation time

Let us finally mention the remarkable demonstration of both SE inhibition and SE enhancement for QDs inserted in GaAs/AlAs micropillars coated with metal [90]. Thanks to a strong reduction of the density of leaky modes, the radiative lifetime of off-resonance QDs is increased by an order of magnitude, whereas QDs in resonance experience a clear SE enhancement ($\tau_{\text{free}}/\tau_{\text{on}} \sim 2.2$).

Clear signatures of the Purcell effect can also be obtained under CW excitation [96,97]; such experiments are very helpful when the photon collection efficiency does not permit time-resolved investigations. Figure 10 displays the PL data obtained at $8\,\text{K}$ as a function of the excitation powers $P_{\text{ex}}$ for a low-$Q$ microdisk ($R = 1.5\,\mu\text{m}$, $Q \sim 2500$, $F_p = 18$), which contains three QD arrays. When $P_{\text{ex}}$ increases, one observes a clear saturation of the background emission for energies due to the state filling of the ground states of the QDs. As a result, WGM show up much more clearly when $P_{\text{ex}}$ is raised. For microdisks containing QWs, a similar behavior is observed due to lasing. Here, however, optical losses – estimated by $\alpha = 2\pi n/\lambda Q$ – are of the order of $75\,\text{cm}^{-1}$, so that the saturated gain of the 3 QD arrays is not sufficient to sustain lasing. For this microdisk, lasing is only observed for higher energy WGMs, when enough additional gain is brought by the optical transitions involving excited QD states. For non-lasing WGMs, the peak intensity first increases linearly and then saturates when $P_{\text{ex}}$ is raised. This behavior is thus similar to that of the background, but for the onset of the saturation, which is observed at a much higher ($\times 5$) critical excitation power as shown in Fig. 10b. Thanks to the Purcell effect they experience, in-resonance QDs are less subject to state-filling than out-of-resonance QDs. Assuming that in-resonance QDs and out of-resonance QDs are pumped the same way, the ratio of the saturation powers gives an estimate of $\tau_{\text{free}}/\tau_{\text{on}}$, which is of the order of 5 in this experiment. Much larger effects are observed for high-$F_p$ microdisks obtained by wet chemical etching (see Fig. 10a). For a $1.8\,\mu\text{m}$ diameter disk ($Q = 5300$ and $F_p = 68$ for the WGM under study) we obtain here $\tau_{\text{free}}/\tau_{\text{on}} \sim 14$ for the QD which are on-resonance with this WGM.

It is clear that very complex phenomena lie behind this apparently simple saturation phenomenon. For instance, the exciton–biexciton splitting is much larger than the linewidth of the WGMs, so that a QD, which is in resonance with a WGM under weak excitation conditions, is out of resonance when it contains more than one exciton. On the other hand, multiexcitonic emission from some other QDs can be coupled to the WGM for high pumping rates. Finally, some broadening of the QD emission lines is expected under high excitation conditions. The validity of a quantitative estimate of $\tau_{\text{free}}/\tau_{\text{on}}$ from these CW PL data is therefore hard to justify from a theoretical point of view, and has to be checked experimentally. A good agreement between time-resolved data and CW data has been reported for instance for pillar microcavities [96]. For the data of Fig. 10, we can easily check that the estimated magnitude of the Purcell effect is reasonable. One expects indeed

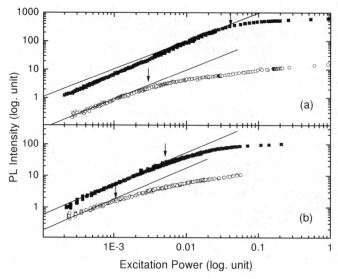

**Fig. 10.** Input–output curves for two microdisks: (**a**) $d = 1.8\,\mu m$, $Q = 5300$, $F_p = 84$; (**b**) $d = 3\,\mu m$, $Q = 3000$, $F_p = 24$. *Full squares* correspond to a non-lasing WGM in resonance with the low-energy emission tail of the QD array, and *empty dots* to reference out-of-resonance QDs. The *arrows* mark the onset of the PL intensity saturation

to observe here an average enhancement factor $\tau_{free}/\tau_{on} \sim 2F_p/(2.2.2) + 1$, where the various corrective factors (from left to right) come from the WGM two-fold degeneracy, the spectral and spatial averagings, the random orientation of the QD dipole and the emission into leaky modes. We get in this way $\tau_{free}/\tau_{on} \sim 17$ for the high-$F_p$ disk and $\tau_{free}/\tau_{on} \sim 5.5$ for the low-$F_p$ disk, which is close to the ratio of the saturation powers.

Apart from micropillars and microdisks, this approach is a convenient tool to characterize 2D photonic crystal microcavities. A ten-fold SE rate enhancement of in-resonance InAs QDs has been reported by *Happ* et al. [98].

## 6.3   Purcell Effect for Single QDs in 0D Microcavities

The "single QD in a microcavity" system is particularly interesting since it allows one to probe the full magnitude of CQED effects, at least in principle. On a more practical side, this system constitutes also the core of an efficient single photon source.

Several groups have demonstrated since 2000 the coupling of a single QD (or a few QDs) to a single cavity mode using different approaches. By depositing a quantity of InAs very close to the critical thickness, very dilute QD arrays can be grown [33,99] (areal density around $10^7\,cm^{-2}$). For such a low areal density, micropillars or microdisks in the few micron diameter

range contain a single QD (or less) on average. Because of QD size fluctuations, however, the probability to get a spectral resonance of the QD with
a given cavity mode is rather low. One can also start from a dense QD array
and use a spectral detuning of the cavity mode toward the low energy tail
of the inhomogeneously broadened line of the QD array in order to reduce
in a controlled way the average number of QDs on resonance with the cavity
mode [100]. Controlling both the QD location within the cavity and its spectral resonance is very desirable, but obviously challenging. In a pioneering
experiment, *Zwiller* et al. have controlled the QD location in a microdisk using a detailed mapping of the QD sample by μPL to drive the processing of
the microdisk [79]. Concerning spectral tuning, temperature control can be
used to some extent but the development of appropriate tunable microcavities
would be extremely useful.

Several groups have reported the observation of SE enhancement for single
QDs in microcavities. As we shall see, the analysis of the experimental data
is usually not straightforward.

The most direct (and convincing) evidence has been presented by *Kiraz* et al., who have studied a single InAs QD placed in a microdisk, using
temperature to tune this QD in and out of resonance with a WGM [15].
A crossing behavior of the WGM and the QD exciton line is observed, which
confirms that the QD is in the weak-coupling regime. Since the experiment
is conducted in the low temperature range (4 to 50 K), the QD exhibits the
same exciton radiative lifetime for positive and negative spectral detunings
when it is out of resonance with the WGM. Therefore, lifetime modifications
can be safely attributed to a microcavity effect. A strong enhancement of the
intensity of the WGM emission and a 6-fold reduction of the QD radiative
lifetime are observed when the QD is tuned in resonance with the WGM
($d = 5\,\mu m$, $Q = 6500$, $F_p = 17$).

This experiment raises however some open questions, which highlight the
difficulties related to a quantitative interpretation of experiments on single QDs. Here, the radiative lifetime of the QD under study is around 3.4 ns
when it is out of resonance, whereas similar QDs in the unprocessed sample
exhibit a 1.1 ns lifetime. This difference has been initially attributed to the
different photonic environments surrounding the QDs [101]. Though SE inhibition is indeed expected for excitons embedded in a sub-wavelength thick
dielectric slab in air, calculations show that this effect should be rather small
in the present case, due to the in-plane orientation of the QD dipole and
to the fact that the microdisk thickness is close to $\lambda/n$ [102]. By the way,
this effect has not been observed for collections of out-of-resonance QDs in
a similar disk geometry [12]. Therefore, the unusually long lifetime of the QD
under study could be related to a very specific location within the microdisk
(e.g. very close to the sidewalls) or to strong fluctuations of the "free-space"
exciton radiative lifetime from one QD to the other. In any case, this discussion highlights the need for a good reference to discuss the magnitude of the

Purcell effect. In the present experiment, this magnitude will be around 2 or 6, depending on the assumption made for the radiative lifetime of the QD in bulk GaAs, 1.1 or 3.4 ns.

The CNRS/LPN [16,103] and Stanford [14,104] groups have studied single QDs in micropillars by time-resolved µPL. Some typical data taken from [16] are shown on Fig. 11. µPL experiments are first conducted at different excitation powers in order to identify QD exciton lines; under very strong excitation conditions, single QDs exhibit several meV broad emission bands which permit one to probe the mode structure of the pillar microcavity. As for experiments on collection of QDs (see Sect. 6.2), those QDs, which are out-of-resonance with respect to the cavity mode, exhibit lifetimes rather close to the free-space value ($\sim 1.3$ ns). This suggests in particular that non-radiative recombination can be neglected. QDs in spectral resonance exhibit much shorter lifetimes (400 ps in Fig. 11, limited by the spectral resolution of the photon correlation set-up used in this study) due to the Purcell effect. Here $F > 3$ for this micropillar ($d = 1$ µm, $Q = 500$, $F_\mathrm{p} = 9$). *Solomon* et al. have observed even stronger effects ($F \sim 6$ for $d = 0.6$ µm, $Q = 630$, $F_\mathrm{p} = 26$) for InAs QDs displaying amazingly long "free-space" lifetimes ($\sim 25$ ns) [104].

Considering a specific QD on resonance, it is not possible to rule out *a priori* some contribution of non-radiative processes to the observed lifetime shortening. Since the location of the QD within the micropillar is not known, a quantitative estimate of the expected radiative lifetime (such as the one done in [14]) is by no means possible. Furthermore, the observation of a stronger PL signal for QDs in resonance than for other QDs is not a proof of fully radiative exciton recombination, since this effect is mostly due to the better collection by the µPL setup of the light emitted into the cavity mode. By contrast, the direct measurement of the efficiency of the triggered single photon source formed by this "QD in micropillar" system permits us to address this issue safely. It has been shown in two independent experiments [103,104] that this efficiency is well understood assuming negligible non-radiative recombination process.

As for collections of QDs, clear signatures of the Purcell effect are also observed under CW excitation. Figure 12 shows data obtained for three QDs in a pillar microcavity, the first two being resonant with the fundamental mode of the pillar, whereas the third emits at an energy higher than the stop-band of the DBRs. Their exciton lines present qualitatively the same excitation power dependence, with a linear increase for low pumping powers, followed by a saturation and decrease for large pumping rates. This behavior reflects the variation of the probability to have a single exciton in the QD, which drops when the probability to have $N$ excitons in the QD ($N > 1$) becomes large. A remarkable feature in Fig. 12 is the large difference in saturation pumping powers of these three QDs. Due to their smaller exciton radiative lifetime, QDs in resonance experience state filling for larger critical pumping rates at a later stage. This behavior can be modeled by simple

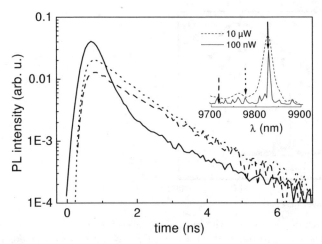

**Fig. 11.** Time-resolved spectra obtained at 4 K for three QDs in a pillar microcavity ($d = 1\,\mu m$, $Q = 500$, $F_p = 9$). The PL spectra obtained under low and high excitation conditions is shown in the *insert*

rate equations. Using typical radiative lifetimes measured for the exciton and the biexciton for similar QDs in GaAs, it is possible to extract estimates of the Purcell SE enhancement factor. Here $F = 10$ for the QD which is well matched spectrally with the cavity mode, which is pretty close to the maximum SE enhancement achievable in this micropillar ($F_p + 1 = 14$ here). As expected, a stronger Purcell effect can be observed for a single QD compared to collections of randomly distributed QDs, provided this specific QD is well matched to the cavity mode.

Obviously, the analysis of this CW experiment relies on two main assumptions: (1) the "free space" exciton radiative lifetime of these QDs is the same and (2) the pumping rate is the same for all three QDs. Assumption (1) is reasonable in the present case since reference QDs in GaAs mesas display small fluctuations of their radiative lifetime ($\pm 10\%$ typically). The validity of the second assumption can be checked in a very simple way by studying the time-integrated PL intensity of the same lines for a *pulsed* excitation. For such pumping conditions, the intensity of the exciton line saturates when more than one electron–hole pair on average per pulse are injected in the QD. The onset of saturation only depends on the pumping rate, and not on the radiative lifetime of the QD. In the present experiment, the pumping rates were similar within experimental accuracy ($\sim 20\%$).

### 6.4 "Nearly" Single-Mode Spontaneous Emission

It has been recognized as early as 1991 that small, high-$Q$ micropillars can potentially collect a large fraction $\beta$ of the spontaneous emission (SE) of a monochromatic emitter into a single confined mode ($\beta \to 1$) [57,58]. This

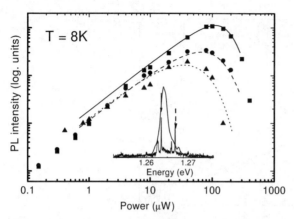

**Fig. 12.** Excitation power dependence of the integrated intensity of the "exciton" line of three different QDs in a micropillar ($d = 0.85\,\mu$m, $F_\mathrm{p} = 13$). Fits are obtained assuming $F = 10$ for the QD in resonance (*solid line*), $F = 2$ for the detuned QD at $1.267\,$eV (*dashed line*) and $F = 1$ for a QD emitting above the DBR stopband at $1.34\,$eV (*dotted line*)

result can be viewed as a consequence of the selective enhancement of the SE into the cavity mode (Purcell effect), which provides a dynamic funnelling of the photons into this specific mode [11]. Let us consider an isolated QD in a 0D microcavity, which supports a continuum of "leaky" modes and a set of discrete resonant cavity modes. If the emitter is coupled to a single resonant mode, the fraction of the SE which is emitted into this mode, $\beta$, can be written as

$$\beta = \frac{F}{F + \gamma}, \tag{21}$$

where $\gamma/\tau_\mathrm{free}$ is the SE rate into leaky modes and $F/\tau_\mathrm{free}$ its SE rate into the cavity mode. As shown in the previous sections, $\gamma$ is experimentally of the order of one, so that very large values of $\beta$ are obtained when the Purcell effect is strong (e.g. $\beta > 0.8$ for $F > 5$).

When the QD is similarly coupled to two resonant cavity modes, $\beta = F/(F + \gamma)$ approaches 50% for both modes. For InAs QDs – whose dipole is randomly oriented in the plane of the sample – this situation is encountered in circular micropillars when the QD is coupled to the polarization-degenerate fundamental mode. It is also the case in microdisks due to the degeneracy of counterpropagating WGMs. Such degeneracies can be lifted by breaking the symmetry of the microcavity. Elliptical micron-sized micropillars display for instance a well resolved fundamental doublet of modes, which are linearly polarized along the long and short axis of the pillar cross-section [105]. Due to their preferential coupling to one of these modes, single QDs on resonance exhibit a very high degree of linear polarization ($> 90\%$). This effect has been implemented to realize a single-mode source of polarized single photons based on a single QD [16], as discussed in Sect. 7.2.

The combination of SE enhancement into the cavity mode and inhibition of SE into leaky modes is a very efficient way for further improving $\beta$. Considering the data published by *Bayer* et al. ($F + \gamma = 2.2$ and $\gamma = 0.1$) [90], we see that the fraction of the SE coupled to the (2-fold degenerate) fundamental mode of the metal-coated micropillar is around 0.96, which is a very impressive figure of merit.

To conclude this section, let us emphasize that these values of $\beta$, which are *experimentally* deduced from measurements of the SE enhancement factor $F$, are much larger than previous values obtained for quantum wells in GaAs/AlAs VCSELs ($\beta \sim 0.04$ [106]) or microdisks ($\beta \sim 0.2$ [67,69]) and determined from the study of the input/output characteristic curves of these microcavity lasers. Due to its large homogeneous linewidth (especially under the high excitation conditions corresponding to the onset of lasing), a QW does not experience a strong Purcell effect. Although a clear $\beta$ increase is observed with respect to standard QW laser diodes ($\beta \sim 10^{-5}$), this improvement is mostly related to the reduction of the number of modes to which this broadband emitter is coupled. For a comparable microcavity size, much larger $\beta$s are obtained for QDs. Their very small emission linewidth (for a low $T$ and weak excitation conditions) is the key to observe a very efficient dynamic funnelling of the photons into a single mode.

# 7   Application Prospects

## 7.1   QD-Microcavity Lasers: Toward Single QD-Lasers

Until now, microcavity lasers did not benefit much from the implementation of QD arrays instead of QWs as active medium, mostly because of the important homogeneous and inhomogeneous broadenings of their gain curve.

For instance, the lowest reported threshold current for QD-VCSELs is around 200 μA at 300 K [107], to be compared with 36 μA for the best QW-VCSELs [8]. However, carrier localization in QDs can be very helpful by preventing non-radiative carrier recombination at microcavity sidewalls [34,35]; this effect has been highlighted experimentally by studies of the differential efficiency of 1 μm diameter oxide-apertured light emitting diodes [108], as well as by the operation of QD-PC lasers under pulsed optical pumping with a relatively low threshold (120 μW at 300 K [109]). In these few experiments, experimental conditions – and first of all the temperature – were not well suited to the observation of lasing in a genuine CQED regime. We discuss briefly possible routes toward novel QD lasers, which use a single QD as active medium and rely on CQED for their operation. Unique characteristics, such as an extremely low threshold current or thresholdless behavior are predicted.

Single QD microcavity lasers are appealing for several reasons. On one hand, working with a single QD alleviates all limitations of QD lasers related to inhomogeneous broadening. On the other, this system is, at least at first sight, a solid-state analog of micromasers [110] and single-atom lasers [111], which operate with one atom or less on average in their cavity.

Micromasers are based on a high-$Q$ single mode cavity, pumped by a weak beam of excited atoms crossing the cavity one by one. When crossing the cavity mode, the atoms experience a strong coupling regime and undergo a coherent evolution during an interaction time defined by the mode size and the atom velocity [112]. If $Q$ is high enough for cavity relaxation to be negligible between two successive atom transits, a field can build up in the cavity. Since the frequency of the atom/cavity Rabi oscillation depends [as $(n+1)^{1/2}$] on the initial number of photons in the cavity, micromasers display an oscillatory gain curve and very interesting non-linear characteristics, such as switching between different stable operating points. Well above threshold, micromasers can also operate within domains of negative differential gain, which results in a strong squeezing of the intensity fluctuations of the emitted field.

Is it possible to build a single QD laser operating, as the micromaser does, in the strong coupling regimes? To adress this point, it is important to notice first major differences between these two quantum systems. Micromasers are pumped by an atomic beam, each atom having a finite interaction time with the cavity. By contrast, the single QD is fixed in the cavity of the microlaser and remains coupled to the cavity mode as long as it contains one exciton or less. This puts in practice very severe contraints on the QD repumping scheme. To illustrate this point, let us consider a single QD describing a Rabi oscillation between the two states $|X,0\rangle$ and $|g,1\rangle$, and let us inject in the QD an additional electron–hole pair. The coupled system behaves very differently depending on its state just before this event occurs: if it is $|g,1\rangle$, this state becomes $|X,1\rangle$ and the system experiences a Rabi oscillation at angular frequency $\Omega\sqrt{2}$; if it is $|X,0\rangle$, it becomes $|XX,0\rangle$, and the Rabi oscillation is quenched due to the spectral detuning of the cavity mode and QD optical transition. Since expected Rabi periods for QDs are typically in the ps range, i.e. faster than relaxation processes, controlled non-resonant pumping is not possible. Tuning the QD transition energy using, e.g. electro-optical effects to control the QD-cavity interaction on such a short time scale seems also very difficult. Therefore, single QD lasers will most likely operate in the weak coupling regime.

The weak coupling regime results from a too fast photon escape or emitter dephasing. Let us consider briefly the first case ($\Omega < \omega/4Q$). We know from Sect. 3 that the enhanced SE rate of the emitter in the cavity mode is at most

$\omega/2Q$ (value achieved for an ideally matched emitter at critical coupling), so that the average photon number in the cavity at saturation $\langle n \rangle_{\text{sat}}$ satisfies

$$\langle n \rangle_{\text{sat}} = \frac{\tau_{\text{c}}}{\tau_{\text{cav}}} < \frac{\omega}{2Q} \frac{Q}{\omega} < \frac{1}{2} . \tag{22}$$

Lasing is therefore not possible in this regime, since emitted photons leave the cavity too fast to have the possibility to stimulate the emission of a second photon by the (repumped) QD.

More interesting is the second case, for which the emitter dephasing is too strong to permit a coherent evolution of the QD-cavity system, but the cavity has very small losses ($\omega/Q_{\text{e}} > \Omega \gg \omega/Q$). Under such conditions, standard rate equations can be used to describe the evolution of the electron and photon populations [113,114]. The threshold condition $\langle n \rangle = 1$ can be written as

$$\beta \frac{p}{\tau_{\text{cav}}} \tau_{\text{ph}} = 1 \qquad \text{or} \qquad p \frac{F}{\tau_{\text{free}}} \tau_{\text{ph}} = 1 , \tag{23}$$

where the inversion parameter $p$ ($p < 1$) is the probability that the QD contains one electron–hole pair and $\tau_{\text{ph}} = Q/\omega$ is the photon lifetime in the cavity. Here, the magnitude of the Purcell effect is much smaller than $F_{\text{p}}$ since the emitter linewidth is much broader than the mode linewidth. The SE rate in the cavity mode is given by

$$F = \frac{4\Omega^2 Q_{\text{e}} \tau_{\text{free}}}{\omega} = \left( \frac{\Omega Q_{\text{e}}}{\omega} \right) (4\Omega \tau_{\text{free}}) . \tag{24}$$

We see therefore that lasing is possible only if $4 \left( \Omega/\Delta \omega_{\text{e}} \right) \left( \Omega/\Delta \omega_{\text{c}} \right) > 1$ where the first (resp. second) term which appears in this product is smaller (larger) than 1.

It is also interesting to write the lasing condition as $F \left( \tau_{\text{ph}}/\tau_{\text{free}} \right) > 1$, which shows clearly that two different types of single QD lasers can be built, depending on the fact they rely or not on a strong Purcell effect ($F > 1$ or $F < 1$).

### 7.1.1 Single QD Lasing Without Purcell Effect ($F < 1$)

In this case, $\tau_{\text{ph}}/\tau_{\text{free}} > 1$. Since InAs QDs have a free-space lifetime in the 1 ns range, the microcavity $Q$ must be larger than $10^6$, a value which can be obtained only with microspheres. The system formed by a single QD coupled to a microsphere has been proposed as an ultralow threshold laser by *Pelton* and *Yamamoto* [115].

For this system, the total QD SE rate is weakly modified compared to its free space value, and most of the SE goes into the continuum of leaky modes. In spite of this low $\beta$, lasing can be obtained with a remarkably low threshold thanks to the very low losses of the WGM. An important parameter appears to be the QD to microsphere distance: bringing the GaAs bulk or mesa structure containing the QD closer to the microsphere improves the QD/WGM coupling but spoils the $Q$ of WGMs, down to $10^6$ typically when the two systems are in contact [115].

In order to estimate the typical threshold of this single QD laser and highlight the crucial temperature issue (which is not adressed in [115]), let us consider a QD which is optimally located in the vicinity of a 60 μm diameter microsphere (QD-microsphere distance $\sim 0.3$ μm, $Q \sim 10^7$, $\tau_{ph} \sim 6$ ns). The effective volume of a WGM is given by $\pi^2 \lambda d^2/60$, where all lengths are measured in μm [45], so that $V \sim 2000(\lambda/n)^3$ and $F_p \sim 400$. Let us consider that $\Delta E_{em}$ is around 10 μeV, a realistic value for a single QD at low temperature ($\sim 10$ K). We get $F = 0.1\, F_p Q_{em}/Q \sim 0.4$, where the factor 0.1 accounts for the relative intensity of the vacuum field at the QD location referenced to its maximum value. Since $\tau_{free} \sim 1$ ns, lasing is obtained for $p > 0.4$. Assuming a perfect pumping efficiency, this corresponds to a threshold current $I_{th} = ep(1 + F)/\tau_{free} \sim 100$ pA, which is five orders of magnitude smaller than the best reported value for semiconductor lasers [8].

### 7.1.2   Single-QD Lasing With a Strong Purcell Effect

When the QD experiences a strong Purcell effect (in spite of its linewidth being larger than the mode linewidth), the requirement on the cavity $Q$ is less stringent and other microcavities – such as microdisks – could be used to build a single QD laser.

Let us consider for instance a single QD optimally located in a 2 μm diameter microdisk ($\hbar\Omega = 50$ μeV) and let us assume that $Q \sim 30\,000$. (It is reasonable to think that such a value, larger than the best $Q$s reported to date for microdisks, will be achieved in the future.) According to Sect. 5, this QD would be in the strong coupling regime at low temperature, since both emitter and mode linewidth are small compared to the Rabi energy. When the temperature is raised gradually, the QD emission line entails an increasing homogeneous broadening, so that the system finally enters a weak coupling regime. Let us assume that $\Delta E_e$ is around 200 μeV, which is a typical value around 90 K in good quality samples [26,27]. We have then $\Delta E_e \gg \hbar\Omega$ (weak coupling regime), $\tau_{ph} \sim 18$ ps, $F \sim 80$ from (24) and $\tau_{cav} \sim 12$ ps. The lasing threshold is reached for $p \sim 2/3$ and the threshold current is of the order of 10 nA.

Due to the strong enhancement of the SE rate, a large fraction of the SE is coupled to the cavity mode ($\beta > 0.98$). Such a laser can be regarded in a very good approximation as a thresholdless laser. Thresholdless lasers are ideal converters of electrical signals into optical signals: above as well as below threshold, all photons are emitted into a single cavity mode. Their input/output and input/noise characteristics do not display any peculiar feature at threshold ($\langle n \rangle = 1$), hence their name. (Experimentally, the onset of lasing can be detected however thanks to a spectral narrowing of the emission above threshold.)

Apart from practical problems (related in particular to the pumping issue), single QD lasers suffer from several intrinsic drawbacks, which constitute severe limitations in view of possible applications. First of all, raising the temperature results in a broadening of the QD exciton line, as well as in a drastic reduction of $F$ and $\beta$. The operation of a single QD-laser at 300 K ($\Delta E_{\mathrm{em}} \gg 1\,\mathrm{meV}$) is therefore unlikely. Secondly, the QD-WGM system is decoupled as soon as two electron–hole pairs or more are trapped in the QD. This effect leads to several original properties under high injection conditions and limits *in fine* the maximum output power. When the pumping rate is raised under non-resonant excitation, such a laser would exhibit first conventional lasing, then a blinking behavior, before it switches off due to carrier accumulation in the QD [116]. Resonant carrier injection using resonant tunneling, as proposed in [115], or resonant optical pumping might however alleviate this problem.

To conclude, microcavity effects on single QDs open a route toward single QD lasers displaying unique characteristics (ultralow threshold, thresholdless behavior), but the practical interest of such devices is not yet clear. In the near and mid-term, the main applications of the Purcell effect will concern devices relying on (enhanced) SE, such as the single photon source based on a single QD in a microcavity, which is presented in the next section.

## 7.2   The Single-Mode Single Photon Source and the Photon Collection Issue

As discussed by Michler in his contribution to this volume, the development of efficient solid-state sources of single photons (S4Ps) is a major challenge in the context of quantum communications [117,118] and quantum information processing [119]. S4Ps implement a single emitter able to emit photons one by one, such as a molecule [120], a colour center [121], a semiconductor nanocrystal [122] or self-assembled QD [16,17,103,104,123,124,125,126]. Collecting efficiently these single photons is essential for a practical use of the S4P. Considering firstly quantum key distribution, the preparation of a secret key using error correction and privacy amplification is possible only if the probability of detecting a photon in a given time slot is larger than the detector's dark count probability [118]. A low efficiency $\varepsilon$ of the S4P (defined as the probability to emit one photon instead of none for a light pulse) puts therefore a severe constraint on the length of the transmission line, besides its obvious influence on the transmission rate itself. Secondly, a single mode operation of the S4P as well as good efficiency ($\varepsilon \sim 1$) are required to ensure the operation of quantum optical gates based on single photon interference and the scalability of related quantum computing schemes [119].

A very promising route toward efficient single-mode S4Ps is based on the combination of a single QD and of a pillar microcavity, as first proposed [96] and demonstrated by CNRS/LPN [16]. Following the experimental observation of the strong Purcell effect, it has been suggested that photon collection efficiencies close to unity would be achievable for single InAs QDs in high-$F_p$ GaAs/AlAs micropillars [11,14]. Unfortunately, such a statement, which is implicitly based on an identification of $\beta$ and $\varepsilon$, is not perfectly valid. In some cases, the S4P efficiency $\varepsilon$ is much smaller than the SE coupling factor $\beta$, due to several intrinsic and extrinsic loss mechanisms whose role has been overlooked until recently [127].

The first, intrinsic, loss mechanism is related to the finite reflectivity of the bottom distributed Bragg reflector (DBR). For a micropillar built with balanced DBRs, confined photons would have for instance equal probabilities to escape the pillar microcavity from its top or from its bottom. In practice, mirror losses can be easily engineered by adjusting the number of periods of both DBRs. Though rather obvious, this first point highlights that $\varepsilon$ and $\beta$ are both qualitatively and quantitatively different.

Some extrinsic losses such as residual absorption or scattering by interface roughness already exist in planar microcavities and will be present as well in micropillars. As discussed in Sect. 4, novel extrinsic and intrinsic losses affect small diameter micropillars ($d < 3\,\mu\text{m}$) [53] and first of all, the scattering induced by the roughness of the pillar sidewalls. All these effects tend to reduce the S4P efficiency. We estimate in the following $\varepsilon$ as the fraction of the light, which is emitted into the cavity mode and exits the cavity due to the finite reflectivity of the top DBR. This light forms a well collimated beam and can be entirely collected using basic optical components.

In order to estimate $\varepsilon$, we write that in the low-loss regime, $1/Q$ can be written as a sum of terms related to the various loss mechanisms: intrinsic losses due to finite DBR reflectivity ($1/Q_{\text{int}}$), and extrinsic losses already present in the planar cavity ($1/Q_{\text{ext}}$). For micropillars we add a term ($1/Q_{\text{scat}}$) accounting for additional losses, such as sidewall scattering, diffraction effects for partially etched micropillars, as well as the intrinsic losses appearing for sub-micron diameter micropillars. We thus get

$$\frac{1}{Q} = \frac{1}{Q_{\text{2D}}} + \frac{1}{Q_{\text{scat}}} = \frac{1}{Q_{\text{int}}} + \frac{1}{Q_{\text{ext}}} + \frac{1}{Q_{\text{scat}}}. \tag{25}$$

Assuming a negligible transmission for the bottom DBR, we can estimate $\varepsilon$ as:

$$\varepsilon = \beta \frac{1/Q_{\text{int}}}{1/Q} = \beta \left( 1 - \frac{Q}{Q_{\text{scat}}} - \frac{Q}{Q_{\text{ext}}} \right). \tag{26}$$

$Q_{\text{scat}}$ and $Q_{\text{ext}}$ can be estimated from the experimental study of the cavity $Q$s of the planar and pillar microcavities, as illustrated here using the data presented on Fig. 5. First, some modelling using the standard transfer matrix approach show that $Q_{\text{ext}}$ is around $30\,000$, for the planar microcavity as well as for other GaAs/AlAs planar microcavities displaying record cavity

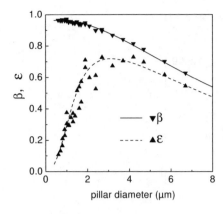

**Fig. 13.** Dependence of $\beta$ and $\varepsilon$ as a function of the pillar radius for a series of high-$Q$ micropillars. $\beta$ and $\varepsilon$ are estimated from the experimental data presented Fig. 5

$Q$s [53,82]. Secondly, $1/Q_{\text{scat}}$ is estimated from the decrease of the micropillar $Q$ with respect to $Q_{2D}$. Simple algebra gives

$$\varepsilon = \beta \left( \frac{Q}{Q_{2D}} - \frac{Q}{Q_{\text{ext}}} \right) . \tag{27}$$

As previously discussed in Sect. 6, $\beta$ can be estimated from the magnitude of the Purcell effect using (21). We present in Fig. 13 the estimates of $\beta$ and $\varepsilon$ for the collection of micropillars of Fig. 5. We assume here that the QD is located at the antinode of the fundamental mode and in perfect spectral resonance ($F = F_p/2$ due to the random in-plane orientation of the QD dipole). Due to their large $F_p$s, very high $\beta$ values are expected for the small diameter micropillars. Quite remarkably, $\beta$s close to one are obtained in spite of the strong degradation of $Q$ induced by the sidewall scattering. By contrast, these losses affect $\varepsilon$ much more seriously. Since intrinsic cavity losses are very weak (large $Q_{2D}$), additional losses due to sidewall scattering become dominant for diameters below $3\,\mu$m. It is thus very detrimental to reduce further the pillar diameter beyond this limit. In the present case, the maximum value for $\varepsilon$ (70 %) is obtained around $d = 3\,\mu$m for such state-of-the-art, high-$Q$ micropillars.

Until recently, most efforts on QDs in pillar microcavities have been concentrated on a maximization of the magnitude of the Purcell effect (and therefore of $\beta$). It is clear that this approach is far from being optimal as far as the S4P efficiency $\varepsilon$ is concerned, for two reasons. Firstly, it tends to favor low-$V$ cavities, for which losses due to sidewall scattering are larger. Secondly, high-finesse planar cavities are also preferred, since a large $Q_{2D}$ helps reduce total losses; this again increases the relative role of scattering losses and absorption losses compared to intrinsic (mirror) losses. $\varepsilon$ displays a maximum which results from a trade-off between the favorable reduction of the cavity volume and the detrimental degradation of $Q$, which are both entailed by a decrease of the pillar diameter. The optimum diameter corresponds roughly to the onset of $Q$ degradation due to sidewall scattering.

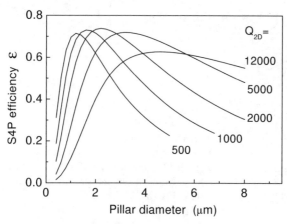

**Fig. 14.** Theoretical dependence of $\beta$ and $\varepsilon$ as a function of the pillar diameter $d$, for various values of the planar cavity losses

We present in Fig. 14 a theoretical estimate of $\varepsilon$ as a function of the pillar diameter for various values of the intrinsic cavity losses (or $Q_{2D}$). We assume constant absorption losses ($Q_{ext} = 30\,000$) and a constant sidewall roughness, similar to the one observed on our two series of micropillars. Note that these assumptions correspond to the best-published results not only for the GaAs/AlAs system, but also for all semiconductor pillar microcavities. For all cases, $\varepsilon$ exhibits qualitatively the same dependence as a function of $d$, but the optimum $d$ depends strongly on $Q_{2D}$; it increases as expected as a function of $Q_{2D}$, since high-finesse microcavities are more sensitive to additional losses. The highest value of $\varepsilon(0.73)$ is obtained for $Q_{2D} \sim 2000$ and $d \sim 2\,\mu m$. Quite remarkably, the maximum value of $\varepsilon$ depends rather weakly on $Q_{2D}$ and remains above 0.7 when $Q_{2D}$ is in the 500–5000 range. This behaviour allows one to choose (if necessary) the pillar diameter without compromising the collection efficiency $\varepsilon$ of the S4P. This is very interesting in practice, since a reduction of the pillar diameter is the most common approach for reducing the number of emitters present in the microcavity down to unity [14,16].

The efficiency of some QD-S4Ps based on this optimized micropillar design ($Q_{2D} = 1000$, $d = 1\,\mu m$) has been recently studied experimentally by microPL [103]. For that purpose, a pulsed excitation is used, with a pump intensity corresponding to the onset of the saturation of the QD exciton line. Since $\eta \sim 1$ at low temperature, the QD emits one photon per pulse under such excitation conditions. After calibration of the detection efficiency of the microPL setup, a comparison of the well-known photon flux delivered by the QD to the microPL signal permits us to estimate $\varepsilon$. Efficiencies as large as 40% have been obtained at CNRS/LPN [16,103] as well as Stanford [104]. This represents typically a 50-fold improvement with respect to single QDs in mesa structures.

QD in high Q PC cavity

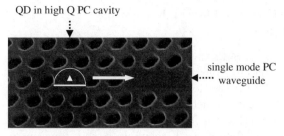

single mode PC
waveguide

**Fig. 15.** Schematic view of a single-mode S4P based on a single QD in a 2D photonic crystal microcavity

Since the QD location within the micropillar is not yet controlled, $\varepsilon$ is significantly smaller than its 70% upper limit for state-of the art processing. However, its experimental value can be fairly well understood by considering the actual magnitude of the Purcell effect. For the QD1 of Fig. 11 for instance, we have $3 < F + 1 < F_{\mathrm{p}} + 1$, $Q_{\mathrm{2D}} = 1000$, $Q = 500$, and $F_{\mathrm{p}} = 9$, so that $0.66 < \beta < 0.9$ and $0.33 < \varepsilon < 0.45$ in good agreement with the experimental estimate ($\varepsilon = 0.39 \pm 0.04$ for QD1).

Because they combine a large $\varepsilon$ and a small residual probability to emit more than one photon, such S4Ps are already much better suited to quantum key distribution than attenuated laser sources. By contrast, their limited efficiency would be a severe hindrance to the scalability of optical quantum computers. Though further improvements of the quality of micropillars can be reasonably expected, alternative SP4 designs should be considered as well. We present in Fig. 15 a single-mode S4P based on a single QD in a 2D PC microcavity, coupled to a photonic waveguide. Such cavities display very high Purcell factors, so that excellent $\beta$s can be expected. By placing the waveguide very close to the cavity, photon escape is dominated by their coupling, and most photons are coupled to the waveguide. Let's consider for instance a state of the art isolated 2D PC cavity: $Q_{\mathrm{iso}} = 2800$ and $F_{\mathrm{p}}^{\mathrm{iso}} = 490$. Let us now assume that its $Q$ drops down to 500 when coupled to a nearby PC waveguide ($Q$ should not be too small, in order to permit the spectral filtering of the $X$-line of the QD versus multi-exciton lines). For this S4P, we have then $F_{\mathrm{p}} \sim 80$, $\beta \sim 0.99$ and $\varepsilon = \beta(1 - Q/Q_{\mathrm{iso}}) \sim 0.83$. PC cavities are therefore already better suited than micropillars to the development of high-efficiency QD S4Ps. Besides the monolithic integration of the S4P and waveguide, an additional asset is related to the possibility of using other emitters than QDs, e.g. molecules or semiconductor nanocrystals. The geometry of 2D PC cavities is well suited to the coupling of one such emitter to the cavity mode. Thanks to the ultimately small volume of the PC cavity, this emitter could then experience a strong Purcell effect even at room temperature. This approach might thus well lead to the development of integrated single-mode S4Ps, exploiting a single molecule or nanocrystal and operating at 300 K.

## 7.3    Generation of Indistinguishable Single Photons and Entangled Photon Pairs

Several applications suggested for S4Ps in quantum information processing and communications (QIPC) require consecutive photons to have the same wave packet. This is in particular the case for the optical quantum computing scheme proposed by *Knill* et al. [119], which is based on a two-photon interference effect. (Let us recall here that the Young's slits experiment is based on the interference of impinging photons with themselves: this experiment can be done by sending photons one by one, or by using incoherent (but monochromatic) light.) Two-photon interference relies on the bosonic nature of photons. When two *identical* single photons enter simultaneously opposite input ports of a 50/50 beam splitter, they should always exit the beamsplitter by the same output port [128]. Here, *identical* means that the spatial and temporal overlaps of their wavepackets should be perfect. This behavior has been first demonstrated using photon pairs generated by parametric down-conversion [128]. In a very elegant experiment, *Santori* et al. have recently demonstrated two-photon interference from consecutive single photons generated by a QD-in-micropillar S4P at around 4 K [17].

The Purcell effect plays a key role in this experiment for two main reasons: 1) single photons are prepared by the pillar microcavity in the same spatial mode; 2) the radiative recombination of the QD exciton is strongly enhanced, so that it becomes much faster than decoherence processes (such as spectral diffusion due to electron–electron interaction or electron–phonon scattering), which is not the case for InAs QDs in bulk GaAs [26,28] at 4 K.

In the same spirit, the Purcell effect might permit the generation of polarization-entangled photon pairs from a single QD using the sequential radiative recombination of a biexciton as proposed by *Benson* et al. [129]. This challenging open issue is discussed by Michler in this volume. Though clear temporal [125] and polarization [130] correlations have been observed for such photon pairs, entanglement has not been yet observed either due to decoherence processes or to the splitting of the bright exciton states due to the QD anisotropy. Let us now consider a QD in a micropillar, whose exciton and biexciton lines are both in resonance with a (two-fold polarization degenerate) cavity mode. Here again, a faster radiative recombination decreases the relative impact of decoherence processes and the possible loss of entanglement. When decoherence is negligible, the Purcell effect also increases the lifetime-induced broadening of the bright exciton lines. When the bright exciton splitting is of the order of a few μeV (which is usually the case for InAs QDs), entanglement is absent for a QD in bulk GaAs because it is possible to determine the polarization of the photons from a measurement of their energy. The intrinsic broadening of the bright exciton lines would partly restore entanglement for the QD in the cavity, provided it is larger than the bright exciton splitting.

# 8    Conclusion

To conclude, a spectacular development of solid-state CQED has been witnessed in the last five years, which results from the implementation of single semiconductor QDs as "artificial atoms" as well as from a spectacular improvement of the figures of merit of semiconductor microcavities. This combination already led to several important results such as the observation of a strong Purcell effect and the development of the first solid-state single mode single photon source. This source is the first optoelectronic device to be based on a real CQED effect. Ongoing progress for 0D microcavities as well as QDs opens several challenging avenues toward novel physics and applications.

**Acknowledgements**

This article is mostly based on a work performed by the author at France Télécom R&D and CNRS/LPN labs in Bagneux until 2001. The essential contributions of B. Gayral, B. Legrand, E. Moreau, I. Robert, B. Sermage, I. Abram (optical studies), V. Thierry-Mieg (MBE growth), C. Dupuis and L. Ferlazzo-Manin (processing) are very gratefully acknowledged, as well as very fruitful interactions with J. Bloch, R. Kuszelewicz, J.Y. Marzin, L.C. Andreani (U. Pavia), E. Costard (Thalès-TRT) and J. Bleuse (CEA Grenoble).

References [1] to [4] provide good introductory overviews of solid-state cavity quantum electrodynamics.

# References

1. H. Benisty, J.M. Gérard, R. Houdré, J. Rarity, C. Weisbuch (Eds.): *Confined Photon Systems: Fundamentals and Applications*, Lect. Notes Phys. **531** (Springer, Berlin, Heidelberg 1999)
2. C. Weisbuch, J. Rarity (Eds.): *Microcavities and Photonic Bandgaps: Physics and Applications*, NATO ASI Ser. E **324** (Kluwer, Dordrecht 1996)
3. E. Burstein, C. Weisbuch (Eds.): *Confined Electrons and Photons: New Physics and Applications*, NATO ASI Ser. B **340**, (Plenum, New York 1995)
4. M. Ducloy, D. Bloch (Eds.): *Quantum Optics of Confined Systems*, NATO ASI Ser. E **314**, (Kluwer, Dordrecht 1996)
5. S. Haroche, D. Kleppner: Phys. Today **42**, 24 (1989)
6. C. Weisbuch, M. Nishioka, A. Ishikawa, Y. Arakawa: Phys. Rev. Lett. **69**, 3314 (1992)
7. H. De Neve, J. Blondelle, P. Van Daele, P. Demeester, R. Baets: Appl. Phys. Lett. **70**, 799 (1997)
8. D.L. Huffaker, D.G. Deppe: Appl. Phys. Lett. **71**, 1449 (1997)
   A.E. Bond, P.D. Dapkus, J. O'Brien: IEEE Photonics Technol. Lett. **10**, 1362 (1998)
9. V. Sandoghdar, F. Treussart, J. Hare, V. Lefèvre-Seguin, J.-M. Raimond, S. Haroche: Phys. Rev. A **54**, R1777 (1996)

10. E. M. Purcell: Phys. Rev. **69**, 681 (1946)
11. J. M. Gérard, B. Sermage, B.Gayral, E. Costard, V. Thierry-Mieg: Phys. Rev. Lett. **81**, 1110 (1998)
12. B. Gayral, J. M. Gérard, B. Sermage, A. Lemaître, C. Dupuis: Appl. Phys. Lett. **78**, 2828 (2001)
13. L. A. Graham, D. L. Huffaker, D. G. Deppe: Appl. Phys. Lett. **74**, 2408 (1999)
14. G. Solomon, M. Pelton, Y. Yamamoto: Phys. Rev. Lett. **86**, 3903 (2001)
15. A. Kiraz, P. Michler, C. Becher, B. Gayral, A. Imamoglu, L. Zhang, E. Hu: Appl. Phys. Lett. **78**, 3932 (2001)
16. E. Moreau, I. Robert, J. M. Gérard, I. Abram, L. Manin, V. Thierry-Mieg: Appl. Phys. Lett. **79**, 2865 (2001)
17. C. Santori, D. Fattal, J. Vuckovic, G. Solomon, Y. Yamamoto: Nature **419**, 594 (2002)
18. L. Besombes, K. Kheng, D. Martrou: Phys. Rev. Lett. **85**, 425 (2000)
19. V. D. Kulakovskii, G. Bacher, R. Weigand, T. Kümmell, A. Forchel, E. Borovitskaya, K. Leonardi, D. Hommel: Phys. Rev. Lett. **82**, 1780 (1999)
20. K. Brunner, G. Abstreiter, G. Böhm, G. Tränkle, G. Weimann: Phys. Rev. Lett. **73**, 1138 (1994)
21. D. Gammon, E. S. Snow, B. V. Shanabrook, D. S. Katzer, D. Park: Science **273**, 87 (1996)
22. M. V. Artemyev, U. Woggon: Appl. Phys. Lett. **76**, 1353 (2000)
23. X. Fan, P. Palginis, S. Lacey, H. Wang, M. C. Lonergan: Opt. Lett. **25**, 1600 (2000)
24. L. Goldstein, F. Glas, J. Y. Marzin, M. N. Charasse, G. Le Roux: Appl. Phys. Lett. **47**, 1099 (1985)
25. J. Y. Marzin, J. M. Gérard, A. Izraël, D. Barrier, G. Bastard: Phys. Rev. Lett. **73**, 716 (1994)
26. M. Bayer, A. Forchel: Phys. Rev. B **65**, 041308 (2002)
27. C. Kammerer, C. Voisin, G. Cassabois, C. Delalande, Ph. Roussignol, F. Klopf, J. P. Reithmaier, A. Forchel, J. M. Gerard: Phys. Rev B. **66**, R041306 (2002)
28. P. Borri, W. Langbein, S. Schneider, U. Woggon, R. L. Sellin, D. Ouyang, D. Bimberg: Phys. Rev. Lett. **87**, 157401 (2001)
29. K. Matsuda, K. Ikeda, T. Saiki, H. Tsuchiya, H. Saito, K. Nishi: Phys. Rev. B **63**, 121304 (2001)
30. L. Landin, M. Miller, M. E. Pistol, C. E. Pryor, L. Samuelson: Science **280**, 262 (1998)
31. J. M. Gérard, A. Lemaître, B. Legrand, A. Ponchet, B. Gayral, V. Thierry-Mieg: J. Crystal Growth **201/202**, 1109 (1999)
32. R. J. Warburton, C. S. Dürr, K. Karrai, J. P. Kotthaus, G. Medeiros-Ribeiro, P. M. Petroff: Phys. Rev. Lett. **79**, 5282 (1997)
33. J. M. Gérard, J. B. Génin, J. Lefebvre, J. M. Moison, N. Lebouché, F. Barthe: J. Crystal Growth **150**, 351 (1995)
34. J. M. Gérard, O. Cabrol, B. Sermage: Appl. Phys. Lett. **68**, 1113 (1996)
35. J. M. Gérard: in [2]
36. See e.g. J. M. Raimond in [3] or J. M. Raimond and S. Haroche in [4]
37. L. C. Andreani, G. Panzarini, J. M. Gérard: Phys. Rev. B. **60**, 13276 (1999)
38. I. Wilson-Rae, A. Imamoglu: Phys. Rev. B **65**, 235311 (2002)
39. P. W. Milonni, P. L. Knight: Opt. Commun. **9**, 119 (1973)
40. S. D. Brorson, H. Yokoyama, E. Ippen: IEEE J. Quant. Electron. **26**, 1492 (1990)

41. G. Bourdon, I. Robert, R. Adams, K. Nelep, I. Sagnes, J. M. Moison, I. Abram: Appl. Phys. Lett. **77**, 1345 (2000)
42. S. M. Barnett, P. M. Radmore: Opt. Commun. **68**, 364 (1988)
43. H. Yokoyama in [3]
44. V. B. Braginsky, M. L. Gorodetsky, V. S. Ilchenko: Phys. Lett. A **137**, 393 (1989)
45. L. Collot, V. Lefèvre-Seguin, M. Brune, J. M. Raymond, S. Haroche: Europhys. Lett. **23**, 327 (1993)
46. D. K. Armani, T. J. Kippenberg, S. M. Spillane, K. J. Vahala: Nature **421**, 925 (2003)
47. S. L. Mc Call, A. F. J. Levi, R. E. Slusher, H. H. Houch, N. A. Whittaker, A. C. Gossard, J. H. English: Appl. Phys. Lett. **60**, 289 (1992)
48. E. Yablonovitch: J. Opt. Soc. Am. B **10**, 283 (1993)
49. J. Jewell, A. Scherer, S. L. Mc Call, Y. H. Lee, S. Walker, J. P. Harbison, L. T. Florez: Electron. Lett. **25**, 1123 (1989)
50. J. M. Gérard, D. Barrier, J. Y. Marzin, R. Kuszelewicz, L. Manin, E. Costard, V. Thierry-Mieg, T. Rivera: Appl. Phys. Lett. **69**, 449 (1996)
51. O. J. Painter, A. Husain, A. Scherer, J. D. O'Brien, I. Kim, D. Dapkus: J. Lightwave Technol. **17**, 2082 (1999)
52. S. Noda, K. Tomoda, N. Yamamoto, A. Chutinan: Science **289**, 604 (2000)
53. T. Rivera, J. P. Debray, B. Legrand, J. M. Gérard, L. Manin-Ferlazzo, J. L. Oudar: Appl. Phys. Lett. **74**, 911 (1999)
54. J. Bloch, R. Planel, V. Thierry-Mieg, J. M. Gérard, D. Barrier, J. Y. Marzin, E. Costard: Superlat. Microstruc. **22**, 371 (1997)
55. K. Iga, F. Koyama, S. Kinoshita: IEEE J. Quant. Electron. **24**, 1845 (1988)
56. A. Scherer, J. L. Jewell, Y. H. Lee, J. P. Harbison, L. T. Florez: Appl. Phys. Lett. **55**, 2724 (1989)
57. T. Baba et al.: IEEE J. Quant. Electron. **27**, 1347 (1991)
58. Y. Yamamoto, S. Machida, G. Björk: Phys. Rev. A **44**, 657 (1991)
59. J. P. Reithmaier, M. Röhner, H. Zull, F. Schäfer, A. Forchel, P. A. Knipp, T. L. Reinecke: Phys. Rev. Lett. **78**, 378 (1997)
60. T. Tezuka, S. Nunoue, H. Yoshida, T. Noda: Jpn. J. Appl. Phys. **32**, L54 (1993)
61. M. Pelton, Y. Yamamoto: Phys. Rev. A **59**, (1999) pp. 2418–2421
62. J. Vuckovic, M. Pelton, A. Scherer, Y. Yamamoto: Phys. Rev. A **66**, 023 808 (2002)
63. A. L. Holmes, M. R. Islam, R. V. Chelakara, F. J. Ciuba, R. D. Dupuis, M. J. Ries, E. I. Chen, S. A. Maranowski, N. Holonyak Jr.: Appl. Phys. Lett. **66**, 2831 (1995)
64. Hyun-Eoi Shin, Young-Gu Ju, Hyun-Woo Song, Dae-Sung Song, Il-Young Han, JungHoon Ser, Han-Youl Ryu, Yong-Hee Lee, Hyo-Hoon Park: Appl. Phys. Lett. **72**, 2205 (1998)
65. J. M. Gérard in [1]
66. L. A. Graham, D. L. Huffaker, D. Deppe: Appl. Phys. Lett. **74**, 2408 (1999)
67. R. E. Slusher, A. F. J. Levi, U. Mohideen, S. L. McCall, S. J. Pearton, R. A. Logan: Appl. Phys. Lett. **63**, 1310 (1993)
68. U. Mohideen, W. S. Hobson, S. J. Pearton, F. Ren, R. E. Slusher: Appl. Phys. Lett. **64**, 1911 (1994)
69. T. Baba, T. Hamano, F. Koya: IEEE Quant. Electron. **27**, (1991) p. 1347

70. B. Gayral, J. M. Gérard, A. Lemaître, C. Dupuis, L. Manin, J. L. Pelouard: Appl. Phys. Lett. **75**, 1908 (1999)
71. P. Michler, A. Kiraz, L. Zhang, C. Becher, E. Hu, A. Imamoglu: Appl. Phys. Lett. **77**, 184 (2000)
72. J. S. Foresi, P. Villeneuve, J. Ferrera, E. Thoen, G. Steinmeyer, S. Fan, J. D. Joannopoulos, L. Kimerling, H. Smith, E. Ippen: Nature **390**, 143 (1997)
73. D. Labilloy, H. Benisty, C. Weisbuch, T. F. Krauss, V. Bardinal, U. Oesterle: Electron. Lett. **33**, 1978 (1997)
74. D. Labilloy, H. Benisty, C. Weisbuch, T. F. Krauss, C. J. M. Smith, R. Houdre, U. Oesterle: Appl. Phys. Lett. **73**, 1314 (1998)
75. J. Vuckovic, M. Lonkar, H. Mabuchi, A. Scherer: Phys. Rev. E **65**, 016 608 (2001)
76. T. Yoshie, J. Vuckovic, A. Scherer, H. Chen, D. Deppe: Appl. Phys. Lett. **79**, 4289 (2001)
77. A. Imamoglu, D. Awschalom, G. Burkard, D. P. Di Vicenzo, D. Loss, M. Sherwin, A. Small: Phys. Rev. Lett. **83**, 4204 (1999)
78. See e.g. C. J. Hood, T. W. Lynn, M. S. Chapman, H. Mabuchi, J. Ye, H. J. Kimble in [1]
79. V. Zwiller: private communication
80. M. V. Artemyev, U. Woggon, R. Wannemacher, H. Jaschinski, W. Langbein, Nano Lett. **1**, 309 (2001)
81. H. Wang: Cavity quantum electrodynamics of quantum dots with dielectric microspheres, in T. Takagahara (Ed.): *Quantum coherence, correeiation and decoherence in semiconductor nanostructures*, (Elsevier, Amsterdam 2003) in press
82. R. P. Stanley, R. Houdré, U. Oesterle, M. Gailhanou, M. Ilegems: Appl. Phys. Lett. **65**, 1883 (1994)
83. A. Kastler: Appl. Opt. **1**, 17 (1962)
84. For a review, see H. Benisty in [1]
85. C. Begon, H. Rigneault, P. Johnsson: Single Mol. **1**, 207 (2000)
86. H. Rigneault, C. Amra, C. Begon, M. Cathelinaud, C. Picard: Appl. Opt. **38**, 3602 (1999)
87. G. Björk, S. Machida, Y. Yamamoto, K. Igeta: Phys. Rev. A **44**, 669 (1991)
88. F. de Martini, M. Marocco, P. Mataloni, L. Crescentini, R. Loudon: Phys. Rev. A **43**, 2480 (1991)
89. I. Abram, I. Robert, R. Kuszelewicz: IEEE J. Quant. Electron. **34**, 71 (1998)
90. M. Bayer, T. L. Reinecke, F. Weidner, A. Larionov, A. Mc Donald, A. Forchel: Phys. Rev. Lett. **86**, 3168 (2001)
91. A. M. Vredenberg, N. E. J. Hunt, E. F. Schubert, D. C. Jacobson, J. M. Poate, G. J. Zydzik: Phys. Rev. Lett. **71**, 517 (1993). See also Hunt et al. in [3].
92. H. Yokoyama, K. Nishi, T. Anan, H. Yamada, S. D. Brorson, E. P. Ippen: Appl. Phys. Lett. **57**, 2814 (1991)
93. M. Suzuki, H. Yokoyama, S. D. Brorson, E. P. Ippen: Appl. Phys. Lett. **58**, 998 (1991)
94. K. Tanaka, T. Nakamura, W. Takamatsu, M. Yamanishi, Y. Lee, T. Ishihara: Phys. Rev. Lett. **74**, 3380 (1995)
95. S. Cortez, O. Krebs, P. Voisin, A. Lemaître, J. M. Gérard: Phys. Rev. B **63**, 23 3306 (2001)
96. J. M. Gérard, B. Gayral: J. Lightwave Technol. **17**, 2089 (1999)

97. B. Gayral, J. M. Gérard: Physica E **7**, 641 (2000)
98. T. D. Happ, I. Tartakovskii, V. D. Kulakovslii, J. P. Reithmaier, M. Kamp, A. Forchel: Phys. Rev. B **66**, R041 303 (2002)
99. D. Leonard, K. Pond, P. M. Petroff: Phys. Rev. B **50**, 11 687 (1994)
100. J. M. Gérard, Le Si Dang, B. Gayral: French patent no. 0006824 (2000)
101. C. Becher, A. Kiraz, P. Michler, A. Imamoglu, W. V. Schoenfeld, P. M. Petroff, L. Zhang, E. Hu: Phys. Rev. B **63**, R12 1312 (2001)
102. S. T. Ho, S. L. Mc Call, R. E. Slusher: Opt. Lett. **18**, 909 (1993)
103. E. Moreau, I. Robert, L. Manin, V. Thierry-Mieg, J. M. Gérard, I. Abram: Physica E **13**, 418 (2002)
104. M. Pelton, C. Santori, J. Vuckovic, B. Zhang, G. S. Solomon, J. Plant, Y. Yamamoto: Phys. Rev. Lett. **89**, 23 3602 (2002)
105. B. Gayral, J. M. Gérard, B. Legrand, E. Costard, V. Thierry-Mieg: Appl. Phys. Lett. **72**, 1421 (1998)
106. D. K. Kuksenkov, H. Temkin, K. L. Lear, H. Q. Lou: Appl. Phys. Lett. **70**, 13 (1997)
107. Z. Zou, D. L. Huffaker, S. Csuzak, D. G. Deppe: Appl. Phys. Lett. **75**, 22 (1999)
108. Z. Zou, H. Chen, D. G. Deppe: Appl. Phys. Lett. **78**, 3067 (2001)
109. T. Yoshie, O. Shchekin, H. Chen, D. G. Deppe, A. Scherer: Electron. Lett. **38**, 967 (2002)
110. D. Meschede, H. Walther, G. Müller: Phys. Rev. Lett. **54**, 551 (1985)
111. K. An, J. J. Childs, R. R. Dasari, M. S. Feld: Phys. Rev. Lett. **73**, 3375 (1994)
112. See e.g. J. M. Raimond in [3]
113. G. Björk, Y. Yamamoto: IEEE J. Quant. Electron. **27**, 2386 (1991)
114. H. Yokoyama: Science **256**, 66 (1992)
115. M. Pelton, Y. Yamamoto: Phys. Rev. A **59**, 2418 (1999)
116. J. Bleuse, J. M. Gérard: unpublished
117. C. H. Bennett, G. Brassard, A. K. Eckert: Sci. Am. **267**, 50 (1992)
118. G. Brassard, N. Lütkenhaus, T. Mor, B. C. Sanders: Phys. Rev. Lett. **85**, 1330 (2000)
    N. Lütkenhaus: Phys. Rev. A **61**, 052 304 (2000)
119. E. Knill, R. Laflamme, G. Millburn: Nature **409**, 46 (2001)
120. Th. Basché, W. E. Moerner, M. Orrit, H. Talon: Phys. Rev. Lett. **69**, 1516 (1992)
    C. Brunel, B. Lounis, Ph. Tamarat, M. Orrit: Phys. Rev. Lett. **83**, 2722 (1999)
    B. Lounis, W. E. Moerner: Nature **407**, 491 (2000)
121. C. Kurtsiefer, S. Mayer, P. Zarda, H. Weinfurter: Phys. Rev. Lett. **89**, 290 (2000)
    R. Brouri, A. Beveratos, J.-Ph. Poizat, Ph. Grangier: Opt. Lett. **25**, 1294 (2000)
    A Beveratos, S. Kühn, R. Brouri, T. Gacoin, J.-Ph. Poizat, P. Grangier: Eur. J. Phys. D **18**, 191 (2002)
122. P. Michler, A. Imamoglu, M. Mason, P.Carson, G. Strouse, S. Buratto: Nature **406**, 968 (2000)
123. P. Michler, A. Kiraz, C. Becher, W. Schoenfeld, P. M. Petroff, L. Zhang, E. Hu, A. Imamoglu: Science **290**, 2282 (2000)
124. C. Santori, M. Pelton, G. Solomon, Y. Dale, Y. Yamamoto: Phys. Rev. Lett. **86**, 1502 (2001)

125. E. Moreau, I. Robert, L. Manin, V. Thierry-Mieg, J. M. Gérard, I. Abram: Phys. Rev. Lett. **87**, 18 3601 (2001)
126. Z. Yuan, B. E. Kardynal, R. M. Stevenson, A. J. Shields, C. J. Lobo, K. Cooper, N. S. Beattie, D. A. Ritchie, M. Pepper: Science **295**, 102 (2002)
127. W. L. Barnes, G. Björk, J. M. Gérard, P. Jonsson, J. Wasey, P. Worthing, V. Zwiller: Eur. J. Phys. D **18**, 197 (2002)
128. C. K. Hong, Z. Y. Ou, L. Mandel: Phys. Rev. Lett. **59**, 2044 (1987)
129. O. Benson, C. Santori, M. Pelton, Y. Yamamoto: Phys. Rev. Lett. **84**, 2513 (2000)
130. C. Santori, D. Fattal, M. Pelton, G. S. Solomon, Y. Yamamoto: Phys. Rev. B **66**, 045 308 (2002)

# Nonclassical Light
# from Single Semiconductor Quantum Dots

Peter Michler

University of Bremen, P. O. Box 330440,
28334 Bremen, Germany
pmichler@physik.uni-bremen.de

**Abstract.** In this article the recent progress in the generation of nonclassical light using semiconductor quantum dots is reviewed. Photon antibunching and triggered single photon emission for both optical and electrical pumping has been observed. The coupling of a single self-assembled quantum dot to a high-quality factor cavity mode is obtained. This gives access to the study of cavity-quantum electrodynamics (QED) effects, e.g., the Purcell effect in an all-semiconductor nanostructure. The positive impact of the Purcell effect on single photon emission is presented. Triggered correlated pairs of photons and the prospects for the generation of entangled photons will also be discussed.

## 1  Introduction

During the last few years the study of quantum optics in semiconductors has become an emerging field of fundamental research [1,2,3,4,5,6]. This interest is mainly triggered by the development of high-quality quantum dot (qd) structures (see the Chapter by Petroff in this volume). They combine atom-like properties such as a discrete energy spectrum and sharp lines in photoluminescence with the advantage that they can naturally be embedded in solid-state systems. Moreover, they possess rich possibilities of generating nonclassical light with tunable photon statistics. These properties make them very attractive for novel device applications in the fields of quantum cryptography, quantum teleportation and quantum computation. Consequently, increasing effort is being devoted towards an understanding of the quantum optical properties of these systems.

The goal of this contribution is to give an introduction to this new and fascinating research topic. We review recent experiments of nonclassical light generation with single semiconductor quantum dots. The chapter is organized as follows: in Sect. 2 we give a brief review of the photon statistics of different light sources and introduce the second-order correlation function $g^{(2)}(\tau)$. We explain how a photon cascade can be produced with a single QD after pulsed optical or electrical excitation. The quantum optical properties of the QD radiation were studied by photon statistics measurements and the experimental technique is described in Sect. 3. The photon statistics of the QD photoluminescence is studied under continuous wave excitation in Sect. 4. Section 5

P. Michler (Ed.): Single Quantum Dots, Topics Appl. Phys. **90**, 315–347 (2003)
© Springer-Verlag Berlin Heidelberg 2003

gives a detailed discussion of the generation of triggered single photons with a single QD both under optical and electrical excitation. Single photon emission from an isolated QD on resonance with a whispering gallery mode and with a fundamental mode of a micropillar is presented. In Sect. 6 cross correlation measurements between the biexciton and the exciton emission as well as between the exciton and the charged exciton emission are analyzed and the prospect for the generation of entangled photons is discussed.

## 2    Theoretical Background

Nonclassical light possesses properties which cannot be explained by the classical electromagnetic theory of waves. Quantization of the radiation field is required to explain the nonclassical properties of light, e.g., the correlations of light generated by a single two level quantum emitter, such as an atom, ion, single molecule or an exciton in a quantum dot. One of the central goals in quantum optics is the generation of light fields with suppressed photon number fluctuations. In the following we discuss this aspect for different light fields.

### 2.1    Photon Statistics

The coherent state (*Glauber* state) represents the best approximation of a classical light source with well defined amplitude and phase in the quantum mechanical description of the electromagnetic field. An example for such a source is a laser which is operated well above the threshold.

The coherent state $|\alpha\rangle$ is an eigenstate of the annihilation operator $\hat{a}$: $\hat{a}|\alpha\rangle = \alpha|\alpha\rangle$. The complex number $\alpha = |\alpha|\exp(i\phi)$ corresponds to the classical complex amplitude of an electromagnetic wave, where $|\alpha|^2 = \langle\alpha|\hat{n}|\alpha\rangle = N$ is the expectation value of the photon number operator $|\hat{n}\rangle$. Thus, $N$ is identical with the average number of photons in a mode and $\phi$ corresponds to the quantum mechanical expectation value of the phase.

$|\alpha\rangle$ can be represented in the basis of the eigenstates of the photon number operator [7]

$$|\alpha\rangle = \sum_{n=0}^{\infty} \exp\left(-\frac{|\alpha|^2}{2}\right) \frac{\alpha^n}{\sqrt{n!}} \, |n\rangle \, . \tag{1}$$

The photon number distribution in this state is a Poisson distribution, i.e., the probability to find $n$ photons in the coherent state with the mean photon number $N = \alpha^2$ is given by

$$P_\alpha(n) = |\langle n|\alpha\rangle|^2 = \frac{N^n}{n!} \exp(-N) \, . \tag{2}$$

States which possess a narrower or a wider photon number distribution are called *sub-* or *super-Poissonian* states where classical light sources can only generate the latter.

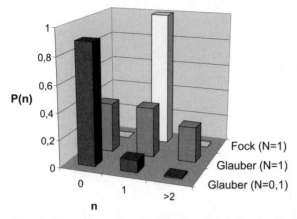

**Fig. 1.** Photon number distribution $P(n)$ for the Fock state with a mean photon number $N = 1$ and for Glauber states with $N = 1$ and 0.1 [8]

A nonclassical light state is the *photon number* or *Fock state* $|n\rangle$. Such a state can be generated, e.g., by a single photon source. The Fock state is the eigenstate of the photon number operator: $\hat{n}|n\rangle = N|n\rangle$, i.e., a mode which is excited in this state is occupied by exactly $N$ photons and the variance $\Delta N = 0$. Figure 1 displays the difference in the photon number distribution for the Glauber and Fock states. If the system is in a Fock state with $N = 1$ a photon number of exactly *one* ($n = 1$) will be measured whereas the coherent state shows a broad distribution of probabilities $P(n)$ for measuring $n$ photons of $P(n = 0) = 0.37$, $P(n = 1) = 0.37$ and $P(n \geq 2) = 0.26$. The high probability $P$ of the emission of two or more photons prevents the application of weak laser pulses of $N = 1$ for quantum cryptography. To reduce $P(n \geq 2)$ significantly ($\leq 1\%$) for a coherent light source one has to adjust the average number of photons as low as $N = 0.1$. But this will result that 90% of the pulses will contain no photons at all.

A characteristic feature of a light source is the temporal sequence of the emitted photons. The photons from conventional light sources, e.g., from a spectral lamp or a thermal light source, arrive in bunches, i.e., the probability for photons to arrive together, in coincidence, is large for short delay time intervals $\tau$. This phenomenon is called *photon bunching* (see Fig. 2a). In contrast, for coherent light emitted from a laser there is no preferred time interval between photons, i.e., the photons are completely uncorrelated. In other words, the probability of detecting two photons with a certain delay $\tau$ is equal for all values of $\tau$. The photon statistics is described by a Poissonian distribution which classically describes random processes (see Fig. 2b), whereas the bunching behavior shows super-Poissonian statistics. A driven single anharmonic quantum system, such as an atom or a molecule, emits *antibunched* photons, i.e., given a first photon is detected it is very unlikely to detect another one immediately after. The physics of photon antibunching

a) bunched, super-Poissonian

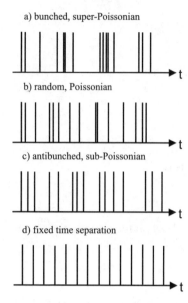

b) random, Poissonian

c) antibunched, sub-Poissonian

d) fixed time separation

**Fig. 2.** Schematic sketch of the photon sequence of (**a**) bunched photons, (**b**) random photons, (**c**) antibunched photons, (**d**) photons with fixed time separations, e.g., from a single photon source. Each *vertical line* represents a photon emission event

is easy to understand: if a two level atom emits a photon at time $\tau = 0$, it is impossible for it to emit another one immediately after, since it is necessarily in the ground state. The next photon can only be emitted after a waiting time which is determined by the effective incoherent pump and recombination rates of the nonresonant optically pumped QD. The result is that there is a deadtime between successive photon emission events, and the generated photon stream, in the absence of other sources of fluctuation, is sub-Poissonian (see Fig. 2c) [9]. A single photon source is able to generate photons on demand. Such a source allows the ultimate quantum control of the photon generation process, i.e., single photons can be generated within short time intervals and a deterministic dwell time between successive photon generation events (see Fig. 2d). This makes it possible to encode information on a single photon level. Such a source is of interest for future applications in quantum computing [10] and quantum cryptography [11]. The generation of single photons on demand from a single QD have been recently demonstrated for the first time by us [2] and will be discussed in detail below.

One method to measure the photon number distribution consists of measuring the second-order correlation function which is defined as

$$g^{(2)}(\tau) = \frac{\langle : I(t)I(t+\tau) : \rangle}{\langle I(t) \rangle^2}, \tag{3}$$

where :: indicates normal ordering and $I(t)$ is the measured intensity. $g^{(2)}(\tau)$ describes the probability to measure a photon at time $\tau$ on condition that a first photon has been detected at time $\tau = 0$. The second-order or intensity correlation function is usually measured by a Hanbury-Brown and Twiss

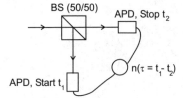

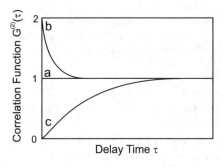

**Fig. 3.** Hanbury-Brown and Twiss experiment to measure the photon statistics

**Fig. 4.** Photon correlation function $g^{(2)}(\tau)$ for a coherent light source $a$, a super-Poissonian source $b$, and a sub-Poissonian source $c$

(HBT) setup which basically consists of a beamsplitter and two photodetectors (see Fig. 3, details in Sect. 3). The signal of one of the photodetectors starts a clock until a detection event of the second detector results in a stop. Then a histogram of the arrival time separations of photon pairs $n(\tau)$ will be produced which is proportional to $g^{(2)}(\tau)$ as long as the measured time separation $\tau$ between photon pairs is much smaller than the mean time between detection events.

Photon correlation measurements under *CW excitation* are an important tool to study the photon statistics and to decide whether the emitted light stems from a single quantum emitter. For a coherent light source $g^{(2)}(\tau) = 1$ (see Fig. 4a) which means that the photons are completely uncorrelated. In contrast, a light source with super-Poissonian statistics shows a clear excess of coincidences ($1 < g^{(2)}(\tau) < 2$) for times shorter than the coherence time $t_{coh}$ of the light source (see Fig. 4b). On the other hand a sub-Poissonian light source satisfies $g^{(2)}(\tau) < 1$, this indicates that two photons are unlikely to appear simultaneously on the detectors (see Fig. 4c). Photoluminescence collected from a single quantum emitter has the property $g^{(2)}(\tau = 0) \approx 0$. This strong quantum correlation is easily washed away with increasing number of emitters. Thus, strong photon antibunching $g^{(2)}(\tau = 0) < 0.5$ can be considered as direct evidence that the radiation source is a single quantum emitter.

Photon correlation measurements carried out under *pulsed excitation* yield signatures for turnstile operation, by discriminating between one and two-photon (Fock-state) pulses as well as coherent-state pulses [14]. The information about the photon number correlation between pulses results from the area of the correlation peaks. In the case of a coherent pulse all the peak areas are identical and the probabilities for the detection of the 'second pho-

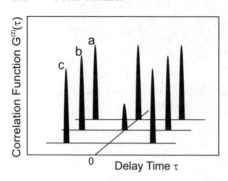

**Fig. 5.** Photon correlation function $G^{(2)}(\tau)$ for a pulsed coherent state $a$, a sub-Poisson state $b$, and a pulsed single-photon-Fock state $c$

ton' are equal within the same pulse or for a succeeding pulse, i.e., Poisson distributed statistics (see Fig. 5a). Sub-Poisson statistics occur if the peak area of the central pulse at $\tau = 0$ is smaller than the peak areas of the other peaks (see Fig. 5b). For perfect single photon emission the peak at $\tau = 0$ is absent since it is not possible to measure a 'second photon' within the first pulse (see Fig. 5c).

## 2.2  Photon Cascade from a Quantum Dot

The three dimensional carrier confinement in a QD results in discrete atomic-like energy spectra. The lens shaped self-assembled InAs QDs approximately possess a parabolic confinement potential in the plane of the QD. Thus, the single particle states can be approximated by a quantum well function in the growth direction ($z$-direction) and by harmonic oscillator functions in the plane of the dot ($xy$-plane). Due to the small thickness of the QD in the $z$-direction the vertical quantization energy is large and it is sufficient to consider the first quantized wavefunction in the $z$-direction. The degeneracies of the electronic shells in the $xy$-plane are given by $2n$, where $n$ is the principal quantum number in the system ($n = 1, 2, 3, \ldots = \text{s}, \text{p}, \text{d}, \ldots$). The occupation of the shells after optical excitation in the GaAs barriers depends strongly on the excitation power and follows Poisson statistics. Due to the three dimensional carrier confinement and the strong Coulomb correlation bound electron–hole states, e.g., the exciton ($1X$), the biexciton ($2X$) or higher multiexction states ($NX$) build up (see Fig. 6a).

The emission spectrum of the QDs depends on the pump power and on the excitation process, i.e., whether pulsed or steady-state excitation schemes are used. After the latter a characteristic PL spectrum is emitted which reflects the steady-state population of the QD, determined by the excitation and recombination rates (see the Chapters by Hawrylak and Korkusiński and Bayer in this volume). The emitted photon statistics can be varied by the excitation power from a sub-Poissonian one, where the photons are temporally antibunched, to super-Poissonian, where they are temporally bunched. This will be discussed in more detail in Sect. 4.

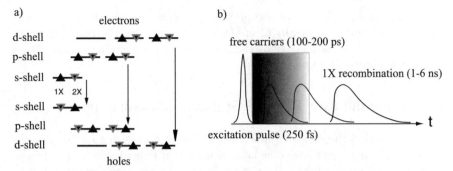

**Fig. 6.** (a) Schematic sketch of the quantum dot energy levels, their population with electron and holes and the respective optical transitions. The *open* and *filled triangles* represent the electrons and holes in different spin configurations. (**b**) Schematic sketch of the emission cascade of an optically excited QD. The *coloured curves* describe the temporal decay of the excitons in the different shells [8]

Even more exciting is the generation of radiative quantum cascades with single QDs. After *pulsed* optical excitation of the semiconductor above the GaAs bandgap electron–hole pairs were mainly generated in the GaAs barriers and subsequently captured by the QDs and relax to the lowest energy levels within a short time scale ($\sim$ 1–100 ps). The recombination of this multiexcitonic state occurs in a cascade process, i.e., in sequential optical transitions of the multiexciton states $NX, (N-1)X, \ldots$ the biexciton $2X$, and the exciton $1X$. The $1X$ exciton possesses the longest lifetime of all multiexcitons and decays as the last one. Figure 6b displays this scenario in the time regime. The energy of the photons emitted during relaxation depends significantly on the number of excitons that exist in the QD, due to Coulomb interactions enhanced by strong carrier confinement (Fig. 7). If the recombination times of the multiexcitonic states are longer than the recombination time of the free electron–hole pairs in the barriers, each excitation pulse can lead to at most one photon emission event at the corresponding $NX$ transition. Therefore, regulation of photon emission process can be achieved due to a combination of Coulomb interactions creating an anharmonic multiexciton spectrum and slow relaxation of highly-excited QDs leading to vanishing reexcitation probability following the corresponding photon emission event at

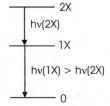

**Fig. 7.** Biexciton–exciton cascade in the *s*-shell

the $NX$-transition. Thus, specific photons from the cascade process, e.g., the $2X$ and $1X$, can be spectrally filtered out and can be used to generate single photons, correlated pairs of photons and even the production of polarization entangled photons is expected.

## 3    Experimental: Photon Statistic Measurements

Investigations of the photon statistics have become a major tool in quantum optics since they reveal the nonclassical nature of light. Experimentally, the photon statistics of single QDs is investigated using a combination of a low-temperature diffraction-limited scanning optical microscopy (SOM) for spatially resolved PL spectroscopy and an ordinary HBT setup for photon correlation measurements. The experimental arrangement is schematically shown in Fig. 8. The system provides spectral resolution of $70\,\mu eV$, spatial resolu-

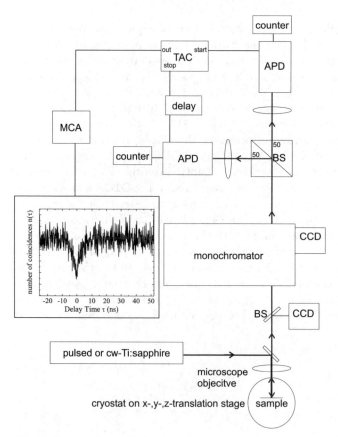

**Fig. 8.** Sketch of a typical experimental setup for photon correlation measurements on quantum dots

tion of 1.7 µm, and temporal resolution of 420 ps. The samples are mounted in a He gas flow cryostat. The cryostat is moved by computer-controlled translation stages, thus allowing for scanning across the sample. Optical excitation is performed either with a continuous wave (CW) or a mode-locked Ti:sapphire laser (82 MHz). Both laser systems are operated at a wavelength where electron–hole pairs are generated in the barriers which are subsequently captured to the QDs. A microscope objective is used to focus the excitation laser onto the sample and to collect the emitted PL from the QDs. The collected light is spectrally filtered by a monochromator and then split with a 50/50 beamsplitter. The resulting two light beams are focused onto two single-photon-counting avalanche photodiodes (SPAD). The pulses from the two SPADs are used to start and stop a time-to-amplitude converter (TAC) whose output is stored in a multichannel analyzer (MCA). The resulting histograms yield the number of photon pairs $n(\tau)$ with arrival time separation of $\tau = t_{\text{start}} - t_{\text{stop}}$.

For continuous wave experiments the measured photon count distribution $n(\tau)$ is usually normalized to the expectation value for counting completely uncorrelated photons obeying a Poissonian arrival time distribution. This is simply done by dividing $n(\tau)$ by the constant $C = R_{\text{start}} R_{\text{stop}} T_{\text{m}} T_{\text{ch}}$, where $T_{\text{m}}$ is the total measuring time, $T_{\text{ch}}$ is the time interval of each channel of the MCA, and $R_{\text{start}}$ and $R_{\text{stop}}$ are the average detection rates of the 'START' and the 'STOP' SPADs, respectively. The normalized measured distribution $\bar{n}(\tau)$ is equivalent to the correlation function $g^{(2)}(\tau)$ as long as the measured time separation $\tau$ between photon pairs is much smaller than the mean time $\Delta T_{\text{D}}$ between detection events [13]. As $\tau \leq 10^{-7}$ s and $\Delta T_{\text{D}} \geq 1 \times 10^{-6}$ s in typical experiments, this condition is easily fulfilled. For the pulsed experiments normalization presumes the knowledge of the temporal profile of the signal intensities [3]. However, it is possible to obtain a 'concentrated' $g^{(2)}(\tau)$ value by dividing the peak areas by the two count rates $R_{\text{start}}$, $R_{\text{stop}}$, the laser repetition period and the total measuring time $T_{\text{m}}$.

## 4  Photon Antibunching in Single Quantum Dot Photoluminescence

The first observation of photon antibunching in the fluorescence from a single atom [15] has been generally regarded as the first proof of the quantum nature of light, since quantization of the radiation field is required to explain the observed correlations. Photon correlation measurements have recently been performed in both colloidal [16] and self-assembled quantum dots [17]. These studies have triggered a series of further quantum optical experiments on single quantum dots. In this section, we report our recent observation of photon antibunching in the photoluminescence of a single self-assembled InAs/GaAs QD at 4 K [17]. The measurements prove that a single self-assembled QD can be considered as a strongly anharmonic quantum emitter.

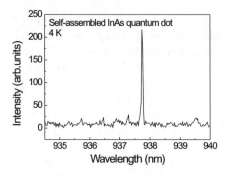

**Fig. 9.** Photoluminescence spectrum of a self-assembled InAs quantum dot, obtained at 4 K at a pump intensity of 66 W/cm² [18]

The emission properties of epitaxially self-assembled QDs can be tailored by their geometrical dimensions and they possess the advantage that they can naturally be embedded in photonic structures, e.g., in micro-resonators, and photonic crystals. These properties make them very attractive for cavity-quantum electrodynamics (QED) experiments and photonic devices.

The self-assembled InAs QDs were grown by molecular beam epitaxy (MBE). The structures are based on a AlAs/GaAs short-period superlattice and a GaAs buffer layer for substrate smoothing. The samples consist of 100 nm GaAs, an InAs QD layer, and 100 nm GaAs. The QDs were grown using the partially covered island technique in order to shift their ground state to higher energies [12] (see also the Chapter by Petroff in this volume). The emission wavelengths of the QDs are in the range from 925 nm to 975 nm. The QDs have a diameter of $\sim$ 400–500 Å and a height of $\sim$ 30 Å. The samples exhibit a gradient in the QD density reaching from $< 10^8 \, \mathrm{cm}^{-2}$ to $\sim 10^{10} \, \mathrm{cm}^{-2}$ across the wafer samples.

The PL spectrum of a self-assembled InAs QD at low temperature (4 K) shows a single, resolution limited (70 μeV) line due to single exciton recombination ($1X$) (Fig. 9). The corresponding photon count distribution $n(\tau)$ for the $1X$ transition is shown in Fig. 10 for two different pump intensities. These intensities correspond to an excitation of the QD well below saturation (a) (see also Fig. 9) and at the onset of saturation (b). Saturation is defined here as the pump intensity at which the $1X$ line reaches its maximum intensity [19]. Traces (a) and (b) exhibit also a clear dip in the correlation counts for a time delay $\tau = 0$, indicating a strong photon antibunching. For comparison, trace (c) of Fig. 10 shows the correlation for multiple emission lines of many QDs in a high-density region of the sample. As expected, this correlation is flat over the complete measurement time and its normalized value of 1 corresponds to Poissonian photon statistics.

In order to correct for the time resolution of our setup the InAs single QD data were fitted with the correlation function

$$g^{(2)}(\tau) = 1 - ae^{-\tau/t_{\mathrm{d}}} \tag{4}$$

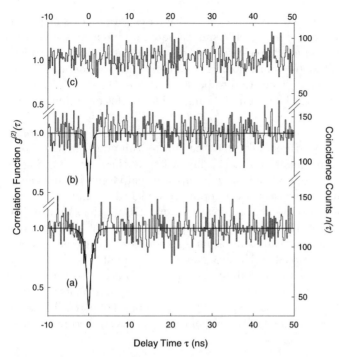

**Fig. 10.** Measured distribution of coincidence counts $n(\tau)$ and fit of the correlation function $g^{(2)}(\tau)$ (*solid line*) for a single InAs QD at 4 K, obtained at two different pump intensities: 66 W/cm² Trace (**a**) and 252 W/cm² Trace (**b**). Trace (**c**) shows the coincidence counts for many QDs in a high density region of the sample [17]

convolved with a Gaussian time distribution with 420 ps full width half maximum. Here, $t_d = 1/(\Gamma + W_p)$ is the antibunching time constant, $\Gamma$ is the spontaneous emission rate, $W_p$ is the effective pump rate and $a$ is introduced to take care of background light. Both $t_d$ and $a$ are fitting parameters. The resulting fitted $g^{(2)}(\tau)$ is shown as the solid line in Fig. 10. The values of $1 - a = g^{(2)}(0)$ obtained from the fit are 0.23 and 0.34 for traces (**a**) and (**b**). The fact that $g^{(2)}(0) < 0.5$ in both traces unambiguously indicates that the measured photon antibunching from the 1$X$ transition stems from a single, anharmonic quantum emitter. However, $g^{(2)}(0)$ does not reach its ideal value of zero due to the presence of background straylight. The correlation function $g_b^{(2)}(\tau)$ expected in the presence of a background radiation is $g_b^{(2)}(\tau) = 1 + \rho^2(g^{(2)}(\tau) - 1)$ where $\rho = S/(S + B)$ is the ratio of signal $S$ to background $B$ counts [20]. From the optical emission spectra of the single QD we can determine $\rho$ for the 1$X$ transition for the two pump intensities to be 0.9 and 0.83, respectively. The resulting values of $g_b^{(2)}(0)$ are 0.18 and 0.30, which are in good agreement with $g^{(2)}(0)$ values determined by the fits. The antibunching time constants $t_d$ obtained from the fit of $g^{(2)}(\tau)$ are 736 ps and 444 ps for traces (**a**) and (**b**). Thus the $t_d$ decreases with increasing pump power in accordance with our model. Using these pump-intensity-dependent time constants and a simple three-level rate equation model taking into account exciton and biexciton transitions we can determine the lifetime of the single exciton ground-state transition to be $900 \pm 100$ ps. Details of the analysis can be found in [17]. This value is comparable to lifetime values found in the literature [21].

More recently, *Regelman* and co-workers [22] and *Kiraz* and co-workers [23] have shown experimentally that the emitted photon statistics from a single semiconductor quantum dot can be varied by the excitation power from a sub-Poissonian one, where the photons are temporally antibunched, to super-Poissonian, where they are temporally bunched at finite delay, in addition to the antibunching at zero delay. To understand this, we need to consider that at high excitation powers the average number of electron–hole pairs occupying the dot in steady state is large. Therefore, the average probability to find the QD occupied with only a single pair is small. The correlation measurement now reflects the probability to find the QD occupied with a single electron–hole pair at time $\tau$ after emission of the previous photon, which actually left the QD empty. Just after the QD was emptied of pairs, the probability that the QD is occupied with only one pair increases, becoming thus larger than the average one.

# 5 A Quantum Dot Single-Photon Source

A single-photon source, which is able to generate photons on demand, has been a major challenge for many years. Such a source allows the ultimate quantum control of the photon generation process, i.e., single photons can be generated within short time intervals and a deterministic dwell time between successive photon generation events. This makes it possible to encode information on a single photon level. Such a source is of interest for future applications in quantum computing [10] and quantum cryptography [11].

The single-photon sources used today either employ highly attenuated laser pulses or rely on parametric down-conversion. Both schemes possess a serious disadvantage since photons are created randomly, i.e., with Poissonian photon statistics. In order to maintain a low multi-photon emission probability the average photon number per pulse has to be kept as low as $\sim 0.1$.

The realization of a practical single photon source requires three main elements: a single quantum emitter, regulation of the excitation and/or the recombination process and an efficient output coupling of the single photons. The active emitter needs to possess a high quantum efficiency $\eta \sim 1$ and has to emit one photon after each other. Though working until now at low temperature, epitaxially self-assembled InAs quantum dots are ideal candidates since they exhibit quantum efficiencies of $\eta \sim 1$ and show strong photon antibunching (see Sect. 4). Photon antibunching is a necessary but not sufficient condition for a single photon turnstile device as photons are emitted randomly and not at deterministic times. Thus, an additional mechanism for regulation of the emission process is required to realize single photon pulses on demand. The triggered single-photon emission at the single exciton ground state transition of a self-assembled QD is ensured by a *pulsed excitation*, the *anharmonicity* of the multi-exciton spectrum in combination with *slow relaxation* of highly excited QD states (see Sect. 2.2). Spontaneous emission of a quantum emitter is generally emitted in all directions (full solid angle) and therefore hard to capture efficiently. For practical use, however, the emitter should be coupled to a cavity mode with directional field profile. Self-assembled QDs can be easily embedded into an appropriate microcavity, e.g., into a micropillar. This allows efficient collecting of the single photons and thus using them for external applications.

In the following, we will first demonstrate triggered single photon emission from the exciton transition. Second, we will discuss two different experiments where resonant coupling of the QD transition to a high-$Q$ cavity mode has been achieved. Third, an electrically driven single-photon source is presented.

## 5.1   Triggered Single Photons from a Single Quantum Dot

In this section, we will discuss triggered single photon emission from a single quantum dot which is embedded into a microdisk but whose transition energy $1X$ is far detuned from all resonator modes. For identification of the QD $1X$ transition and the microdisk whispering gallery modes (WGM) we first recorded power dependent PL spectra from the microdisks containing QDs.

Figure 11 shows power dependent PL spectra for a $5\,\mu m$ diameter disk in the range between 1.310 and 1.348 eV. For this measurement the sample was excited with a continuous-wave Ti:sapphire laser at 760 nm. At low excitation power ($1\,W/cm^2$), a single sharp line (1.3222 eV) due to single exciton recombination ($1X$) is observed. With increasing excitation power two lines at 1.3208 and at 1.3196 eV appear below the single exciton line. The line at 1.3196 eV shows a superlinear increase with excitation intensity and originates from a biexciton decay ($2X$) whereas the line at 1.3208 eV (M) is due to background emission which is coupled into a whispering gallery mode (WGM). The correlation of the M line to a WGM will become clear in the discussion below (Sect. 5.2.1). The inset of Fig. 11 shows the measured CW correlation function $g^{(2)}(\tau)$ for the $1X$ transition of the single QD in the microdisk at the onset of saturation. Saturation is defined here as the pump intensity where the $1X$ line reaches its maximum intensity. The dip at $\tau = 0$ arises from photon antibunching [15] and the fact that $g^{(2)}(\tau) < 0.5$ proves that the emitted light from the $1X$ transition stems from a single, anharmonic quantum emitter.

Figure 12 shows the measured unnormalized correlation function $G^{(2)}(\tau)$ for (A) the pulsed Ti:sapphire laser, and (B) the $1X$ transition of a QD that is far detuned from all WGMs ($T = 4\,K$). The pump intensity in this experiment

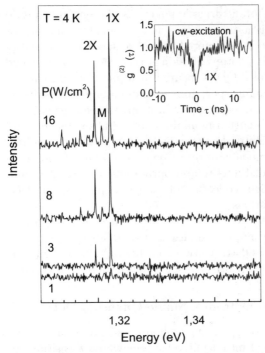

**Fig. 11.** Power dependent PL spectra from a single InAs QD embedded in a 5 μm diameter microdisk. Contributions from the excitonic ground transition ($1X$), higher excited states [e.g. biexciton ($2X$)], and a whispering gallery mode (M) are visible. *Inset*: Measured CW correlation function $g^{(2)}(\tau)$ for the single QD $1X$ transition. The time bin is 195 ps and the excitation power is 160 W/cm² [24]

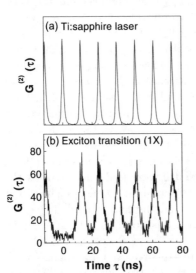

**Fig. 12.** Measured unnormalized correlation function $G^{(2)}(\tau)$ of (**a**) a mode-locked Ti:sapphire laser (FWHM = 250 fs), and (**b**) a single QD excitonic ground state ($1X$) emission under pulsed excitation conditions (82 mHz). The QD $1X$ transition was out of resonance with the microdisk modes [2]

corresponds to an excitation of the QD where the $1X$ emission is well into the saturation regime [19]. As expected, the measured $G^{(2)}(\tau)$ of the pulsed Ti:sapphire laser exhibits peaks at integer multiples of $T_{rep} = 12.27$ ns with negligible signal in between the peaks. The measured $G^{(2)}(\tau)$ of the QD $1X$ emission at $T = 4$ K (Fig. 12b) also shows peaks at integer multiples of $T_{rep}$, indicating the locking of the photon emission to the pulsed excitation. But in contrast to the mode-locked laser, the peak at $\tau = 0$ is no longer present, i.e., the probability of finding a second photon following the detection of the first photon at $\tau = 0$ vanishes. The absence of the peak at $\tau = 0$ provides strong evidence for an ideal single photon turnstile operation.

The lifetime of the single exciton ground-state transition $(1X)$ was determined from CW antibunching experiments to be 2.2 ns, which is the shortest possible total recombination time for a multiply excited QD. As the recombination time in the GaAs barrier and the wetting layer is considerable faster (100–200 ps), the probability of re-excitation of the QD and emission of a second photon is very small ($< 5 \times 10^{-3}$). To ensure that a single photon is indeed emitted for each excitation pulse, the pump power of the excitation laser should be adjusted so that the probability of having no injected electron–hole pair in the QD is negligible. The fact that the photon correlation measurement depicted in Fig. 12b was obtained well in the saturation regime ensures that QD is multiply excited in our experiments. In addition, the quantum efficiency of the QD has to be high $\eta \sim 1$ to avoid nonradiative recombination processes. Recent experiments reported in [17] have shown that for our samples the dominant recombination mechanism is radiative. These facts allow us to conclude that the generated light at the excitonic ground-state transition energy $1X$ is a stream of single photons with a repetition rate of 82 mHz.

It is interesting to note that *Thompson* and coworkers [25] have recently shown that biexciton and trion recombination in single QDs also allow the realization of triggered single photons. Since the radiation lifetime of the biexciton is approximately a factor of two shorter than that of the exciton, the maximum possible emission rate from the biexciton state can be higher. In addition, the time jitter between successive photons is reduced.

## 5.2 Single Photons from Quantum Dots Coupled to a Microcavity

For practical applications the emission energy of the QD should be in resonance with a fundamental mode of a microcavity to achieve a high photon collection efficiency. Single photon emission from an isolated quantum dot on resonance with a whispering gallery mode (WGM) [2] and with a fundamental mode of a micropillar [26] will be discussed in the following.

The transition energy of an exciton in a QD is only defined within the relatively broad inhomogeneous linewidth ($\sim 30$–$50$ meV) of the corresponding ensemble. Thus, a tuning possibility of the exciton transition vs. the mode

**Fig. 13.** The microdisk structure that consists of a 5 μm diameter disk and a 0.5 μm post. The GaAs disk area that supports high-quality factor whispering gallery modes is 200 nm thick and contains InAs quantum dots [2]

resonance is necessary. The stronger temperature dependence of the QD single-exciton resonance allows us to change the relative energy of the mode and the 1X transition by varying the sample temperature [27]. An alternative method could be electric field tuning via the quantum-confined Stark effect which is discussed in the Chapter by Bacher in this volume.

### 5.2.1  Single Photon Emission from a Microdisk

Figure 13 shows a microdisk structure which consists of a 5 μm diameter disk and a 0.5 μm $Al_{0.65}Ga_{0.35}As$ post. The disk area consists of 100 nm GaAs, an InAs QD layer, and 100 nm GaAs. Details of the microdisk processing can be found in [29].

Temperature tuning has been applied to shift the 1X transition shown in Fig. 11 into resonance with the cavity mode M ($Q \sim 6500$). The crossing between the WGM and the QD 1X-transitions is shown in Fig. 14a where we plot the energies of the two lines versus temperature. The WGM appears at an energy of 1320.7 meV at 4 K and shifts only slightly to an energy of 1319.6 meV at 54 K. On the other hand, the QD 1X-transition shifts strongly with temperature, over 3 meV within a 50 K temperature difference. The different energy shifts of the 1X-transition and the WGM with temperature give rise to a crossing of the two resonances. The temperature dependence of the energy of the WGM can be attributed to the change in the refractive index of GaAs with temperature. On the other hand, the temperature dependence of the energy of the 1X-transition is caused by the changes in the bandgaps of InAs and GaAs with temperature.

Figure 14b shows the change in the intensity of the WGM emission as a function of the 1X-WGM detuning. At a temperature of 44 K (zero detuning) the intensity of the WGM luminescence increases by a factor of 29 compared to its value at 4 K, strongly indicating a resonance between the QD 1X-transition and the WGM. The observed crossing together with resonant

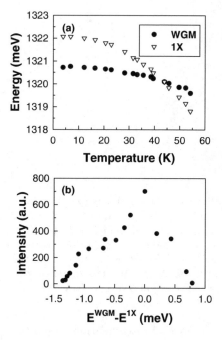

**Fig. 14.** (a) Change of the WGM and the 1$X$-transition emission energy with temperature (excitation power = 13 W/cm$^2$). (b) Change in the intensity of the WGM luminescence with detuning (excitation power = 13 W/cm$^2$) [27]

enhancement of luminescence are evidence for the weak coupling between the WGM and the single QD.

In the weak coupling regime enhancement of the spontaneous emission rate of the QD 1$X$-transition due to the *Purcell* effect [28] is expected (see the Chapter by Gérard in this volume). To quantify the magnitude of the Purcell effect we have carried out pump-power dependent CW photon correlation measurements; this method has been previously shown to be a reliable alternative to standard time-resolved measurements for determining recombination times [17]. Moreover, this method is able to discriminate between single QD emission ($g^2(0) < 0.5$) and emission from several QDs ($g^2(0) > 0.5$). Figure 15 shows photon correlation measurements performed at 4 K (out of resonance) and 44 K (in resonance) with excitation powers of 36 W/cm$^2$ and 5 W/cm$^2$, respectively. After normalization, the measured correlation functions show clear dips at zero time delay ($g^{(2)}(0) = 0.08$ at 4 K, $g^{(2)}(0) = 0.38$ at 44 K) indicating strong photon antibunching. Since $g^{(2)}(0) < 0.5$ in our measurements, we can state that the observed emission lines stem from the 1$X$-transition of a single QD [17]. The observation of $g^{(2)}(0) < 0.5$ at 44 K also supports that, in resonance, the QD 1$X$-transition is the main emission feeding the WGM luminescence.

From photon correlation measurements at two different pump powers, we deduced the lifetime of the 1$X$-transition at 4 K. In these measurements decay times of 2.7 ns and 1.5 ns were observed at excitation levels below saturation (36 W/cm$^2$, Fig. 3a) and at the onset of saturation (92 W/cm$^2$) of

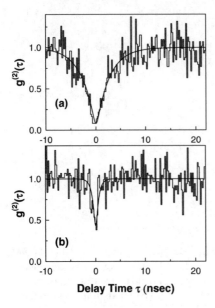

**Fig. 15.** Measured photon correlation function of the $1X$-transition of the single QD: out of resonance with the WGM, at 4 K, under an excitation power of 36 W/cm$^2$ (trace (**a**)), and in resonance with the WGM, at 44 K, under an excitation power of 5 W/cm$^2$ (trace (**b**)) [27]

the $1X$-transition, respectively. From a three-level rate-equation model that includes $1X$ and biexcitonic ($2X$) transitions and omits any higher multiexcitonic recombinations or any other population decay channels (e.g. Auger processes) [17], a lifetime of 3.4 ns is determined for the $1X$-transition at 4 K. A conventional time-correlated single photon counting (TCSPC) measurement on this QD exhibits a decay time of 2.8 ns, showing reasonable agreement with the lifetime deduced from the photon correlation measurements. These values are larger than previously reported lifetimes for the $1X$-transition of a single InAs QD ($\sim 1$ ns) [19]. The lifetime difference can be explained by the different photonic environments created by the microdisk that partially inhibits spontaneous emission [17].

Our photon correlation measurements at resonance, 44 K, revealed decay times of 560 ps and 370 ps at pump powers of 5 W/cm$^2$ (Fig. 3b), and 45 W/cm$^2$ respectively, corresponding to excitation levels below the saturation of the $1X$-transition. By using the pump power dependent method described in the previous paragraph, a lifetime of 590 ps is determined for the $1X$-transition at 44 K. This provides a strong indication of lifetime reduction caused by the Purcell effect. A more detailed temperature dependent TCSPC study that demonstrates the Purcell effect is published in [27].

If a quantum emitter is spectrally matched with a single cavity mode, located at a maximum of the electric field, and its dipole is aligned with the local electric field, the Purcell factor is given by [28]

$$F_{\mathrm{p}} = \frac{3Q\lambda_{\mathrm{c}}^3}{4\pi^2 V}, \tag{5}$$

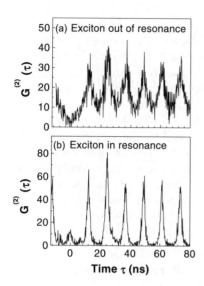

**Fig. 16.** Measured unnormalized correlation function $G^{(2)}(\tau)$ of a single QD excitonic ground state emission (**a**) out of resonance, and (**b**) at resonance with a cavity mode ($Q \sim 6500$), under pulsed excitation conditions (82 mHz). The average pump intensity in both cases was $\sim 22\,\mathrm{W/cm^2}$ [2]

where $Q$ is the quality factor of the cavity, $\lambda_c$ is the wavelength of the emission in the cavity, and $V$ is the effective mode volume. For a microdisk, the ideal spontaneous emission enhancement can be estimated to be $[(2/2)F_p+1]$, where the various terms account, respectively for WGM degeneracy (2), the random dipole orientation in the plane of the QD (1/2), and the contribution of the emission into leaky modes (1) [30]. Taking the parameters of our microdisk an enhancement in the spontaneous emission rate of 17 is estimated. The fact that the measured value ($\sim$ 5–6) is much smaller than the predicted value 17 for ideal coupling can be attributed to the non-ideal spatial overlap of the QD and the WGM. Finally, we want to stress that a thorough theoretical analysis of the Purcell factor in our microdisk would have to account for the special photonic environment of the microdisk which is beyond the scope of this contribution.

Figure 16 shows the measured unnormalized correlation function $G^{(2)}(\tau)$ for the $1X$ transition of Fig. 11a out of resonance ($T = 4\,\mathrm{K}$), and Fig. 11b in resonance ($T = 36\,\mathrm{K}$) with the WGM. The resonance condition is achieved at slightly lower temperature due to the higher pump power used in the pulsed experiment ($22\,\mathrm{W/cm^2}$). We emphasize that the photon correlation signals shown in Figs. 12b and 16 are obtained for different QDs; the $1X$ recombination time for the QD analyzed in Fig. 16 is 3.4 ns, which explains the appearance of broader peaks. When the QD is in resonance with the WGM, the FWHM of the photon correlation peaks are narrower (factor 3.4) than the out of resonance case, i.e., the time jitter between successive photon generation events is reduced. This is a direct consequence of the Purcell effect which causes a reduction of the ground state transition lifetime $\tau_{1X}$ and ensures that photons are primarily emitted into the cavity mode.

A small peak at $\tau = 0$ is observed in the resonance case (see Fig. 16b). The intensity ratio of this peak to the peaks at $iT_{\mathrm{rep}}$ is directly related to the fraction of pulses having two or more photons [14]. An experimental ratio $R = 0.2$ is deduced from Fig. 16b. The fact that $R$ is larger than the ideal value of zero could be due to the Purcell effect, which increases the probability of capturing a second electron–hole pair from the wetting layer after the $1X$ recombination process has occurred. Another possible explanation is the contribution from the background light generated by the wetting layer or by the excited states of other QDs. There are two experimental observations that support the latter explanation: first, even when the ground-state transition of the QD ($1X$) is off resonance the mode emission is still visible, indicating the influence of the background (see Fig. 11). Second, using higher average pump powers $P$ in the resonant case increases $R$ ($R = 0.36$ (0.55) for $P = 56$ (303) $\mathrm{W/cm}^2$).

### 5.2.2  Single Photon Emission from a Micropillar

Recently, high efficiency coupling (78%) of the spontaneous emission from an isolated quantum dot to the fundamental mode of a pillar microcavity has been achieved [31]. This cavity geometry allows an efficient output coupling of the photons due to the directional field profile of the fundamental mode. In addition, photon correlation experiments under pulsed excitation have revealed triggered single photon emission of such a coupled system [26]. The investigated micropillar possesses a $0.9\,\mu\mathrm{m}$ diameter, and a cavity $Q$ close to 400 which results in an estimated Purcell factor of $F_{\mathrm{p}} \sim 8$. Figure 17 shows the μPL spectra obtained for different excitation powers together with a transmission electron microscope picture of the investigated micropillar. Under weak excitation power (1 μW) the QDs display sharp lines due to single

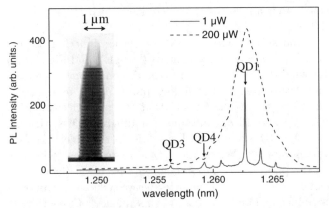

**Fig. 17.** μPL spectra obtained for a $0.9\,\mu\mathrm{m}$ diameter micropillar containing few InAs QDs, in the strong and weak excitation regimes. *Insert*: view of the same pillar by transmission electron microscopy [26]

exciton recombination. At high excitation power $(200\,\mu W)$ the higher QD states are filled and the wetting layer states are populated. Due to Coulomb interaction between the trapped excitons and the surrounding electron–hole plasma the QDs typically display a broad $(5–10\,meV)$ emission line. In this case, the QDs act as a broadband internal light source, and the PL spectrum reflects the modal density of the pillar microcavity.

Figure 18 displays photon correlation measurements under pulsed excitation of QD1. The central peak at $\tau = 0$ is 0.19 which means that the probability of emitting two or more photons is five times smaller than that of a laser delivering the same average number of photons per pulse. The residual unwanted central peak is a result of insufficient rejection of the intense emission from the InAs wetting layer and from additional multiexcitonic recombination lines from neighboring QDs. One possibility to reduce the multi-photon emission is to optically pump the excitons in the excited states instead of pumping above the barrier bandgap. In this case, the contributions of the wetting layer and the other QDs are completely suppressed.

The same group also realized single-mode polarized emission of single QDs placed in *elliptical* micropillars. This is an important issue, e.g. for the application of a single-photon source for polarization-encoded quantum cryptography schemes since for a randomly polarized source, preparation of a specific polarization would result in the loss of half the photons.

It is important to note that in each of the previous experiments accidental matching of a QD resonance with the cavity mode was realized be-

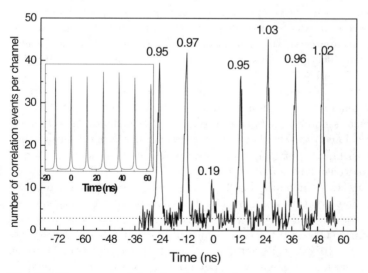

**Fig. 18.** Histograms of the time intervals between detection events on start and stop detectors. The normalized area, calculated after removing the background level due to the dark counts of the APDs (*dashed lines*), is indicated for each correlation peak. A reference histogram obtained for the scattered laser light is shown in the *insert* [26]

cause their transition energies are only defined within the relatively broad inhomogeneous linewidth ($\sim$ 30–50 meV) of the corresponding QD ensemble. From this point of view, single QDs possess a serious disadvantage so far. One improvement could be to use single donor or acceptor bound excitons in a semiconductor as active emitters. This approach is highly attractive with respect to coupling the quantum emitter to a high-quality resonator mode since the emission energies of donor and acceptor-bound excitons are well known and energetically defined within a small range ($\sim$ 1–2 meV), i.e., 1–2 orders of magnitude better than for QDs. Recently, we demonstrated triggered single photon emission using the radiative recombination of single nitrogen bound excitons in a ZnSe quantum well structure [32]. Our results also suggest that single nitrogen bound excitons are well suited for cavity-quantum electrodynamics (QED) experiments and could provide a solution to achieve the challenging goal of the strong coupling regime in a semiconductor microcavity.

## 5.3    Electrically Driven Single-Photon Source

From the viewpoint of practical application one would prefer an electrically driven single-photon source. A single-photon device based on a mesoscopic double barrier p–n heterojunction was proposed in 1994 [33]. An extension of this proposal was recently demonstrated [34] where single as well as multiple photon emission events with a repetition rate of 10 mHz at 50 mK have been reported. This device utilizes Coulomb blockade of tunneling for electrons and holes in a mesoscopic p–n diode structure to regulate the photon generation process. In this scheme, single electron and hole charging energies must be large compared to the thermal background energy to ensure single photon emission. Therefore, this device can only be operated at ultra-low temperatures ($T \leq 1$ K).

The fact that *nonresonant* optical pumping schemes have demonstrated single-photon generation with individual quantum dots suggest that electrical pumping (which is nonresonant) can be readily implemented. In fact, *Yuan* and co-workers have recently shown that electroluminescence from a single quantum dot within the intrinsic region of a p–i–n junction is able to act as an electrically driven single-photon source [35].

Figure 19 shows a schematic sketch of the single-photon-emitting diode in cross-section. The structure was grown by molecular beam epitaxy on a GaAs substrate and consists of a GaAs p–i–n diode with a layer of nanomenter-scale InAs self-organized QDs inserted into the intrinsic region. The dot density was estimated to $\sim 5 \times 10^8$ cm$^{-2}$. Structures with lateral dimensions of $10 \times 10$ μm were fabricated with a small emission aperture in the opaque metal layers on the device surface. The structure shows ideal diode-like behavior with a turn on bias of 1.5 V. Figure 20 displays electroluminescence spectra recorded at 5 K. At low currents only a single line due to exciton recombination is visible. At higher injection currents a second strong line appears at 4.7 meV

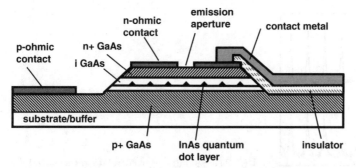

**Fig. 19.** Schematic sketch of the single-photon-emitting diode in cross-section [35]

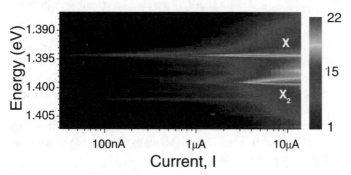

**Fig. 20.** Electroluminescence spectra of the single-photon-emitting diode as a function of emission photon energy (*vertical axis*) and drive current (*horizontal axis*). Sharp line emission (marked $X$ and $X_2$) is seen arising from a single quantum dot in the structure [35]

higher energy. This line strengthens with current as $I^2$ and is ascribed to the biexciton transition of the dot.

Pulsed emission from the diode can be achieved by pulsing of the injection current. Single-photon emission is performed provided that the pulse width is much less than the exciton lifetime. *Yuan* and co-workers [35] used 400 ps rectangular voltage pulses at a repetition rate of 80 MHz, superimposed on a dc bias of 1.5 V. Figure 21 shows the second-order correlation function recorded for the exciton transition $X$ and the wetting layer (WL). As expected the WL exhibits peaks at integer multiples of the repetition time, indicating that a second photon is just as likely to be found in the same emission pulse as the first, as in any other. In contrast to the WL the peak at $\tau = 0$ of the dot correlation is only about 11% of those at finite delays. This indicates an approximate order of magnitude decrease in multiphoton emission pulses as compared to a Poissonian light source.

Though the operation has only been demonstrated at low temperature (5 K) with InAs QDs room temperature operation could in principle be achieved by using QDs with higher confinement potentials to suppress non-

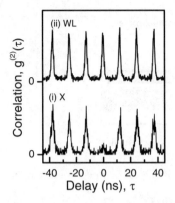

**Fig. 21.** Second-order correlation measured for pulsed electrical injection for ($i$) quantum dot exciton and ($ii$) wetting layer emission [35]

radiative carrier losses into the barriers. CdSe/Zn(S,Se) QDs may provide a solution. We have recently demonstrated the generation of triggered single-photons with epitaxially grown self-assembled CdSe/Zn(S,Se) QDs for temperatures up to 200 K [36]. These results show the high potential of these type of QDs for nonclassical light generation at high temperatures. Furthermore, InAs QDs can be tailored to possess much deeper carrier confinement potentials [37]. These QDs emit in the 1.3 μm wavelength range and are thus well suited for fiber optic communications.

## 6  Correlated Photon Pairs from a Single Quantum Dot

Another major goal in the field of quantum information science is the realization of efficient sources of pairs of correlated photons. Parametric fluorescence setups are generally used as sources for photon pairs. The conversion efficiency of pump to parametric photons is of the order of $10^{-10}$ per pump photon, while at the same time, the probability of emitting two or more pairs of photons is not negligible [3]. This is because the statistics of the nondegenerate parametric photon pair generation follows a Poissonian distribution. In contrast, it has been shown in quantum dots that optical pumping could yield a single photon pair practically for every excitation pulse [38]. Since the incident laser pulses contain $\sim 10^6$ photons the efficiency compared to parametric down-conversion is enhanced by a factor of $\sim 10^4$. In addition, such sources would have a negligible probability of emitting more than one photon pair at a time. Furthermore, it has been proposed that photon pairs emitted by a quantum dot can be expected to display also polarization entanglement [39]. The ideas and problems which are involved with this concept will be discussed in detail below.

## 6.1   Cross-Correlation Measurements

Cross-correlation measurements provide a powerful tool for investigating the QD multiexciton features. Identification of specific spectral lines, the time order of emission pairs as well as polarization correlation studies on photon pairs can be performed [3,23,40].

The second-order cross-correlation function is expressed as

$$g_{i,j}^{(2)}(\tau) = \frac{\langle I_j(t)I_i(t+\tau)\rangle}{\langle I_i(t)\rangle\langle I_j(t)\rangle}, \tag{6}$$

where $I_i(t)$ $(I_j(t))$ is the measured intensity of the $i$th ($j$th) multiexciton transition. $g_{i,j}^{(2)}(\tau)$ describes the probability to measure a photon at time $\tau$ from the $i$th multiexciton on condition that a first photon from the $j$th multiexciton has been detected at time $\tau = 0$. The experimental setup for cross-correlation measurements is shown in Fig. 22. In contrast to the conventional HBT setup of Fig. 8, the luminescence is now sent first to a nonpolarizing beamsplitter which splits the light into two beams. Each arm then includes spectral filters (monochromators or interference filters) to select specific luminescence lines, e.g., photons from biexciton and exciton recombination processes. The polarization relationship between the different components can be studied by inserting additional polarization optics ($\lambda/2-$, $\lambda/4$ plates, and polarizers) into each of the arms.

### 6.1.1   CW Biexciton–Exciton Correlation

In the following we will first discuss the biexciton–exciton transition. The function $g_{1X,2X}^{(2)}(\tau)$ therefore represents the conditional probability that a photon from an exciton recombination $(1X)$ will be emitted at time $\tau$ after a biexciton recombination process $(2X)$ has previously occurred. The measurement of $g_{1X,2X}^{(2)}(\tau)$ is realized by illuminating the start APD by the $2X$ emission and the stop APD by the $1X$ emission. Figure 23 shows the second-order cross-correlation function obtained in a continuous wave experiment from *Moreau* et al. [3]. The cross-correlation function of the two photons exhibits asymmetric bunching ($\tau > 0$) and antibunching ($\tau < 0$) features. The asymmetric cross-bunching behavior indicates that the photon of the biexcitonic recombination is always emitted first and is followed by the photon of the excitonic recombination after an exponentially distributed time interval, which,

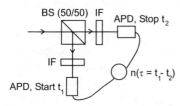

**Fig. 22.** Schematic sketch of the modified HBT setup to perform cross-correlation measurements. The photoluminescence is separated into two beams by a beamsplitter and each arm is sent to an interference filter (IF) tuned, respectively, to the corresponding multiexciton line

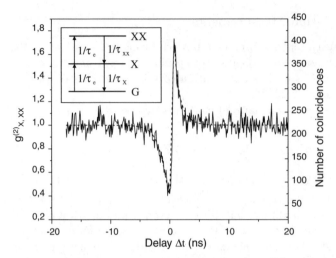

**Fig. 23.** Cross-correlation function of the exciton $(X)$ and biexciton $(XX)$ photons. Time delay is positive when $XX$ arrives before $X$. *Inset*: Schematic of three level model with transition rates with $G$ referring to the ground state and $\tau_e$ to the excitation time [3]

at low incident intensities, corresponds to the exciton lifetime. The cross anti-bunching at negative times can be understood within the same model. When the second photon in the cascade $(1X)$ is detected, the QD has returned to its ground state and thus the probability of detecting the first one $(2X)$ goes to zero until the QD can be re-excited. The measured correlation function $g^{(2)}_{1X,2X}(0)$ does not reach its theoretical minimum of zero because of the presence of background straylight from the wetting layer. When the background is removed from the signal $g^{(2)}_{1X,2X}(0)$ goes down to zero indicating that only *one* photon pair is emitted at a time.

### 6.1.2 CW Trion Correlation with Exciton and Biexciton

The spectral features of single QDs are significantly more complicated if in addition to electron–hole pairs single electrons or holes are captured into the quantum dot. New emission lines due to charged exciton (trion) recombination occur in the spectrum. Cross-correlation spectroscopy between a charged exciton and a biexciton and between a charged exciton and an exciton reveal unique signatures which allows the identification of the charged exciton emission [23].

Figure 24 shows power dependent PL spectra of a single self-assembled InAs QD. At low pump powers, the single-exciton peak $(X1)$ dominates the spectrum. At higher pump powers, two other peaks $(XX$ and $X2)$ become dominant. The $(XX)$ peak is located $3.5$ meV below the $X1$ emission and its intensity has a quadratic dependence on pump power which are characteristic

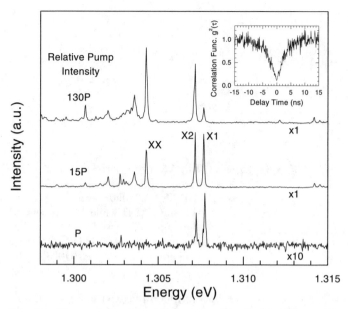

**Fig. 24.** Power dependent PL spectra of a single self-assembled InAs QD. *Inset*: Photon-correlation measurement carried out using the $X1$ emission, showing strong antibunching [23]

features for biexciton emission. The $X2$ peak has been identified to charged exciton recombination due to cross-correlation measurements which will be discussed below.

Figure 25a shows the cross-correlation between $X2$ and $XX$ emissions. The $X2$ emission is sent to the start APD while $XX$ is sent to the stop APD. The cross-correlation only shows antibunching and no bunching signature is observed. The absence of bunching demonstrates that the $XX$ emission does not populate the $X2$ state while the antibunching indicates that those emissions arise from the same QD. This may already suggest an identification of $X2$ as a charged exciton line. To provide further evidence, *Kiraz* et al. [23] have carried out cross-correlation measurements between $X2$ and $X1$ emissions where the start and stop APDs were illuminated by the $X1$ and $X2$ lines, respectively. The resulting $X1–X2$ cross-correlation function (see Fig. 25b) clearly shows asymmetric antibunching which proves once again that the two lines originate from the same dot. The fact that the recovery of the antibunching is faster for $\tau > 0$ than for $\tau < 0$ is expected if $X2$ arises from a charged exciton. This is because the post-measurement state of the charged exciton is a single charged QD. Single-charge injection into the QD is much faster than triple charge injection, which in turn determines the recovery time for $\tau < 0$. Thus cross-correlation measurements provide a pow-

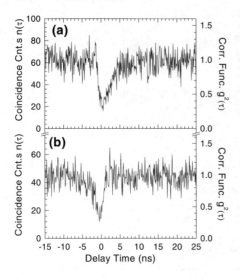

**Fig. 25.** Cross-correlation under CW diode laser excitation at 785 nm. (a) The $X2$ fluorescence is sent to the start APD while $XX$ is sent to the stop APD. (b) The $X2$ emission is sent to the start APD and the $X1$ emission to the stop APD [23]

erful tool for identification of specific lines in a complicated PL spectrum of a single QD.

### 6.1.3 Pulsed Biexciton–Exciton Cascade

Even more exciting is the generation of radiative quantum cascades with single QDs after *pulsed* optical or electrical excitation of a multiexciton state (see Sect. 2.2). For example, the biexciton/exciton cascade can now be used to generate correlated pairs of photons with a definite time order which has been recently shown by *Moreau* and coworkers [3].

Figure 26 presents the second-order cross-correlation function of the exciton $(X)$ and biexciton $(XX)$ photons under pulsed excitation. The $(XX)$ emission is sent to the start APD and the $X$ emission to the stop APD. The zero peak corresponds to the $X$ and $XX$ photons being emitted both after the same laser pulse. It is higher than the others and shows an abrupt rise at delay time $\Delta t = 0$ followed by an exponential decay with a time constant of $T = 1.5$ ns which corresponds to the exciton lifetime of this QD. This is due to the fact that there is an enhanced probability of detecting an $X$ photon after an $XX$ photon. The drop to practically zero at negative time intervals indicates that the probability of emitting a second pair of photons during the same laser pulse is essentially zero. For the other peaks at finite delay times, the two photons are emitted in different periods. They also possess an asymmetric lineshape although their maxima are delayed with respect to the laser pulse by $\sim 1$ ns and their rise times are not as abrupt as those of the central peak. This arises from the jitter introduced in the emission from the $XX$ photon that is used to start the correlation measurement. As a result the rise time reflects essentially the lifetime of the biexciton.

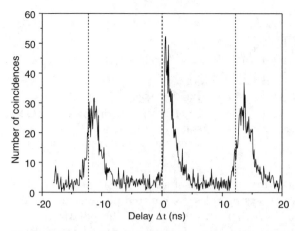

**Fig. 26.** Cross-correlation function of the exciton $(X)$ and biexciton $(XX)$ photons under pulsed excitation. The $(XX)$ emission is sent to the start APD and the $X$ emission to the stop APD. *Full line*: Experimental results. *Cursors* are separated by the repetition rate of the laser (12.2 ns) [3]

An important question is whether the generation of polarization entangled photon pairs is possible with single QDs. Entangled photons are interesting for applications in quantum cryptography and for quantum teleportation experiments where the quantum mechanical state of the system is transferred to another system at a distinct location [11]. Moreover, it has been recently shown that an entangled photon pair beats the classical diffraction limit by a factor of 2 [41].

The biexciton ground state is a spin-singlet state $(S = 0)$; the spins of electrons and holes compensate each other. Thus the polarization of the biexciton transition is controlled by the final state of the recombination, i.e. by the eigenstates of the exciton. Starting from the biexcitonic ground state of a symmetric QD ($D_{2d}$ symmetry), a first electron can recombine with a hole and emit a $\sigma^+$ or a $\sigma^-$ photon (Fig. 27). Due to spin conservation, the decay of the second electron–hole pair has to give a photon of opposite circular polarization. Since the temporal sequence of the two polarizations is not defined, these two photons should display polarization *entanglement*, i.e., a quantum mechanical property which means that though a polarization measurement on an individual photon is always random, the results of the measurements on both photons are always correlated. The quantum mechanical state is given by

$$|\Psi\rangle = \frac{1}{\sqrt{2}}(|\sigma^+\rangle_1|\sigma^-\rangle_2 + |\sigma^-\rangle_1|\sigma^+\rangle_2). \tag{7}$$

An asymmetry of the confinement potential in the QD $xy$-plane ($< D_{2d}$ symmetry) results in an exchange energy splitting and the split states are linear

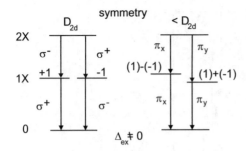

**Fig. 27.** Schematic sketch of the exciton (1*X*) and the biexciton (2*X*) states and the allowed optical transitions in QDs of different symmetry

combinations of the original states. This results in a linear polarization of the QD emission. Due to the selection rules the emitted photons are equally linearly polarized ($\pi_x,\pi_x$ or $\pi_y,\pi_y$). If this is the case, the polarization information is encoded in the energy of the photons and can in principle be inferred, thereby destroying the entanglement.

There are several complications of this simple picture of photon entanglement, like spin relaxation processes and heavy-hole light-hole mixing. Besides carrier–phonon scattering, the dephasing process will critically depend on carrier–carrier scattering, and therefore on the carrier density in the wetting layer. A more fundamental problem could be level mixing between the heavy-hole ($J = 3/2$) and light hole ($J = 1/2$) states due to the strong confinement in the QDs. In this case the two photons are not perfectly anticorrelated with respect to $|\sigma^+\rangle$ and $|\sigma^-\rangle$ polarization.

Recently, *Santori* [5] and *Shields* [6] and colleagues have observed polarization-correlated photon pairs from a single InAs QD in a linear polarization basis, but no entanglement. A polarization degree of $\sim 84\%$ and $\sim 65\%$ between the first and the second photon have been measured, respectively. They explain their findings in terms of biexciton decay in an *asymmetric* QD. More recently, we have obtained a similar result for the biexciton–exciton cascade in a CdSe QD where a polarization of 74% has been achieved in a linear basis [42].

Whether the generation of polarization entangled photons is possible with symmetric QDs has yet to be established and further experiments are presently underway.

## 7   Summary and Outlook

The recent quantum optical experiments on single semiconductor QDs demonstrate the high potential of these systems both for a fundamental understanding of matter–light interaction on a single quantum level and for the realization of unique devices, e.g., triggered single photon or photon pair emitters.

However, a few technological challenges remain for practical devices. The currently investigated QDs emit in the near infrared spectral region (0.85 μm– 1 μm) and possess shallow electronic confinement potentials which prevent a high quantum efficiency of the emission at temperatures above $\sim 50\,\mathrm{K}$. CdSe/Zn(S,Se) or (In,GaN)/GaN QDs which possess higher electronic confinement potentials and higher biexciton binding energies may provide a solution for room temperature operation in the blue-green spectral region. Promising results from CdSe/Zn(S,Se) QDs have been recently reported by us [36]. We observed the generation of triggered single-photons with epitaxially grown self-assembled CdSe/Zn(S,Se) QDs for temperatures up to $200\,\mathrm{K}$. For fiber optic communications one would prefer to use the dispersion and absorption minima of the optical fiber at $\sim 1.3\,\mu\mathrm{m}$ and $\sim 1.55\,\mu\mathrm{m}$, respectively. InAs QDs can be tailored to possess much deeper carrier confinement potentials [37]. Moreover, these QDs emit in the $1.3\,\mu\mathrm{m}$ wavelength range and are thus well suited for fiber optic applications.

A precondition for mass production of devices is the *controlled* coupling of the QD emission to a cavity mode with directional field profile in order to achieve a high photon collection efficiency. For this task the spatial position of the QD and the energy of its transition has to be adjusted with respect to the maximum of the mode profile and the energy of the cavity mode, respectively. The former task is still a challenge although considerable progress has been made on pre-patterning of MBE samples to create nucleation spots on which QD growth takes place (see the Chapter by Petroff in this volume). The energetic resonance condition can be achieved by temperature tuning or by electric field tuning via the quantum-confined Stark effect. However, the QD transiton energies are only defined within a relatively broad inhomogeneous linewidth ($\sim 30\text{--}50\,\mathrm{meV}$) of the corresponding QD ensemble. From this point of view single donor or acceptor bound excitons are advantagous as active emitters in microcavities since their emission energies are at predetermined and well defined energies. Recent quantum optical experiments suggest that single nitrogen bound excitons are well suited for triggered single photon sources and cavity-quantum electrodynamics experiments [32].

Finally, the new and innovative field of *Semiconductor Quantum Optics with Quantum Dots* will further provide us with unexpected interesting physics and will be the basis for future generations of photonic devices operating with single or only a few photons.

### Acknowledgements

I am indebted to Atac Imamoglu, Alper Kiraz, and Christoph Becher for their collaboration in obtaining the experimental and theoretical results during my research stay at the University of California Santa Barbara and W. V. Schoenfeld, P. M. Petroff, Lidong Zhang, and E. Hu for growing and processing the quantum dot and microcavity samples. I thank J.-M. Gérard and A. J. Shields for providing me with several figures from photon correlation measurements and S. Strauf for critical reading of the manuscript. The research has been supported by the Max-Kade-Foundation.

# References

1. M. Kira, F. Jahnke, W. Hoyer, S. W. Koch: Prog. Quant. Electron. **23**, 189 (1999)
2. P. Michler, A. Kiraz, C. Becher, W. V. Schoenfeld, P. M. Petroff, Lidong Zhang, E. Hu, A. Imamoglu: Science **290**, 2282 (2000)
3. E. Moreau, I. Robert, L. Manin, V. Thierry-Mieg, J.-M. Gérard, I. Abram: Phys. Rev. Lett. **87**, 183601 (2001)
4. M. Bayer, T. L. Reinecke, F. Weidner, A. Larionov, A. McDonald, A. Forchel: Phys. Rev. Lett. **86**, 3168 (2001)
5. C. Santori, D. Fattal, M. Pelton, G. S. Solomon, Y. Yamamoto: Phys. Rev. B **66**, 045308 (2002)
6. R. M. Stevenson, R. M. Thompson, A. J. Shields, I. Farrer, B. E. Kardynal, D. A. Ritchie, M. Pepper: Phys. Rev. B **66**(R), 081302 (2002)
7. H.-A. Bachor: *A Guide to Experiments in Quantum Optics* (Wiley-VCH, Weinheim 1998)
8. P. Michler, C. Becher: Phys. Bl. **57**, 55 (2001)
9. D. F. Walls, G. J. Milburn: *Quantum Optics* (Springer, Berlin, Heidelberg 1994)
10. E. Knill, R. Laflamme: Nature **409**, 46 (2001)
11. D. Bouwmeester, A. Ekert, A. Zeilinger: *The Physics of Quantum Information* (Springer, Berlin, Heidelberg 2000)
12. J. M. Garcia, T. Mankad, P. O. Holtz, P. J. Wellman, P. M. Petroff: Appl. Phys. Lett. **72**, 3172 (1998)
    J. M. Garcia, G. Medeiros-Ribeiro, K. Schmidt, T. Ngo, J. L. Feng, A. Lorke, J. Kotthaus, P. M. Petroff: Appl. Phys. Lett. **71**, 2014 (1997)
13. R. Hanbury-Brown, R. Q. Twiss: Nature **178**, 1447 (1956)
14. C. Brunel, B. Lounis, P. Tamarat, M. Orrit: Phys. Rev. Lett. **83**, 2722 (1999)
15. H. J. Kimble, M. Dagenais, L. Mandel: Phys. Rev. Lett. **39**, 691 (1977)
16. P. Michler, A. Imamoglu, M. D. Mason, P. J. Carson, G. F. Strouse, S. K. Buratto: Nature **406**, 968 (2000)
17. C. Becher, A. Kiraz, P. Michler, A. Imamoglu, W. V. Schoenfeld, P. M. Petroff, Lidong Zhang, E. Hu: Phys. Rev. B **63**, 121312(R) (2001)
18. P. Michler, A. Imamoglu, A. Kiraz, C. Becher, M. D. Mason, P. J. Carson, G. F. Strouse, S. K. Buratto, W. V. Schoenfeld, P. M. Petroff: Phys. Stat. Solidi (b) **229**, 399 (2002)
19. E. Dekel, D. V. Regelman, D. Gershoni, E. Ehrenfreund, W. V. Schoenfeld, P. M. Petroff: Phys. Rev. B **62**, 11038 (2000)
20. R. Brouri, A. Beveratos, J.-P. Poizat, P. Grangier: Opt. Lett. **25**, 1294 (2000)
21. E. Dekel, D. V. Regelman, D. Gershoni, E. Ehrenfreund, W. V. Schoenfeld, P. M. Petroff: Solid State Commun. **117**, 395 (2001)
22. D. V. Regelman, U. Mizrahi, D. Gershoni, E. Ehrenfreund, W. V. Schoenfeld, P. M. Petroff: Phys. Rev. Lett. **87**, 257401 (2001)
23. A. Kiraz, S. Fälth, C. Becher, B. Gayral, W. V. Schoenfeld, P. M. Petroff, L. Zhang, E. Hu, A. Imamoglu: Phys. Rev. B **65**, 161303(R) (2002)
24. P. Michler, A. Kiraz, C. Becher, W. V. Schoenfeld, P. M. Petroff, L. Zhang, E. Hu, A. Imamoglu: Adv. Solid State Phys. **41**, 3 (2001) (Springer, Berlin, Heidelberg 2001)
25. R. M. Thompson, R. M. Stevenson, A. J. Shields, I. Farrer, C. J. Lobo, D. A. Ritchie, M. L. Leadbeater, M. Pepper: Phys. Rev. B **64**, 201302 (2001)

26. E. Moreau, I. Robert, J.-M. Gérard, I. Abram, L. Manin, V. Thierry-Mieg: Appl. Phys. Lett. **79**, 2865 (2001)
27. A. Kiraz, P. Michler, C. Becher, B. Gayral, A. Imamoglu, Lidong Zhang, E. Hu, W. V. Schoenfeld, P. M. Petroff: Appl. Phys. Lett. **78**, 3932 (2001)
28. E. M. Purcell: Phys. Rev. **69**, 681 (1946)
29. P. Michler, A. Kiraz, L. Zhang, C. Becher, E. Hu, A. Imamoglu: Appl. Phys. Lett. **77**, 184 (2000)
30. J.-M. Gérard, B. Gayral: J. Lightwave Technol. **17**, 2089 (1999)
31. G. S. Solomon, M. Pelton, Y. Yamamoto: Phys. Rev. Lett. **86**, 3903 (2001)
32. S. Strauf, P. Michler, M. Klude, D. Hommel, G. Bacher, A. Forchel: Phys. Rev. Lett. **89**, 177403 (2002).
33. A. Imamoglu, Y. Yamamoto: Phys. Rev. Lett. **72**, 210 (1994)
34. J. Kim, O. Benson, H. Kan, Y. Yamamoto: Nature **397**, 500 (1999)
35. Z. Yuan, B. E. Kardynal, R. M. Stevenson, A. J. Shields, C. J. Lobo, K. Cooper, N. S. Beattie, D. A. Ritchie, M. Pepper: Science **295**, 102 (2002)
36. K. Sebald, P. Michler, T. Passow, D. Hommel, G. Bacher, A. Forchel: Appl. Phys. Lett. **81**, 2920 (2002)
37. R. Murray, D. Childs, S. Malik, P. Sirerns, C. Roberts, J.-M. Hartmann, P. Starrinou: Jpn. J. Appl. Phys. **38**, 528 (1999)
38. J.-M. Gérard, B. Gayral: Physica (Amsterdam) **9E**, 131 (2001)
39. O. Benson, C. Santori, M. Pelton, Y. Yamamoto: Phys. Rev. Lett. **84**, 2513 (2000)
40. D. V. Regelman, U. Mizrahi, D. Gershoni, E. Ehrenfreund, W. V. Schoenfeld, P. M. Petroff: Phys. Rev. Lett. **87**, 257401 (2001)
41. M. D'Angelo, M. V. Chekhova, Y. Shih: Phys. Rev. Lett. **87**, 013602 (2001)
42. S. Strauf, S. M. Ulrich, K. Sebald, P. Michler, T. Passow, D. Hommel, G. Bacher, A. Forchel: Phys. Stat. Solidi (b) **238**, 321 (2003)

# Index

acoustic phonons, 153
adiabatic approximation, 33
AlAs, 16, 17
AlGaAs, 9, 10
AlGaN, 9
AlInGaAs, 9
AlN, 21
anisotropy, 155
antiferromagnetic interaction, 172
application, 199, 219
artificial dot, 188
– lithography based, 206
"atom-like" shell model, 10
Auger process, 79

biexciton, 58, 147, 157, 168
bimodal dot distribution, 200
binding energy, 157, 158
Bir–Pikus–Bahder Hamiltonian, 31
Bragg mirror, 197
bright exciton, 154
Brillouin function, 172
broadening
– homogeneous, 192, 211
– inhomogeneous, 193, 211

carrier
– capture, 219
– capture time, 214, 217
– lifetime, 215
cascaded decay, 160
catastrophic optical damage (COD), 224
cavity quantum electro dynamics, 269
CdMgTe, 173
CdMnSe/ZnSe, 171
CdMnTe, 171, 173
CdSe/ZnMnSe, 147, 171

CdSe/ZnSe, 150
centre of mass operator, 74
characteristic temperature, 196, 210
charge storage, 16
charged exciton, 11, 12, 57, 58, 99, 157, 161
charged quantum dot, 11
chemical potential, 209
chirp, 222
circular polarization, 154
confinement potential, 95
– angular momentum, 96
– three-dimensional confinement, 94
continuous elasticity
– isotropic strain, 28
– total elastic energy, 29
cross correlation measurements, 339

density matrix, 242
density of states, 209
density of transparency, 209, 214
dephasing time, 243
device, 16, 219
diluted magnetc semiconductor, 172
dipole moment, 164
disk-shaped dot, 33
doping, 161
doping of the dot, 197
dot
– AlInAs, 204
– CdSe, 205
– GaInAs, 194
– GaInAsN, 194, 199
– GaInAsP, 196
– GaInP, 202
– in optical cavity, 61
– InAs, 193
– InP, 200

– laser threshold, 211
– lens-shaped, 36
– material, 26
– neutral, 11
– nitride based, 204
– stacking, 201
– symmetry, 98, 99
– ZnSe based, 205
dry etching, 208
dynamics, 160

effective cavity volume, 273
eigenstate, 167
electric field, 147, 164
– lateral, 164
electrical injection, 161
electron beam lithography, 150, 163,
    165, 206
electron–phonon interaction, 60
electronic coupling, 8, 13, 15
electronic level, 11, 14
electronic state, 208
empirical pseudopotential, 44
energy splitting, 154
etching, 150
exchange field, 174
exchange interaction, 147, 154, 169, 171
excited state, 152
exciton, 49, 147, 168
exciton fine structure, 99
– bright exciton, 115
– dark exciton, 115, 154
– exchange interaction, 115
– Zeeman interaction, 115
exciton in coupled quantum dot, 13
exciton radiative lifetime, 248
exciton storage, 16
exciton–phonon coupling, 112, 150, 152
exciton–phonon interaction, 98, 112,
    250

far-infrared spectroscopy, 85
Faraday geometry, 167
Fermi edge singularity, 82
filling factor, 215
fine structure, 157
fluctuation dissipation theorem, 176
Fock state, 317
Fock–Darwin state, 49

g-factor, 147, 169, 173
GaAs, 2, 6
gain
– compression, 217
– differential, 217, 219
– material, 215
– optical, 210, 219
– resonator, 214
GaP substrate, 203
giant magnetooptic effect, 171
giant Zeeman effect, 171
Glauber, 316
growth mode
– Frank–Van der Merwe, 190
– Stranski–Krastanov, 190
– Volmer–Weber, 190

Hanbury-Brown and Twiss, 319
hidden symmetry, 67
high power laser, 223
homogeneous broadening, 244
Hund's rule, 73

identical, 14
indirect band structure, 204
InGaAs, 8, 9
inhibition, 299
inhomogeneous broadening, 245
interband polarization operator, 50, 77
island nucleation, 6

Jacobi coordinate, 50

$k \cdot p$ Hamiltonian
– Eight-band model, 41
– Four-band model, 38

laser
– damping, 217, 219
– DFB laser, 198, 219
– – complex coupled, 219
– dynamics, 211
– long wavelength, 199
lattice, 2, 7
linear polarization, 154
linewidth, 150, 176
LO phonon, 153
losses, 211
– internal, 212–214

– optical resonator, 213
– outcoupling, 212, 214

magnetic field, 147, 168, 176
magnetic fluctuation, 176
magnetic ion, 171
magnetic localization, 173
magnetic moment, 177
magnetic polaron, 147, 172
magnetic semiconductor, 147, 171
magnetization, 147, 172, 176
magneto-exciton, 53
many-body effect, 1, 10, 12, 16, 21
MBE, 191
microcavity
– laser, 299
– pillar, 279
– planar, 288
microdisk, 283, 293, 295
micropillar, 290, 296
Mn, 171
mode supression, 221
modulation bandwidth, 217
modulation dynamics, 221
modulation response, 215
molecule, 161
MOVPE, 191
multiexciton, 62, 255
– exact diagonalization technique, 64
– excited state, 71

nano-aperture, 173
neutral exciton, 99
– absorption spectrum, 98
– Kohn's theorem for excitons, 107
nucleation, 2, 4, 6–8, 20

optical Bloch equation, 238
optical quantum computing, 308
optical Rabi oscillation, 241, 258
optical transitions, 210
optimum dot density, 218
optimum laser length, 218
ordering, 3, 4, 9

paramagnet, 172
Pauli exclusion principle, 157
Pauli principle, 94
phonon interaction, 147

photocurrent spectroscopy, 86
photoluminescence, 8, 11, 148
photoluminescence property of
    quantum dot, 11
photon
– antibunching, 317
– beat, 156
– bunching, 317
– correlated, 338
– entangled pair, 308
– indistinguishable, 308
– polarized, 299
photonic crystal, 284
piezoelectric potential, 32
Poisson distribution, 316
polarizability, 164
polarization, 154, 158, 167
polarization entanglement, 343
population relaxation, 243, 255
positioning, 4, 9
Purcell
– effect, 289, 294, 304, 333
– factor, 276, 332
– strong, 302
pure dephasing, 243, 251, 255

quantum computation, 178
quantum dot, 6, 10, see dot
quantum dot memory, 16
quantum dot molecule, 99, 178
– electronic state, 45
– entanglement of isospin, 56
– excitonic state, 54
– isospin, 55
quantum efficiency, 215

Rabi frequency, 241
radiation field, 161
radiative lifetime, 160
radiative recombination, 151
Raman spectra, 59
rate equations, 211
recombination, 160
red light emitting laser, 200
relaxation time, 243

S4P efficiency, 304
Schottky diode, 166
SE inhibition, 293

second-order correlation function, 318
selection rule, 160
self assembled dot, 188, 189
single electron injection, 163
single photon source, 303, 317, 326
single photon turnstile, 329
single QD-laser, 299, 303
single quantum dot, 147, 178
single-mode solid state single photon
    source, 307
single-mode spontaneous emission, 297
sp-d exchange interaction, 172
spatially resolved optical spectroscopy,
    147
spectral diffusion, 152
spectral linewidth, 192
spin, 154, 161
– alignment, 172
– flip, 160
– injection, 16, 18
– polarized LED, 18
spintronic, 148, 178
spontaneous lifetime, 211
Stark effect, 164
statistical magnetic fluctuation, 175
strain coupling, 7, 8
strain elements, 27
strained local band edge, 31
strong coupling regime, 272, 273, 285,
    300
structural factor, 212, 218
structurization, 207
sub-Poissonian, 316, 319

super-Poissonian, 316, 319
superposition of state, 156
superradiance effect, 160

telecom wavelength, 195, 198
threshold
– current, 212, 215
– current density, 196
– density, 211, 214, 218
thresholdless laser, 302
tight-binding Hamiltonian, 43
time-resolved spectroscopy, 159
transfer matrix method, 34
trion, 161
tunable laser, 193
tunneling, 165
type II band alignment, 202

upconversion, 13, 14
upconversion process in quantum dot,
    13

valence force field model
– Keating–Martin, 30
– Stillinger–Weber, 30
– Tersoff, 31
VCSEL, 197
vertical electric field, 163

wavefunction, 161
whispering gallery mode, 329

Zeeman effect, 167

# Topics in Applied Physics

73 **Hydrogen in Metals III**
Properties and Applications
By H. Wipf (Ed.) 1997. 117 figs. XV, 348 pages

74 **Millimeter and Submillimeter Wave Spectroscopy of Solids**
By G. Grüner (Ed.) 1998. 173 figs. XI, 286 pages

75 **Light Scattering in Solids VII**
Christal-Field and Magnetic Excitations
By M. Cardona and G. Güntherodt (Eds.) 1999. 96 figs. X, 310 pages

76 **Light Scattering in Solids VIII**
C60, Semiconductor Surfaces, Coherent Phonons
By M. Cardona and G. Güntherodt (Eds.) 1999. 86 figs. XII, 228 pages

77 **Photomechanics**
By P. K. Rastogi (Ed.) 2000, 314 Figs. XVI, 472 pages

78 **High-Power Diode Lasers**
By R. Diehl (Ed.) 2000, 260 Figs. XIV, 416 pages

79 **Frequency Measurement and Control**
Advanced Techniques and Future Trends
By A. N. Luiten (Ed.) 2001, 169 Figs. XIV, 394 pages

80 **Carbon Nanotubes**
Synthesis, Structure, Properties, and Applications
By M. S. Dresselhaus, G. Dresselhaus, Ph. Avouris (Eds.) 2001, 235 Figs. XVI, 448 pages

81 **Near-Field Optics and Surface Plasmon Polaritons**
By S. Kawata (Ed.) 2001, 136 Figs. X, 210 pages

82 **Optical Properties of Nanostructured Random Media**
By Vladimir M. Shalaev (Ed.) 2002, 185 Figs. XIV, 450 pages

83 **Spin Dynamics in Confined Magnetic Structures I**
By B. Hillebrands and K. Ounadjela (Eds.) 2002, 166 Figs. XVI, 336 pages

84 **Imaing of Complex Media with Acoustic and Seismic Waves**
By M. Fink, W. A. Kuperman, J.-P. Montagner, A. Tourin (Eds.) 2002, 162 Figs. XII, 336 pages

85 **Solid–Liquid Interfaces**
Macroscopic Phenomena – Microscopic Understanding
By K. Wandelt and S. Thurgate (Eds.) 2003, 228 Figs. XVIII, 444 pages

86 **Infrared Holography for Optical Communications**
Techniques, Materials, and Devices
By P. Boffi, D. Piccinin, M. C. Ubaldi (Eds.) 2003, 90 Figs. XII, 182 pages

87 **Spin Dynamics in Confined Magnetic Structures II**
By B. Hillebrands and K. Ounadjela (Eds.) 2003, 179 Figs. XVI, 321 pages

88 **Optical Nanotechnologies**
The Manipulation of Surface and Local Plasmons
By J. Tominaga and D. P. Tsai (Eds.) 2003, 168 Figs. XII, 212 pages

89 **Solid-State Mid-Infrared Laser Sources**
By I. T. Sorokina and K. L. Vodopyanov (Eds.) 2003, 263 Figs. XVI, 557 pages

90 **Single Quantum Dots**
Fundamentals, Applications, and New Concepts
By P. Michler (Ed.) 2003, 181 Figs. XII, 352 pages